Lecture Notes on
Topoi and Quasitopoi

Lecture Notes on
Topoi and Quasitopoi

Oswald Wyler
Department of Mathematics
Carnegie-Mellon University

World Scientific
Singapore • New Jersey • London • Hong Kong

Published by

World Scientific Publishing Co. Pte. Ltd.
P O Box 128, Farrer Road, Singapore 9128
USA office: 687 Hartwell Street, Teaneck, NJ 07666
UK office: 73 Lynton Mead, Totteridge, London N20 8DH

Library of Congress Cataloging-in-Publication data is available.

LECTURE NOTES ON TOPOI AND QUASITOPOI

ISBN 981-02-0153-2

Printed in Singapore by JBW Printers & Binders Pte. Ltd.

PREFACE

A topos is a category which can be viewed as a set theory, with an intrinsic logic which is intuitionistic, but for most topoi not Boolean. This means that most of the familiar laws of logic and set theory are valid in a topos, except for the principle that a statement must be true or false and related laws. Quasitopoi generalize topoi and retain most of the basic properties of topoi. Thus quasitopoi are used when set-like behavior of objects is desirable, but topoi are too set-like. At present, quasitopoi are used mainly in Topology and in Fuzzy Set Theory.

With these Notes, I am trying to redeem an old promise. In [104], I stated a number of theorems on quasitopoi, with most proofs sketched or missing, and I promised to discuss quasitopoi more fully, and with complete proofs, in a set of lecture notes. P. JOHNSTONE, in the preface of [58], cited this promise as one of the reasons for not treating quasitopoi in his book. Presenting quasitopoi in lecture notes became less urgent when J. PENON published an account of the theory in [83]. However, many details remained left out. Since there has been renewed interest in quasitopoi in recent years, it seemed appropriate to write down the notes which I promised years ago. The present Notes form the first coherent and reasonably complete account of the basic theory of quasitopoi, with some new results, and with examples and applications.

Many theorems for quasitopoi generalize theorems for topoi, and their proofs often follow closely the proofs of the corresponding theorems for topoi. This is usually the case for elementary constructive proofs, but may fail for more elegant proofs. Thus it seems appropriate to develop the basic theory of topoi, or at least those parts of it which can be generalized to quasitopoi, together with the theory of quasitopoi, using constructive proofs whenever possible. I have done this in the present Notes, but now with the task nearly done, I almost feel like the proverbial journeyman: "Meischter, d'Arbeit isch fertig, sell i si grad flicke?"

For example, a quasitopos **E** has two basic properties: **E** is cartesian closed, and (strong) partial morphisms in **E** are represented. While writing the present Notes, I obtained new results, included in the Notes, which make it clear that these two properties are almost independent of each other, but I have not carried out the reorganization which this independence suggests.

In writing these Notes, I have tried to make proofs as elementary and constructive as reasonably possible, and to give enough details to make the inevitably remaining gaps small. Even so, words like "clearly" or "it is easily

v

seen" occur quite often. The reader who wants to understand fully what is going on should be prepared to fill in the gaps thus signaled. I have tried to provide diagrams in the text when these diagrams seem appropriate and are not too big or too complicated, but the reader may find it helpful to provide additional diagrams or to extend some diagrams in the text.

The reader of these Notes should be reasonably familiar with the basic concepts of category theory, including categories, functors and natural transformations, diagrams, limits and colimits, adjoint functors, monads and comonads, and cartesian closed categories. For the reader's convenience, and in order to establish the notations to be used, these concepts will be reviewed in Chapter 0, with examples and with concise proofs of basic results. Other categorical concepts which may be less familiar will be reviewed before their first use in these Notes. In addition to the brief review of categorical tools in Chapter 0, the following references may be useful. S. MacLane's book [72], written about twenty years ago, is still the most complete and authoritative textbook on category theory. The interested reader may also consult [48] or [10], or one of several other texts on categories. The books [12] and [40] on topoi have useful introductory chapters on categories, and the chapter on categories in [67] is especially helpful. The recent book [4] by Adámek, Herrlich and Strecker has many definitions and results not found in earlier texts, and it is the first book with some results on quasitopoi.

After the review chapter on categorical tools, these Notes are subdivided into eight chapters. Chapter 1 is a prerequisite for everything which follows, and Chapter 2 presents useful examples of topoi and quasitopoi. Chapters 3 – 6 discuss the general theory. These chapters can be read independently although some cross-references occur, especially in Chapter 5. Chapters 7 and 8 deal with applications of quasitopoi to Topology and to Fuzzy Set Theory, with examples.

Every chapter of these Notes starts with a brief synopsis of its contents. References to the literature have been provided in the text where it seemed appropriate. The bibliography includes further references, but it does not aim for completeness. Numbers in brackets refer, as usual, to bibliography items. Sections are numbered consecutively, and subdivided into numbered items. Equation and diagram numbers are subordinate to item numbers, and subitem numbers are used occasionally. A reference $m.n$ points to Item n of Section m, and a reference $m.n.p$ or $m.n.(p)$ to subitem p or equation (p) of Item $m.n$. Every item starts with its item number, which includes the section number, in boldface type. This number may be followed by a title, in boldface.

Standard concepts and notations are used in these Notes whenever this is possible, but some comments about notations still are necessary.

Composition of functions and relations, of morphisms in a category, and of functors and natural transformations, will always be "from right to left", in the traditional manner. Depending on the context, we shall denote compositions formally in the $g \circ f$ style, or less formally in the $g \cdot f$ style, or just by juxtaposition gf.

Traditionally, an argument of a function is enclosed in parentheses, except for some special functions. These parentheses are often unnecessary, but we shall usually provide them. However, an argument of a functor or natural transformation will not be enclosed in parentheses, except if the parentheses are needed for grouping. We use subscripted arguments where this is traditional, e.g. for natural transformations, but we write these arguments on the line if it is more convenient to do so.

There seems to be no standard notation for images and inverse images of subsets by mappings. We denote by PA the powerset of a set A, and we use $S \subset A$ to denote that S is a subset of A, including the case $S = A$. A notation for "proper" subsets, *i.e.* excluding the case $S = A$, will never be needed. If $f : A \to B$ is a mapping, then we denote by

$$f^{\to}(S) = \{f(x) \mid x \in S\}$$

the image by f of a subset S of A, and by

$$f^{\leftarrow}(T) = \{x \in A \mid f(x) \in T\}$$

the inverse image by f of a subset T of B. With this notation, images and inverse images by f become values of mappings

$$f^{\to} : PA \to PB \qquad \text{and} \qquad f^{\leftarrow} : PB \to PA$$

between powersets. We note that always

$$f^{\to}(S) \subset T \quad \Longleftrightarrow \quad S \subset f^{\leftarrow}(T)$$

for $S \subset A$ and $T \subset B$.

Writing these Notes has forced me to take a definite stand on some quite controversial issues. Thus I write about topoi and quasitopoi, not about toposes and quasitoposes. I shall use monads and comonads when this is appropriate, but not triples or cotriples. In general, I have used what I consider standard notations, but I have taken the precaution of defining the categorical concepts used in these Notes, either in Chapter 0 or when I need them.

Foundations seem to be less controversial than some aspects of terminology, but no book on categories is considered complete without a statement about foundations. Category Theory deals with constructs such as the category of all sets or the category of all groups; thus the question whether and how category theory can be based on set theory is not trivial. The consensus, as argued *e.g.* by A. Blass in [13], is clearly that this can be done, probably in more than one way. This does of course not preclude the future use of foundations for Mathematics which are not based on sets and set membership. Experience tells us that the constructions used in category theory are probably safe, and independent of a particular underlying system of foundations, as long as they are true constructions, expressible in a first-order language. Thus if I say *e.g.* that a category $\mathbf{E}$ has finite products, I imply that a definite terminal object of $\mathbf{E}$ is given, and that a definite product $A \xleftarrow{p} A \times B \xrightarrow{q} B$, including the projections p and q, is given for every ordered pair of objects A, B of $\mathbf{E}$.

Several colleagues have helped me with advice and preprints; I want to thank in particular J. ADÁMEK, J. and J.-L. COULON, H. HERRLICH, U. HÖHLE, and L.N. STOUT. Several people, including DAN NESMITH, have helped with TEXnical matters. I also want to thank World Scientific for publishing these Notes, and my editor, H.M. HO, for her help and advice. Last but not least, I wish to thank Carnegie Mellon University and its School of Computer Science for providing the facilities which made this book possible.

CONTENTS

Chapter 0

CATEGORICAL TOOLCHEST

This preliminary chapter reviews the basic notions and results of Category Theory which we shall use throughout these Notes. We shall discuss categories, functors and natural transformations, universal morphisms and adjunctions, limits and colimits, monads and comonads, and cartesian closed categories, with examples and proofs. We also discuss some special topics and results in Category Theory which we shall need repeatedly in these Notes. Other special topics and results will be taken up when needed in later chapters.

1. Categories

1.1. Definition. A *category* **C** consists of two classes, the class of *objects* of **C** and the class of *morphisms* of **C**, with four operations and seven formal laws.

The first two operations assign to every morphism f of **C** two objects of **C**, called the *domain* dom f of f, and the *codomain* cod f of f. We often write $f : A \to B$ or $A \xrightarrow{f} B$ to denote that f is a morphism with domain A and codomain B. The third operation assigns to every object A of **C** the *identity morphism* id_A in **C**.

The fourth operation, *composition*, is a partial binary operation on morphisms. It assigns a morphism $g \circ f$ of **C** to every pair (f, g) of morphisms of **C** such that dom $g = $ cod f. This composition is often denoted by $g \cdot f$ or just by gf.

Four of the seven formal laws assign domains and codomains. We ask that

$$\mathrm{dom}(\mathrm{id}_A) = A = \mathrm{cod}(\mathrm{id}_A)$$

for an object A of **C**, and that

$$\mathrm{dom}(g \circ f) = \mathrm{dom}\, f \quad \text{and} \quad \mathrm{cod}(g \circ f) = \mathrm{cod}\, g$$

for morphisms f and g such that dom $g = $ cod f.

The other three formal laws are more substantial. They state that we must have

$$f \circ \mathrm{id}_A = f = \mathrm{id}_B \circ f$$

1

for $f : A \to B$ in **C**, and that

$$h \circ (g \circ f) = (h \circ g) \circ f,$$

for morphisms f, g, h such that $\operatorname{dom} g = \operatorname{cod} f$ and $\operatorname{dom} h = \operatorname{cod} g$.

1.2. Concrete categories. Many categories have objects which can be viewed as sets with some kind of structure, and morphisms which can be viewed as mappings which "preserve" structure. Categories of this sort are called *concrete*. Identity morphisms in a concrete category are identity mappings, and the composition of morphisms is the composition of mappings.

The prime example of a concrete category is the category of sets, with sets as objects, mappings as morphisms, and with domain, codomain, identity morphisms and composition defined in the usual way.

Other examples abound: groups and group homomorphisms, rings and ring homomorphisms, vector spaces over a field and linear transformations, topological spaces and continuous mappings, ordered sets and order preserving mappings are just a few.

1.3. Monoids and ordered sets. We recall that a *monoid* is a set M with an associative binary operation, and with a *neutral element*. If the operation is denoted by $(x, y) \mapsto x \cdot y$, then the neutral element e satisfies $e \cdot x = x = x \cdot e$, for all $x \in M$. A monoid M can be considered as a category, with a single object, and with the elements of M as morphisms. Composition is the opeation of M, and the neutral element is the identity morphism.

An *ordered set*, also called a *partly ordered set* or *poset*, is a set with a reflexive, antisymmetric and transitive relation. An ordered set S, with order relation $\leqslant$, can be regarded as a category, with the elements of S as objects, with one morphism $x \to y$ if $x \leqslant y$, and with no other morphisms. Conversely, a category $\mathcal{S}$ with at most one morphism $x \to y$ for objects x and y, can be considered as a *preordered class*, with the objects of $\mathcal{S}$ as elements, and with $x \leqslant y$ if there is a morphism $x \to y$. Identity morphisms make the order relation reflexive, and composition makes it transitive.

1.4. Duality. Every category **C** has a *dual category* $\mathbf{C}^{\mathrm{op}}$, with the same objects and morphisms as **C**, but with arrows reversed. This means that domains and codomains are interchanged, and that if $g \circ f = h$ in **C**, then $f \circ g = h$ in $\mathbf{C}^{\mathrm{op}}$.

Duality is symmetric; we have $(\mathbf{C}^{\mathrm{op}})^{\mathrm{op}} = \mathbf{C}$.

Duality assigns to every categorical concept a dual concept, obtained for **C** by formulating the given concept for $\mathbf{C}^{\mathrm{op}}$ and interpreting the result in **C**.

In the same way, we obtain a dual result for every general result in the theory of categories. The dual of the dual of a concept or result is the given concept or result.

1.5. Special morphisms. An *isomorphism* in a category **C** is a morphism of **C** with a two-sided inverse. Thus $u : A \to B$ is an isomorphism iff there is a morphism $v : B \to A$ such that $vu = \text{id}_A$ and $uv = \text{id}_B$. This determines v uniquely; one usually puts $v = u^{-1}$.

A *monomorphism* in a category **C** is a morphism m of **C** such that always $u = v$ for morphisms u, v of **C** such that $m \circ u = m \circ v$ in **C**. Dually, an *epimorphism* in a category **C** is a morphism e of **C** such that always $u = v$ for morphisms u, v of **C** such that $u \circ e = v \circ e$ in **C**.

A *retraction* in a category **C** is a morphism u with a right inverse v, satisfying $u \circ v = \text{id}_B$ if B is the codomain of u. Dually, u is a *coretraction*, or a *section*, if u has a left inverse.

In the category of sets, the monomorphisms are the injective mappings, and the epimorphisms are the surjective mappings. In a concrete category, every injective morphism is monomorphic, and every surjective morphism epimorphic, but the converse statements need not be valid.

1.6. Subcategories. A *subcategory* **B** of a category **C** has as objects a class of objects of **C** and as morphisms a class of morphisms of **C**. Domains and codomains are the same for **B** as for **C**. If A is an object of **B**, then the morphism id_A of **C** is an identity morphism of **B**, and if morphisms f and g of **B** have a composition gf in **C**, then this composition is also the composition gf in **B**.

We say that **B** is a *full subcategory* of **C** if **B** is a subcategory of **C**, and for objects A, B of **B**, every morphism $f : A \to B$ of **C** is a morphism of **B**. A full subcategory is determined by its class of objects.

2. Functors

2.1. Definition. For categories **A** and **B**, a *functor* $F : \mathbf{A} \to \mathbf{B}$ assigns to every object A of **A** an object FA of **B**, and to every morphism f of **A** a morphism Ff of **B**, subject to the following rules.

(1) $\text{dom}(Ff) = F(\text{dom} f)$, and $\text{cod}(Ff) = F(\text{cod} f)$ for every morphism f of **A**.

(2) $F\text{id}_A = \text{id}_{FA}$ for every object A of **A**.

(3) $F(g \circ f) = (Fg) \circ (Ff)$ if $g \circ f$ is defined in **A**.

Thus a functor is a homomorphism of categories.

2.2. Examples. The *forgetful functor* of a concrete category **C** assigns to every object of **C** the underlying set, and to every morphism of **C** the underlying mapping.

The free group functor from sets to groups assigns to every set S the free group FS over S, and to every mapping $f : S \to T$ the unique extension of f to a group homomorphism $Ff : FS \to FT$.

The discrete space functor from sets to topological spaces assigns to every set S the discrete topological space DS with underlying set S, and to every mapping $f : S \to T$ the continuous map $f : DS \to DT$.

If **B** is a subcategory of **C**, then the embedding **B** → **C** is a functor.

Assigning to every set S its powerset PS, and to every mapping $f : A \to B$ the direct image mapping $f^{\to} : PA \to PB$, defines a functor on sets.

If A and B are preordered sets, regarded as categories, then a functor $f : A \to B$ is an order preserving mapping.

If A and B are monoids, regarded as categories with one object, then a functor $f : A \to B$ is a homomorphism of monoids.

2.3. Dual and contravariant functors. Every functor $F : \mathbf{A} \to \mathbf{B}$ induces a *dual functor* $F^{\mathrm{op}} : \mathbf{A}^{\mathrm{op}} \to \mathbf{B}^{\mathrm{op}}$, with the same assignments for objects and morphisms as F.

A functor $F : \mathbf{A}^{\mathrm{op}} \to \mathbf{B}$, or equivalently a functor $F : \mathbf{A} \to \mathbf{B}^{\mathrm{op}}$, is called a *contravariant functor* from **A** to **B**. A contravariant functor switches domains and codomains, and it reverses the order of composition. Functors $F : \mathbf{A} \to \mathbf{B}$ are sometimes called *covariant functors* from **A** to **B**.

The contravariant *powerset functor* P : SET^{op} → SET is an example. It assigns to every set S its powerset PS, and to every mapping $f : A \to B$ the inverse image mapping $Pf = f^{\leftarrow} : PB \to PA$.

A contravariant functor between preordered sets is an order reversing mapping.

2.4. Full and faithful functors. For objects A and B of a category **A**, a functor $F : \mathbf{A} \to \mathbf{B}$ induces a mapping $F_{A,B}$ from morphisms $f : A \to B$ of **A** to morphisms $g : FA \to FB$ of **B**. We say that F is *faithful* if every mapping $F_{A,B}$ is injective, and *full* if every mapping $F_{A,B}$ is surjective.

The forgetful functor from a concrete category to sets is always faithful; so is an order preserving mapping regarded as a functor, and the embedding functor of a subcategory. The embedding functor of a full subcategory is also full.

2.5. Preserve and reflect. We say that a functor $F : \mathbf{A} \to \mathbf{B}$ *preserves* monomorphisms if Ff is a monomorphism of **B** for every monomorphism of **A**, and we say that F *reflects* monomorphisms if f in **A** is a monomorphism of **A** whenever Ff is a monomorphism of **B**.

This terminology is used not only for monomorphisms, but for every categorical concept to which it can be applied.

For example, every functor preserves isomorphisms, and retractions and coretractions. A faithful functor reflects monomorphisms and epimorphisms.

2.6. Locally small categories and hom functors. For objects A and B of a category **C**, we denote by $\mathbf{C}(A, B)$ the class of all morphisms $f : A \to B$ in **C**. We say that **C** is *locally small* if every class $\mathbf{C}(A, B)$ is a set. These sets are called *hom sets* of **C**.

A category **C** is called a *small category* if the morphisms of **C** form a set. This is the case iff **C** is locally small and the objects of **C** form a set.

If **C** is locally small, then every object A of **C** induces a *covariant hom functor* H_A and a *contravariant hom functor* H^A, from **C** to sets, as follows.

The functor $H_A = \mathbf{C}(A, -)$ assigns to every object B of **C** the hom set $H_A B = \mathbf{C}(A, B)$, and to $f : B \to C$ in **C** the mapping

$$H_A f = \mathbf{C}(A, f) : \mathbf{C}(A, B) \to \mathbf{C}(A, C) : u \mapsto f \circ u.$$

The functor $H^A = \mathbf{C}(-, A)$ assigns to every object B of **C** the hom set $H^A B = \mathbf{C}(B, A)$, and to $f : B \to C$ in **C** the mapping

$$H^A f = \mathbf{C}(f, A) : \mathbf{C}(C, A) \to \mathbf{C}(B, A) : v \mapsto v \circ f.$$

2.7. Composition of functors. For every category **A**, there is an *identity functor* Id **A**, with the obvious assignments. If $F : \mathbf{A} \to \mathbf{B}$ and $G : \mathbf{B} \to \mathbf{C}$ are functors, then $(G \circ F)x = G(Fx)$, for an object or morphism x of **A**, clearly defines a functor $G \circ F : \mathbf{A} \to \mathbf{C}$, the *composition* of F and G.

Thus we can construct categories of categories, with categories as objects and functors as morphisms. A "category of all categories" cannot be viewed as a legitimate construct, but we can construct a locally small category CAT of all small categories, and more generally categories of categories with suitable size restrictions.

3. Natural Transformations

3.1. Definition. For functors $F : \mathbf{A} \to \mathbf{B}$ and $G : \mathbf{A} \to \mathbf{B}$, with the same domain and the same codomain, a *natural transformation* $\nu : F \to G$ assigns to every object A of **A** a morphism $\nu_A : FA \to GA$ of **B**, with the property that the square

$$
\begin{array}{ccc}
FA & \xrightarrow{Ff} & FB \\
\downarrow{\nu_A} & & \downarrow{\nu_B} \\
GA & \xrightarrow{Gf} & GB
\end{array}
$$

commutes for every morphism $f : A \to B$ of $\mathbf{A}$.

A natural transformation $\nu : F \to G$ is also called a *morphism of functors*. If every morphism ν_A is an isomorphism of $\mathbf{B}$, then ν is called a *natural isomorphism* or a *natural equivalence*.

3.2. Composition. If F, G, H are functors with the same domain $\mathbf{A}$ and the same codomain $\mathbf{B}$, then natural transformations $\mu : F \to G$ and $\nu : G \to H$ can be composed, by putting

$$(\nu \circ \mu)_A = \nu_A \circ \mu_A$$

for an object A of $\mathbf{A}$, with the composition carried out in $\mathbf{B}$. It is easily verified that this defines a natural transformation $\nu \circ \mu : F \to H$.

For a functor $F : \mathbf{A} \to \mathbf{B}$, there is also an identity natural transformation $\mathrm{id}_F : F \to F$, with $(\mathrm{id}_F)_A = \mathrm{id}_{FA}$ for an object A of $\mathbf{A}$

3.3. Functor categories. For categories $\mathbf{A}$ and $\mathbf{B}$, functors $F : A \to B$ are the objects, and natural transformations between these functors the morphisms of a category, with identity morphisms and composition defined in 3.2. We call this category a *functor category* and denote it by $[\mathbf{A}, \mathbf{B}]$.

It is easily seen that the functor category $[\mathbf{A}, \mathbf{B}]$ is locally small if $\mathbf{A}$ is small and $\mathbf{B}$ locally small.

Examples of functor categories $[\mathbb{C}^{\mathrm{op}}, \mathrm{SET}]$, with $\mathbb{C}$ small, are given in Section 27.

3.4. Calculus of natural transformations. For functors $H : \mathbb{C} \to \mathbf{A}$ and $K : \mathbf{B} \to \mathbf{D}$, we have functors

$$[H, \mathbf{B}] : [\mathbf{A}, \mathbf{B}] \to [\mathbf{C}, \mathbf{B}] \qquad \text{and} \qquad [\mathbf{A}, K] : [\mathbf{A}, \mathbf{B}] \to [\mathbf{A}, \mathbf{D}],$$

with

$$[H, \mathbf{B}]F = F \circ H \qquad \text{and} \qquad [\mathbf{A}, K]F = K \circ F$$

for a functor $F : \mathbf{A} \to \mathbf{B}$. We put

$$[H, \mathbf{B}]\mu = \mu \cdot H \qquad \text{and} \qquad [\mathbf{A}, K]\mu = K \cdot \mu$$

for a morphism μ of $[\mathbf{A}, \mathbf{B}]$. This defines natural transformations

$$\mu \cdot H : FH \to GH, \qquad \text{and} \qquad K \cdot \mu : KF \to KG,$$

if $\mu : F \to G$, with

$$(\mu \cdot H)_C = \mu_{HC} \quad \text{and} \quad (K \cdot \mu)_A = K\mu_A$$

for objects C of $\mathbf{C}$ and A of $\mathbf{A}$. It follows that

$$K \cdot (\mu \cdot H) = (K \cdot \mu) \cdot H,$$

if the compositions are defined. We may write μH for $\mu \cdot H$ and $K\mu$ for $K \cdot \mu$ if the meaning is clear from the context.

The compositions $\mu \cdot H$ and $K \cdot \mu$ satisfy the four formal laws

$$(M \circ K) \cdot \mu = M \cdot (K \cdot \mu), \qquad \mu \cdot (H \circ L) = (\mu \cdot H) \cdot L,$$

$$K \cdot (\nu \circ \mu) = (K \cdot \nu) \circ (K \cdot \mu), \qquad (\nu \circ \mu) \cdot H = (\nu \cdot H) \circ (\mu \cdot H),$$

which are valid whenever the compositions on either side are defined. These laws follow easily from the definitions. In addition, we have

$$(\rho \cdot G) \circ (H \cdot \mu) = (K \cdot \mu) \circ (\rho \cdot F) : H \circ F \to K \circ G,$$

for $\mu : F \to G$ and $\rho : H \to K$, if the compositions on either side are defined. This follows easily from the naturality of ρ.

This "calculus of natural transformations" is due to GODEMENT [36].

3.5. Special cases. For ordered sets S and T, and for order preserving mappings $f : S \to T$ and $g : S \to T$, there is at most one natural transformation from f to g. This natural transformation exists if and only if $f(x) \leqslant g(x)$ in T for all $x \in S$. Thus the functor category $[S, T]$ is an ordered set, consisting of all order preserving mappings $f : S \to T$ with the "pointwise" order.

For monoids M and N, and for homomorphisms $f : M \to N$ and $g : M \to N$, a natural transformation $u : f \to g$ is given by an element u of N which must satisfy

$$u \cdot f(x) = g(x) \cdot u$$

in N, for all $x \in M$. If N is a group and $f : M \to N$ is given, then this determines a unique homomorphism g, for every element u of N.

3.6. The Yoneda Lemma. For every object A of a locally small category $\mathbf{C}$, we have a contravariant hom functor

$$\mathsf{Y}A = \mathbf{C}(-, A) : \mathbf{C}^{\mathrm{op}} \to \mathrm{SET},$$

and every morphism $f : A \to B$ of $\mathbf{C}$ induces a natural transformation

$$\mathsf{Y}f = \mathbf{C}(-, f) : \mathsf{Y}A \to \mathsf{Y}B.$$

The resulting embedding $\mathsf{Y} : \mathbf{C} \to [\mathbf{C}^{\mathrm{op}}, \mathrm{SET}]$ is called the YONEDA embedding.

The following result is called the YONEDA Lemma.

Theorem. *For an object A of a locally small category $\mathbf{C}$ and a functor $F : \mathbf{C}^{\mathrm{op}} \to \mathrm{SET}$, there is a bijection between members s of FA and natural transformations $\mu : YA \to F$, with*

$$s = \mu_A(\mathrm{id}_A) \qquad \text{and} \qquad \mu_B(u) = (Fu)(s)$$

for corresponding s and μ, and for $u : B \to A$ in $\mathbf{C}$.

Corollary. *The Yoneda embedding is full and faithful.*

PROOF. The first displayed equation follows from the second one for $B = A$ and $u = \mathrm{id}_A$. By naturality of μ, we have

$$\mu_B \circ \mathbf{C}(u, A) = Fu \circ \mu_A : \mathbf{C}(A, A) \to FB$$

for $u : B \to A$ in $\mathbf{C}$, and applying this to id_A, we get the second equation of the Theorem from the first one. If μ is defined by the second equation, with $s \in FA$, then it is easily verified that μ_B is natural in B.

Applying the Theorem to the case $F = YB$, we see that the natural transformations $YA \to YB$ are the morphisms Yg for g in $(YB)(A)$, i.e. $g : A \to B$ in $\mathbf{C}$. The Corollary follows from this observation.

3.7. Duality. For every natural transformation $\nu : F \to G : \mathbf{A} \to \mathbf{B}$, there is a *dual natural transformation* $\nu^{\mathrm{op}} : G^{\mathrm{op}} \to F^{\mathrm{op}} : \mathbf{A}^{\mathrm{op}} \to \mathbf{B}^{\mathrm{op}}$, with $(\nu^{\mathrm{op}})_A$ the same as ν_A for an object A of $\mathbf{A}$, but interpreted as a morphism of $\mathbf{B}^{\mathrm{op}}$. Clearly $(\nu^{\mathrm{op}})^{\mathrm{op}} = \nu$, and the assignments $F \mapsto F^{\mathrm{op}}, \nu \mapsto \nu^{\mathrm{op}}$ define an isomorphism of the categories $[\mathbf{A}, \mathbf{B}]^{\mathrm{op}}$ and $[\mathbf{A}^{\mathrm{op}}, \mathbf{B}^{\mathrm{op}}]$.

4. Universal Morphisms and Adjunctions

4.1. Universal and couniversal pairs. A *universal pair* (A, h) for a functor $G : \mathbf{A} \to \mathbf{B}$, at an object B of $\mathbf{B}$, consists of an object A of $\mathbf{A}$ and a morphism $h : B \to GA$ of $\mathbf{A}$, such that for every pair (X, v), with X an object of $\mathbf{A}$ and $v : B \to GX$ in $\mathbf{B}$, there is a unique morphism $u : A \to X$ of $\mathbf{A}$ such that $v = Gu \circ h$. By *abus de langage*, we may speak of $h : B \to GA$ as a *universal morphism* for G at B.

A *couniversal pair* for G, or a *couniversal morphism* $k : GA \to B$ for G, is defined dually; its dual is a universal pair for the functor $G^{\mathrm{op}} : \mathbf{A}^{\mathrm{op}} \to \mathbf{B}^{\mathrm{op}}$.

There will be many examples of universal and couniversal pairs in these Notes; the following example is typical. A free group FS over a set S is constructed as the group of all formal compositions of elements of S and their formal inverses. The universal embedding $h : S \to FS$ maps $x \in S$ into a composition of one element x. A mapping g from S to a group G then has a unique extension to a group homomorphism $f : FS \to G$ such that $g = f \circ h$, at the set level. Thus a free group is a universal pair (FS, h), for the forgetful functor from groups to sets. We omit the details of this construction.

4.2. We note that universal and couniversal pairs are determined up to isomorphisms.

Proposition. *If (A, h) is a is a universal pair for $G : \mathbf{A} \to \mathbf{B}$, at an object B of $\mathbf{B}$, then (A', h') is a universal pair for G at B if and only if $h' = Gu \circ h$ for an isomorphism $u : A \to A'$ of $\mathbf{A}$.*

PROOF. If (A, h) and (A', h') are universal pairs at B, then $h' = Gu \circ h$ and $h = Gv \circ h$ for morphisms $u : A \to A'$ and $v : A' \to A$; with $vu = \mathrm{id}_A$ since $h = Gx \circ h$ only for $x = \mathrm{id}_A$, and similarly $uv = \mathrm{id}_{A'}$. Conversely, if (A, h) is a universal pair for G and $u : A \to A'$ an isomorphism of $\mathbf{A}$, then it is easily seen that $(A', Gu \circ h)$ is also a universal pair for G.

4.3. Definition. An *adjunction* $F \dashv G : \mathbf{A} \to \mathbf{B}$, for functors $G : \mathbf{A} \to \mathbf{B}$ and $F : \mathbf{B} \to \mathbf{A}$, assigns to every pair of objects A of $\mathbf{A}$ and B of $\mathbf{B}$ a bijection, natural in A and B, between morphisms $u : FB \to A$ of $\mathbf{A}$ and $v : B \to GA$ of $\mathbf{B}$. In this situation, we say that F is *left adjoint* to G and G *right adjoint* to F, and that corresponding morphisms $u : FB \to A$ and $v : B \to GA$ are *adjoint* for $F \dashv G$. Naturality of the bijections means that if $u : FB \to A$ is adjoint to to $v : B \to GA$, then fu is adjoint to $Gf \circ v$, and $u \circ Fg$ to vg, for $f : A \to A'$ in $\mathbf{A}$ and $g : B' \to B$ in $\mathbf{B}$.

4.4. An adjunction is already determined by parts of its data.

Theorem. *An adjunction $F \dashv G : \mathbf{A} \to \mathbf{B}$ is fully determined by one of the following collections of data.*

(1) *A functor $G : \mathbf{A} \to \mathbf{B}$, and for every object B of $\mathbf{B}$ a universal pair (FB, η_B) for G at B.*

(2) *A functor $F : \mathbf{B} \to \mathbf{A}$, and for every object A of $\mathbf{A}$ a couniversal pair (GA, ε_A) for F at A.*

(3) *Functors $G : \mathbf{A} \to \mathbf{B}$ and $F : \mathbf{B} \to \mathbf{A}$, and natural transformations $\eta : \mathrm{Id}\,\mathbf{B} \to GF$ and $\varepsilon : FG \to \mathrm{Id}\,\mathbf{A}$, so that always*

(4) $$\varepsilon_{FB} \circ F\eta_B = \mathrm{id}_{FB} \quad \text{and} \quad G\varepsilon_A \circ \eta_{GA} = \mathrm{id}_{GA},$$
for objects A of $\mathbf{A}$ and B of $\mathbf{B}$.

Adjoint morphisms $u : FB \to A$ in $\mathbf{A}$ and $v : B \to GA$ in $\mathbf{B}$ then are given by

(5) $$v = Gu \circ \eta_B \quad \Longleftrightarrow \quad u = \varepsilon_A \circ Fv,$$

for objects A of $\mathbf{A}$ and B.

We call η the *unit* and ε the *counit* of the adjunction, with

$$\varepsilon F \circ F\eta = \mathrm{id}_F \qquad \text{and} \qquad G\varepsilon \circ \eta G = \mathrm{id}_G$$

in the notation of 3.4.

PROOF. An adjunction $F \dashv G$ satisfies (1), with $v = \eta_B$ corresponding to $u = \mathrm{id}_{FB}$ by the adjunction.

If (1) is valid, then $v = Gu \circ \eta_B$ defines a bijection, natural in A, between morphisms $u : FB \to A$ of $\mathbf{A}$ and $v : B \to GA$ of $\mathbf{B}$. To make this natural in B, we need $u = Fg$ corresponding to $v = \eta_B \circ g$, for $g : B' \to B$ in $\mathbf{B}$. Thus we must put

$$GFg \circ \eta_{B'} = \eta_B \circ g.$$

This determines $Fg : FB' \to FB$ uniquely. One sees easily that F with these values Fg is a functor; thus η becomes a natural transformation. If $u : FB \to A$ corresponds to $v : B \to GA$, then $u \circ Fg$ corresponds to

$$Gu \circ GFg \circ \eta_{B'} = Gu \circ \eta_B \circ g = v \circ g;$$

thus the bijections between $u : FB \to A$ and $v : B \to GA$ are natural in B.

Dually, the adjunction $F \dashv G$ is fully determined by (2), with ε a natural transformation. By what we already proved, the adjunction is determined by (5). We get (4) from (5) by putting $u = \varepsilon_A, v = \mathrm{id}_{GA}$ and $u = \mathrm{id}_{FB}, v = \eta_B$. Conversely, if (4) is valid and $v = Gu \circ \eta_B$, then

$$\varepsilon_A \circ Fv = \varepsilon_A \circ FGu \circ F\eta_B = u \circ \varepsilon_{FB} \circ F\eta_B = u.$$

This proves half of (5), and the other half is proved dually.

4.5. Examples. Examples abound. For every finitary algebraic theory over sets, free algebras provide a left adjoint functor of the forgetful functor from algebras to sets. As remarked in 11.5, discrete and trivial topological spaces define a left adjoint functor and a right adjoint functor for the forgetful functor from topological spaces to sets. Other examples will occur throughout these Notes.

The *free category of a graph* is another example. We get a functor G from categories to directed graphs by forgetting about unit morphisms and composition. On the other hand, we define a *path* in a directed graph Γ, from a vertex a to a vertex b, as a string $(a_0, u_1, a_1, \ldots, u_n, a_n)$ of alternating vertices and arrows of Γ, with $a_0 = a, a_n = b$, and with $u_i : a_{i-1} \to a_i$ in Γ for $1 \leqslant i \leqslant n$. Vertices and paths of Γ then are the objects and morphisms of a category $F\Gamma$, with composition by concatenation and with identity paths

$\mathrm{id}_a = (a)$. We obtain a universal pair $(F\Gamma, h_\Gamma)$ for the functor G by putting $h_\Gamma a = a$ for a vertex of Γ, and $h_\Gamma f = (a_0, f, a_1)$ for an arrow $f : a_0 \to a_1$. Thus we have an adjunction $F \dashv G$.

4.6. Discussion. Adjunction is a self-dual concept; we have $F \dashv G$ for functors F and G iff $G^{\mathrm{op}} \dashv F^{\mathrm{op}}$ for the dual functors. If η is the unit and ε the counit of $F \dashv G$, then $\varepsilon^{\mathrm{op}}$ is the unit and η^{op} the counit of the adjunction $G^{\mathrm{op}} \dashv F^{\mathrm{op}}$.

If $F \dashv G : \mathbf{A} \to \mathbf{B}$, then each of the functors F and G determines the other functor up to natural equivalence, by 4.2. If η is the unit and ε the counit of $F \dashv G$, and if $\lambda : F \to F'$ is a natural isomorphism, then it is easily seen that

$$\eta'_B = G\lambda_B \circ \eta_B \quad \text{and} \quad \varepsilon_A = \varepsilon'_A \circ \lambda_{GA},$$

for objects A of $\mathbf{A}$ and B of $\mathbf{B}$, define the unit η' and the counit ε' of the resulting adjunction $F' \dashv G$.

The Adjoint Functor Theorem of N. BOURBAKI [14] and P. FREYD [33] states necessary and sufficient conditions for a functor to have a left or right adjoint. We shall not use this theorem, preferring to obtain adjunctions by constructing them.

4.7. Proposition. *If $F \dashv G : \mathbf{A} \to \mathbf{B}$, then G preserves monomorphisms. Dually, F preserves epimorphisms.*

PROOF. Let $m : A \to A'$ be monomorphic in $\mathbf{A}$. If $Gm \circ v = Gm \circ v'$ in $\mathbf{B}$, for $X \overset{v}{\underset{v'}{\rightrightarrows}} GA$, and if u and u' in $\mathbf{A}$ are adjoint to v and v', then $mu = mu'$. But then $u = u'$, and $v = v'$ follows. Thus Gm is monomorphic in $\mathbf{B}$. The second part is proved dually.

4.8. Proposition. *If $F \dashv G : \mathbf{A} \to \mathbf{B}$, with counit ε, then G is faithful if and only if every morphism ε_A is epimorphic in $\mathbf{A}$, and G is full if and only if every morphism ε_A is a coretraction in $\mathbf{A}$.*

PROOF. The first part follows immediately from the fact that for morphisms $f : A \to A'$ of $\mathbf{A}$, the adjunction induces a bijection between morphisms $f \circ \varepsilon_A$ in $\mathbf{A}$, and morphisms Gf in $\mathbf{B}$. If G is full, then $\eta_{GA} = Gt$ for a morphism $t : A \to FGA$ of $\mathbf{A}$, with

$$G(t \circ \varepsilon_A) \circ \eta_{GA} = Gt = \eta_{GA}$$

by 4.4.(4). But then $t \circ \varepsilon_A = \mathrm{id}_{FGA}$ by the universal property of η_{GA}. Conversely, if ε_A has a left inverse t, and if $v : GA \to GA'$ is adjoint to $u : FGA \to A'$, then $u = ut \circ \varepsilon_A$, and $v = G(ut)$ follows.

4.9. Composition of adjunctions. Adjunctions can be composed, and thus viewed as morphisms of a category, with categories suitably restricted in size as its objects, and with identity adjunctions Id $\mathbf{A}$ $\dashv$ Id $\mathbf{A}$ as identity morphisms.

Proposition. *If $F \dashv G : \mathbf{A} \to \mathbf{B}$ and $F' \dashv G' : \mathbf{B} \to \mathbf{C}$, with units η and η', and with counits ε and ε', then $FF' \dashv G'G$, with unit and counit given by*

$$G'\eta_{F'C} \circ \eta'_C \qquad \text{and} \qquad \varepsilon_A \circ F\varepsilon'_{GA},$$

for objects C of $\mathbf{C}$ and A of $\mathbf{A}$.

PROOF. We have bijections between morphisms

$$u : FF'C \to A, \qquad v : F'C \to GA, \qquad w : C \to G'GA,$$

given by

$$u = \varepsilon_A \circ Fv \qquad \Longleftrightarrow \qquad v = Gu \circ \eta_{F'V}$$

and

$$v = \varepsilon'_{GA} \circ F'w \qquad \Longleftrightarrow \qquad w = G'v \circ \eta'_C;$$

thus $FF' \dashv G'G$. If $u = \mathrm{id}_{FF'C}$, then w is the unit of the composed adjunction. Dually, u is the counit for $w = \mathrm{id}_{G'GA}$.

5. Special Adjunctions

5.1. Contravariant adjunctions. Adjunction becomes symmetric for contravariant functors. For functors $F : \mathbf{B}^{\mathrm{op}} \to \mathbf{A}$ and $G : \mathbf{A}^{\mathrm{op}} \to \mathbf{B}$, we say that F is *adjoint to G on the right* if $F^{\mathrm{op}} \dashv G$. Dually, we say that F is *adjoint to G on the left* if $G \dashv F^{\mathrm{op}}$. These concepts are self-dual. An adjunction on the right is given by bijections, natural in A and B, between morphisms $u : A \to FB$ of $\mathbf{A}$ and $v : B \to GA$ of $\mathbf{B}$. An adjunction on the right has two units, Id $\mathbf{B} \to GF^{\mathrm{op}}$ and Id $\mathbf{A} \to FG^{\mathrm{op}}$, instead of a unit and a counit. The dual of the second unit is the counit of $F^{\mathrm{op}} \dashv G$.

5.2. Self-adjoint functors. We say that a contravariant endofunctor $G : \mathbf{A}^{\mathrm{op}} \to \mathbf{A}$ is *self-adjoint on the right* if there is a natural transformation $h : \mathrm{Id}\,\mathbf{A} \to GG^{\mathrm{op}}$ such that always

$$v = Gu \circ h_B \qquad \Longleftrightarrow \qquad u = Gv \circ h_A,$$

for $u : A \to GB$ and $v : B \to GA$ in $\mathbf{A}$. Thus a self-adjunction on the right is an adjunction $G^{\mathrm{op}} \dashv G$ for which the two units of the adjunction coincide.

The contravariant powerset functor P on sets, with $Pf = f^{\leftarrow}$ for a mapping f, is self-adjoint on the right, with $u : A \to PB$ and $v : B \to PA$ adjoint if $y \in u(x) \iff x \in v(y)$, for $x \in A$ and $y \in B$. The unit $h_A : A \to PPA$ assigns to $x \in A$ the *point filter* on A at x, consisting of all $S \subset A$ with $x \in S$.

Dual vector spaces over a field provide another classical example of self-adjunction on the right; we omit the details.

5.3. Equivalent categories.

An *isomorphism of categories* **A** and **B** is a functor $G : \mathbf{A} \to \mathbf{B}$, with an inverse functor $F : \mathbf{A} \to \mathbf{B}$, with $G \circ F = \text{Id } \mathbf{A}$ and $F \circ G = \text{Id } \mathbf{B}$.

More generally, an *equivalence of categories* **A** and **B** is a full and faithful functor $G : \mathbf{A} \to \mathbf{B}$ such that every object B of **B** is isomorphic in **B** to an object GA.

If we denote such an object of **A** by FB, and the isomorphism by $\eta_B : B \to GFB$, then the pair (FB, η_B) clearly is universal for the full and faithful functor G. Thus the equivalence $G : \mathbf{A} \to \mathbf{B}$ is part of an adjunction $F \dashv G$, by 4.4. By 4.8 and its dual, the counit ε of this adjunction is a natural isomorphism, and the functor F is full and faithful. Thus $F : \mathbf{B} \to \mathbf{A}$ is also an equivalence of categories.

Isomorphic categories are equivalent; it is easily seen that an isomorphism of categories is the same as an equivalence which is bijective for objects. Equivalence of categories is reflective, and symmetric by the preceding paragraph. It is easily verified that equivalence of categories is also transitive.

A contravariant equivalence $G : \mathbf{A}^{\text{op}} \to \mathbf{B}$ is called a *duality* between the categories **A** and **B**.

5.4. Reflective and coreflective subcategories.

We define a *reflective subcategory* of a category **C** as a subcategory **B** of **C** for which the inclusion functor $I : \mathbf{B} \to \mathbf{C}$ has a left adjoint. If $F \dashv I$, then F or the functor $R = IF$ is called a *reflector* for **B** in **C**, and the universal morphism $\eta_A : A \to RA$, for an object A of **C**, is called a *reflection* for **B** at A.

Dually, we say that **B** is a *coreflective* subcategory of **C** if the inclusion functor I has a right adjoint G. This right adjoint, or the composite functor IG, is then called a *coreflector* for **B**, and a couniversal morphism $\varepsilon_A : IGA \to A$ for I is called a *coreflection* for **B** at A.

For a full reflective subcategory **B** of **C**, with reflections $\eta_A : A \to RA$, the counit of the adjunction is a natural isomorphism by 4.8, with ε_B and η_{IB} inverse isomorphisms for an object B of **B**, by 4.4.(4). In this situation, $R\eta_A$ and ε_{RA} are also inverse isomorphisms for an object A of **C**, and $R\eta_A = \eta_{RA}$ follows easily.

5.5. Galois connections. We have seen in 1.3 that ordered and pre-ordered sets and classes can be regarded as categories, with at most one morphism $x \to y$ for objects x and y. Functors for ordered and preordered sets become order-preserving mappings. An adjunction between order preserving mappings is called a *Galois connection.* Thus if A and B are preordered sets, then a Galois connection $f \dashv g : A \to B$ is given by order preserving mappings $f : B \to A$ and $g : A \to B$ such that

$$(1) \qquad\qquad f(y) \leqslant x \quad \Longleftrightarrow \quad y \leqslant g(x),$$

for $x \in A$ and $y \in B$. The unit and counit then are replaced by inequalities

$$(2) \qquad\qquad y \leqslant g(f(y)) \quad \text{and} \quad f(g(x)) \leqslant y,$$

for all $x \in A$ and $y \in B$.

If we order powersets by set inclusion, then direct images $f^{\to}(S)$ and inverse images $f^{\leftarrow}(T)$ for a mapping $f : A \to B$ provide an example, since

$$f^{\to}(S) \subset T \quad \Longleftrightarrow \quad S \subset f^{\leftarrow}(T)$$

for $S \subset A$ and $T \subset B$. Thus $f^{\to} \dashv f^{\leftarrow}$.

Restricted to ordered and preordered sets and classes, contravariant functors become order-reversing mappings, and adjunctions on the right become contravariant Galois connections, with $f^{\mathrm{op}} \dashv g : A^{\mathrm{op}} \to B$ iff

$$x \leqslant f(y) \quad \Longleftrightarrow \quad y \leqslant g(x),$$

or equivalently

$$x \leqslant f(g(x)) \quad \text{and} \quad y \leqslant g(f(y)),$$

for all $x \in A$ and $y \in B$.

5.6. Adjoint natural transformations. In their general form, adjoint natural transformations are defined for a *frame*, a noncommutative square

$$
\begin{array}{ccc}
A & \xrightarrow{\;P\;} & A' \\
F \uparrow\downarrow G & & F' \uparrow\downarrow G' \\
B & \xrightarrow{\;Q\;} & B'
\end{array}
$$

(1)

consisting of functors P and Q, often identity functors, and adjunctions — not just adjoint functors — $F \dashv G$ and $F' \dashv G'$, as shown.

We say that natural transformations $\lambda : F'Q \to PF$ and $\mu : QG \to G'P$ are *adjoint* for the frame (1), and we may write $\lambda \dashv \mu$, if for morphisms $f : FB \to A$ and $g : B \to GA$, adjoint for $F \dashv G$, the morphisms $Pf \circ \lambda_B$ and $\mu_A \circ Qg$ always are adjoint for $F' \dashv G'$.

Proposition. *For a frame* (1), *adjunction of natural transformations defines a bijection between natural transformations* $\lambda : F'Q \to PF$ *and* $\mu :$ $QG \to G'P$, *with* λ_B *adjoint to* $\mu_{FB} \circ Q\eta_B$ *for* $F' \dashv G'$, *and* $P\varepsilon_A \circ \lambda_{GA}$ *to* μ_A, *for objects* A *of* **A** *and* B *of* **B**, *and the unit* η *and counit* ε *of* $F \dashv G$.

PROOF. A morphism id_{FB} is adjoint to η_B; thus λ_B must be adjoint to $\mu_{FB} \circ Q\eta_B$. This determines λ if μ is given. Now $Pf \circ \lambda_B$ is adjoint to

$$G'Pf \circ \mu_{FB} \circ Q\eta_B = \mu_A \circ QGf \circ Q\eta_B = \mu_A \circ Qg$$

if $f : FB \to A$ is adjoint to $g = Gf \circ \eta_B : B \to GA$ for $F \dashv G$.

Thus we have obtained a unique left adjoint λ of μ; we must show that λ_B is natural in B. For $v : B \to B'$ in **B**, both $\lambda_{B'} \circ F'Qv$ and $PFv \circ \lambda_B$ are adjoint to

$$\mu_{FB'} \circ Q\eta_{B'} \circ Qv = \mu_{FB'} \circ QGFv \circ \eta_B = G'PFv \circ \mu_{FB} \circ \eta_B ;$$

thus they are equal.

The other half of the Proposition is dual to this, and proved dually.

5.7. Composition of adjoint natural transformations. Frames can be composed in two ways. For frames

$$\mathbf{A'} \overset{P'}{\longrightarrow} \mathbf{A''} \qquad\qquad \mathbf{B} \overset{Q}{\longrightarrow} \mathbf{B'}$$

$$F' \big\uparrow\big\downarrow G' \quad F'' \big\uparrow\big\downarrow G'' \quad \text{and} \quad T\big\uparrow\big\downarrow U \quad T'\big\uparrow\big\downarrow U' \quad ,$$

$$\mathbf{B'} \overset{Q'}{\longrightarrow} \mathbf{B''} \qquad\qquad \mathbf{C} \overset{R}{\longrightarrow} \mathbf{C'}$$

we can compose frame 5.6.(1) laterally with the frame at left, and transversally with the frame at right, by putting the frames together along the common piece, taking this piece out, and composing side pieces. These composition can be extended to adjoint natural transformations as follows.

Proposition. *Let* $\lambda \dashv \mu$ *for the frame* (1) *of 5.6. If* $\lambda' : F''Q' \to P'F'$ *and* $\mu' : Q'G' \to G''P'$ *are adjoint for the frame above at left, then* $P'\lambda \circ \lambda'Q$ *and* $\mu'P \circ Q'\mu$ *are adjoint for the lateral composition of* (1) *with this frame. If* $\rho : T'R \to QT$ *and* $\sigma : RU \to U'Q$ *are adjoint for the frame above at right, then* $\lambda T \circ F'\rho$ *and* $U'\mu \circ \sigma G$ *are adjoint for the transversal composition of* (1) *and this frame.*

PROOF. Let $f : FB \to A$ and $g : A \to GB$ be adjoint for $F \dashv G$. Then $Pf \circ \lambda_B$ and $\mu_A \circ Qg$ are adjoint for $F' \dashv G'$, and hence $P'Pf \circ P'\lambda_B \circ \lambda'_{QB}$ and $\mu'_{PA} \circ Q'\mu_A \circ Q'Qg$ for $F'' \dashv G''$. This proves the first part.

For the second part, let $f : FTC \to A$ and $h : C \to UGA$ be adjoint for $FT \dashv GU$, with f adjoint to $g : TC \to GA$ for $F \dashv G$, and g to h for $T \dashv U$. Then $Pf \circ \lambda_{TC} \circ F'\rho_C$ and $\mu_A \circ Qg \circ \rho_C$ are adjoint for $F' \dashv G'$, and the latter is adjoint to $U'\mu_A \circ \sigma_{GA} \circ Rh$ for $T' \dashv U'$.

5.8. Remark. The compositions just defined, for frames and for pairs of adjoint natural transformations, satisfy the associative and identity laws one expects, and in addition "middle interchange laws" of the form

$$(u \circ v) * (x \circ y) = (u * x) \circ (v * y),$$

for a two by two matrix of frames or pairs of adjoint natural transformations. Proofs are straightforward; we omit all details.

6. Limits and Colimits

6.1. Diagrams. A *diagram* D in a category $\mathbf{C}$, with a directed graph Δ as domain, is a homomorphism from Δ to the underlying graph of $\mathbf{C}$. By 4.5, we can also view a diagram with domain Δ as a functor, from the free category $F\Delta$ to $\mathbf{C}$. This allows us to generalize: we shall view a diagram in $\mathbf{C}$ as a *functor* $D : \Delta \to \mathbf{C}$, where Δ can be any category. We shall say that D is *finite* or *small* if Δ is finite or small. With diagrams as functors, a *morphism of diagrams* is simply a natural transformation.

Diagrams in a category $\mathbf{C}$, with a given domain Δ, are the objects of a category, with morphisms of diagrams as its morphisms. We denote this category by $[\Delta, \mathbf{C}]$, or by $\mathbf{C}^\Delta$.

We shall often use the language of graphs for diagrams, saying that Di is a *vertex* of a diagram $D : \Delta \to \mathbf{C}$ if i is a vertex of Δ, and that Du is an *arrow* of D, with *source* Di and *target* Dj, if $u : i \to j$ in Δ.

6.2. Cones. For categories Δ and $\mathbf{C}$, we associate with every object A of $\mathbf{C}$ a *constant diagram* $K_A : \Delta \to \mathbf{C}$, with $K_A i = A$ for every vertex i of Δ, and $K_A u = \mathrm{id}_A$ for every arrow u of Δ. Every morphism $f : A \to B$ of $\mathbf{C}$ then induces a morphism $K_f : K_A \to K_B$ of constant diagrams. Thus constant diagrams determine a functor, from $\mathbf{C}$ to the category $\mathbf{C}^\Delta$ of all Δ-diagrams in $\mathbf{C}$. By *abus de langage*, we often denote a diagram K_A just by A.

We define a *cone* in $\mathbf{C}$ as a morphism $D_1 \to D_2$ of diagrams in $\mathbf{C}$ with a constant domain or a constant codomain. Morphisms of diagrams with constant codomain are also called *cocones*.

6.3. Limits and colimits. We define a *limit* of a diagram $D : \Delta \to \mathbf{C}$ in a category $\mathbf{C}$ as a couniversal pair (L, λ) for the constant diagram functor

$K : \mathbf{C} \to \mathbf{C}^{\Delta}$. Thus a limit of a diagram D consists of an object L of $\mathbf{C}$ and a cone $\lambda : L \to D$, with the universal property that for every cone $\alpha : A \to D$ with codomain D, there is a unique morphism $f : A \to L$ in $\mathbf{C}$ such that $\alpha = \lambda \circ f$, i.e. $\alpha_i = \lambda_i \circ f$ for every vertex i of the domain Δ of D. Dually, a *colimit* of D is a universal pair (R, ρ) for the constant diagram functor for Δ-diagrams in $\mathbf{C}$. Colimits in $\mathbf{C}$ are limits in $\mathbf{C}^{\mathrm{op}}$.

The components λ_i of a limit cone are called the *projections* of the limit, and the components of a colimit cone are called the *injections* of the colimit.

6.4. Products and coproducts. A family $(A_i)_{i \in I}$ of objects of a category $\mathbf{C}$ can be regarded as a diagram $A : I \to \mathbf{C}$ in $\mathbf{C}$, with a discrete domain. A limit P of this diagram is called a *product* of the objects A_i of $\mathbf{C}$, and the components π_i of a limit cone $\pi : P \to A$ are called the *projections* of the product. The product has the universal property that for every family $f_i : X \to A_i$ in $\mathbf{C}$, with a common domain X, there is a unique morphism $f : X \to P$ in $\mathbf{C}$ such that $f_i = \pi_i \circ f$ for every $i \in I$. A product of objects A_i is often denoted by $\prod A_i$, and a product of two objects A and B by $A \times B$.

A product of an empty family is called a *terminal object*. A terminal object T of $\mathbf{C}$ has the universal property that for every object A of $\mathbf{C}$, there is exactly one morphism $A \to T$ in $\mathbf{C}$.

If $A \times B$ is a product with projectiosn $A \xleftarrow{p} A \times B \xrightarrow{q} B$, then we denote by $\langle f, g \rangle : X \to A \times B$ the morphism determined by

$$ p \circ \langle f, g \rangle = f, \qquad q \circ \langle f, g \rangle = g, $$

for morphisms $f : X \to A$ and $g : X \to B$ with common domain X.

Coproducts and their injections are dual to products and their projections. The coproduct of an empty family is called an *initial object*.

Product sets $A \times B$ and $\prod A_i$ are products in the categorical sense. For a concrete category $\mathbf{C}$, products in $\mathbf{C}$ are usually constructed as products of underlying sets, with a suitable structure. The coproduct $\coprod_{i \in I} A_i$ of a family of sets is their disjoint union, constructed as the set of all pairs (i, x) with $i \in I$ and $x \in A_i$.

6.5. Pullbacks and pushouts. A commutative square

$$
\begin{array}{ccc}
P & \xrightarrow{v} & B \\
\downarrow{\scriptstyle u} & & \downarrow{\scriptstyle g} \\
A & \xrightarrow{f} & Q
\end{array}
$$

(1)

in a category $\mathbf{C}$ is called a *pullback square* in $\mathbf{C}$ if the morphisms u, v, and $fu = gv$ form a limit cone for the diagram $A \xrightarrow{f} Q \xleftarrow{g} B$ in $\mathbf{C}$. This means that for every commutative square $fx = gy$ in $\mathbf{C}$, there is exactly one morphism t in $\mathbf{C}$ such that $x = ut$ and $y = vt$ in $\mathbf{C}$.

For sets and mappings, we construct a pullback square (1) by letting P be the set of all pairs (x, y) in $A \times B$ such that $f(x) = g(y)$, with $u(x, y) = x$ and $v(x, y) = y$.

Dually, we say that (1) is a *pushout square* if the morphisms f, g and $fu = gv$ form a colimit cone for the diagram $A \xleftarrow{u} P \xrightarrow{v} B$ in $\mathbf{C}$.

A product $P = A \times B$ can be constructed as a pullback (1), with Q a terminal object, and a coproduct $Q = A \amalg B$ as a pushout (1), with P an initial object.

6.6. Equalizers and coequalizers. If a diagram $A \underset{g}{\overset{f}{\rightrightarrows}} B$ of parallel arrows in a category $\mathbf{C}$ has a limit E in $\mathbf{C}$, then the morphism $e : E \to A$ of the limit cone is called an *equalizer* of f and g in $\mathbf{C}$. This means that $fe = ge$, and that for every morphism $x : X \to A$ in $\mathbf{C}$ with $fx = gx$, there is a unique morphism $y : X \to E$ in $\mathbf{C}$ such that $x = ey$. A coequalizer of f and g is defined dually.

For sets, an equalizer can always be constructed as a subset inclusion $S \to A$, with S consisting of all $x \in A$ such that $f(x) = g(x)$.

6.7. Other limits and colimits. If a category $\mathbf{C}$ has a terminal object and products $A \times B$, then it is easily seen that $\mathbf{C}$ has all finite products, i.e. every finite family of objects of $\mathbf{C}$ has a product in $\mathbf{C}$. Limits of finite diagrams can be constructed from finite products (including a terminal object) and equalizers, or from a terminal object and pullbacks. If $\mathbf{C}$ has small products, and either pullbacks or equalizers, then $\mathbf{C}$ has limits for all small diagrams. We omit the proofs, and the dual statements for colimits.

7. Properties of Limits and Colimits

7.1. Limit and colimit functors. Every property of limits obtained in this Section has a dual property for colimits. We do not always state these dual properties, and we do not prove them.

Since a limit cone for a diagram is couniversal for a constant diagram functor, a category $\mathbf{C}$ has limits for all diagrams with a given domain Δ if and only if the constant diagram functor, from $\mathbf{C}$ to the diagram category $\mathbf{C}^\Delta$, has a right adjoint. The counit for this adjunction then consists of limit cones, and the unit of isomorphisms. Dually, every diagram with domain Δ has a colimit in **C** if and only if the constant diagram functor $K : \mathbf{C} \to \mathbf{C}^\Delta$ has a left adjoint, with a unit consisting of colimit cones.

7.2. The following result is important and very useful.

Theorem. *Right adjoint functors preserve all limits. Dually, left adjoint functors preserve all colimits.*

PROOF. Let $F \dashv G : \mathbf{A} \to \mathbf{B}$. We must show that if $\lambda : L \to D$ is a limit cone in **A**, then the cone $G\lambda : GL \to GD$ in **B** is also a limit cone. Let Δ be the domain of D.

A cone $\sigma : B \to GD$ consists of morphisms $\sigma_i : B \to GDi$, one for every vertex i of Δ. By naturality of the adjunction, the adjoint morphisms $\rho_i : FB \to Di$ of **A** form a cone $\rho : FB \to D$. Thus $\rho = \lambda \circ u$ for a unique morphism $u : FB \to L$. Since $\rho_i = \lambda_i \circ u$ iff $\sigma_i = G\lambda_i \circ v$ for the adjoint morphism $v : B \to GL$, the morphism v adjoint to u is the unique morphism v of **B** such that $\sigma = G\lambda \circ v$, and $G\lambda$ is indeed a limit cone.

7.3. Proposition. (i) *If*

$$
\begin{array}{ccc}
\cdot & \xrightarrow{\;f'\;} & \cdot \\
{\scriptstyle m'}\downarrow & & \downarrow{\scriptstyle m} \\
\cdot & \xrightarrow{\;f\;} & \cdot
\end{array}
$$

is a pullback square with m monomorphic, then m' is monomorphic.

(ii) *If the righthand square of a commutative diagram*

$$
(1) \qquad
\begin{array}{ccccc}
\cdot & \xrightarrow{\;f'\;} & \cdot & \xrightarrow{\;g'\;} & \cdot \\
{\scriptstyle u}\downarrow & & \downarrow{\scriptstyle v} & & \downarrow{\scriptstyle w} \\
\cdot & \xrightarrow{\;f\;} & \cdot & \xrightarrow{\;g\;} & \cdot
\end{array}
$$

is a pullback square, then the lefthand square is a pullback square if and only if the "outer rectangle"

$$
(2) \qquad
\begin{array}{ccc}
\cdot & \xrightarrow{\;g'f'\;} & \cdot \\
{\scriptstyle u}\downarrow & & \downarrow{\scriptstyle w} \\
\cdot & \xrightarrow{\;gf\;} & \cdot
\end{array}
$$

of (1) is a pullback square.

PROOF. For (i), if $m'u = m'v$, then also $mf'u = mf'v$, hence $f'u = f'v$. But then $u = v$ by the definition of a pullback, and m' is monomorphic.

For (ii), if the righthand square in (1) is a pullback, and if fp is defined in $\mathbf{C}$, then $q = g'r$ gives a bijection between morphisms q such that $wq = gfp$, and morphisms r with $vr = fp$. In this situation, if $p = ut$, then $q = g'f't$ iff $r = f't$. It follows that (2) is a pullback square iff the lefthand square in (1) is a pullback square.

7.4. Proposition. *If a morphism $\mu : D_1 \to D_2$ of diagrams in a category* $\mathbf{C}$ *has all components monomorphic in* $\mathbf{C}$*, and if $\lambda_1 : L_1 \to D_1$ and $\lambda_2 : L_2 \to D_2$ are limit cones, then the morphism $m : L_1 \to L_2$ for which $\mu \circ \lambda_1 = \lambda_2 \circ m$ is monomorphic in* $\mathbf{C}$.

PROOF. The morphism m exists because $\mu\lambda_1$ is a cone and λ_2 a limit cone. If $mu = mv$ in $\mathbf{C}$, then $\mu\lambda_1 u = \mu\lambda_1 v$, and hence $\lambda_1 u = \lambda_1 v$. But then $u = v$ by the definition of a limit; thus m is monomorphic.

7.5. Limits and colimits in functor categories. Limits and colimits in a functor category $[\mathbf{A}, \mathbf{B}]$ are usually constructed "object by object" as follows. For $D : \Delta \to [\mathbf{A}, \mathbf{B}]$ and an object A of $\mathbf{A}$, we have an object $(Di)A$ of $\mathbf{B}$ for every vertex i of Δ, and a morphism $(D\alpha)_A : (Di)A \to (Dj)A$ for every arrow $\alpha : i \to j$ of Δ. These objects and morphisms define a diagram $DA : \Delta \to \mathbf{B}$. For $f : A \to B$ in $\mathbf{A}$, the morphisms $(Di)f$ of $\mathbf{B}$, for the vertices i of Δ, define a morphism $Df : DA \to DB$ of diagrams in $\mathbf{B}$ with domain Δ.

If every diagram DA admits a limit cone $\lambda_A : LA \to DA$ in $\mathbf{B}$, then for every morphism $f : A \to B$ in $\mathbf{A}$, there is a unique morphism $Lf : LA \to LB$ in $\mathbf{B}$ such that $Df \circ \lambda_A = \lambda_B \circ Lf$. In this situation, the objects LA and morphisms Lf define a functor $L : \mathbf{A} \to \mathbf{B}$, and the cones λ_A are components of a cone $\lambda : L \to D$. This cone is a limit cone for the given diagram D.

Colimits for $[\mathbf{A}, \mathbf{B}]$ are constructed dually; we omit the details.

7.6. Limits in CAT. A diagram of categories and functors can be regarded as a diagram in a suitable category CAT (see 2.7). Thus limits and colimits for such diagrams can be defined.

To construct the limit in CAT of a small diagram D of categories and functors, we observe first that the forgetful functor from categories to directed graphs has a left adjoint by 4.5; thus it preserves limits by 7.2. Directed graphs can be regarded as diagrams with a domain $\cdot \rightrightarrows \cdot$; thus we can use 7.5 to construct a limit cone $\lambda : \mathrm{Lim}\, D \to D$ for graphs. This is also a limit cone for categories, with identity morphisms and composition in $\mathrm{Lim}\, D$ determined uniquely by the fact that the projections λ_i of the limit cone must be functors.

Product categories are an example. The *terminal category* **1** has one object, one identity morphism, and no other morphisms. For categories **A** and **B**, the *product category* $\mathbf{A} \times \mathbf{B}$ has as objects all pairs (A, B) with A an object of **A** and B an object of **B**. Morphisms $(f, g) : (A, B) \to (A', B')$ are pairs, consisting of a morphism $f : A \to A'$ of **A**, and a morphism $g : B \to B'$ of **B**. Identity mosprhisms are pairs $(\mathrm{id}_A, \mathrm{id}_B)$, and composition is given by $(f', g') \circ (f, g) = (f' \circ f, g' \circ g)$.

7.7. Product functors. If a category **C** has products $A \times B$ of objects, then these products define a *product functor* from $\mathbf{C} \times \mathbf{C}$ to **C**. This functor assigns to a pair (A, B) of objects of **C** their product $A \times B$, and to a pair of morphisms $f : A \to A'$ and $g : B \to B'$ the morphism

$$f \times g : A \times B \to A' \times B'$$

of **C** determined by the equations

$$p' \circ (f \times g) = f \circ p, \qquad q' \circ (f \times g) = f \circ q,$$

for the projections $A \xleftarrow{p} A \times B \xrightarrow{q} B$ and $A' \xleftarrow{p'} A' \times B' \xrightarrow{q'} B'$ of the products $A \times B$ and $A' \times B'$.

Using product functors in a category of categories and in functor categories, we get products of functors and of natural transformations.

7.8. Dense functors. If $G : \mathbf{A} \to \mathbf{B}$ is a functor and B an object of **B**, then pairs (A, u), with A an object of **A** and $u : GA \to B$ in **B**, are objects of a category $G \downarrow B$, with morphisms $f : (A', u') \to (A, u)$ for $f : A' \to A$ in **A** with $u' = u \cdot Gf$ in **B**. Putting $D_{G \downarrow B} f = f : A' \to A$ for $f : (A', u') \to (A, u)$ then defines a *domain functor* for $G \downarrow B$. We say that G is a *dense functor*, or that G is *colimit-dense*, if for objects B, B' of **B** and a cone $\lambda : GD_{G \downarrow B} \to B'$, there is always a unique morphism $g : B \to B'$ in **B** such that $\lambda_{A,u} = gu$ in **B** for every object (A, u) of $G \downarrow B$.

Proposition. *A full and faithful dense functor preserves all limits.*

PROOF. Let $G : \mathbf{A} \to \mathbf{B}$ be the functor, and $\mu : M \to D$ a limit cone in **A**. For a cone $\rho : B \to GD$ in **B** and an object (A, u) of $G \downarrow B$, we have $\rho \cdot u = G\sigma$ for a unique cone $\sigma : A \to D$ in **A**, and $\sigma = \mu \cdot \lambda_{A,u}$ for a unique morphism $\lambda_{A,u} : A \to M$ in **A**. Since G is full and faithful and μ a limit cone, we have $\rho = G\mu \cdot g$ for $g : B \to GM$ in **B** iff $gu = G\lambda_{A,u}$ for every object (A, u) of $G \downarrow B$. Since G is dense and the $G\lambda_{A,u}$ form a cone with domain $GD_{G \downarrow B}$, this determines g uniquely.

8. Monads and Comonads

8.1. Definitions. A *monad* on a category $\mathbf{C}$ is a triple (T, η, μ) consisting of a functor $T : \mathbf{C} \to \mathbf{C}$ and natural transformations $\eta : \mathrm{Id}\,\mathbf{C} \to T$ and $\mu : T \circ T \to T$ which satisfy the conditions

$$T\eta \circ \mu = \mathrm{id}_T = \eta T \circ \mu,$$

and
$$\mu \circ \mu T = \mu \circ T\mu,$$

in the notation of 3.4. The natural transformations η and μ are called the *unit* and the *multiplication* of the monad.

A *comonad* on $\mathbf{C}$ is the dual of a monad on $\mathbf{C}^{\mathrm{op}}$, a triple (G, ε, ψ) consisting of a functor $G : \mathbf{C} \to \mathbf{C}$ and natural transformations $\varepsilon : G \to \mathrm{Id}\,\mathbf{C}$ and $\psi : G \to G \circ G$ which satisfy

$$G\varepsilon \circ \psi = \mathrm{id}_G = \varepsilon G \circ \psi,$$

and
$$G\psi \circ \psi = \psi G \circ \psi.$$

The natural transformations ε and ψ are called the *counit* and the *comultiplication* of the comonad.

Monads and comonads are called *triples* and *cotriples* by some authors.

8.2. Example. The data and formal laws for a monad are analogous to the data and formal laws for a monoid, hence the name "monad". This analogy is strengthened by the following example.

For a monoid M, we have a functor $T = M \times -$ on sets, with $TA = M \times A$ for a set A, and $Tf = \mathrm{id}_M \times f$ for a mapping f. We define $\eta_A : A \to TA$ by putting $\eta_A(x) = (e, x)$ for $x \in A$, with e the neutral element of M, and we define $\mu_A : TTA \to TA$ by putting $\mu_A(a, (b, x)) = (ab, x)$ for a, b in M and $x \in A$, using the composition ab in M. We omit the easy verifications that this defines a monad on sets.

8.3. Theorem. *If* $F \dashv U : \mathbf{A} \to \mathbf{B}$ *is an adjunction, with unit* η *and counit* ε, *then* $(UF, \eta, U\varepsilon F)$ *is a monad on* $\mathbf{B}$, *and* $(FU, \varepsilon, F\eta U)$ *is a comonad on* $\mathbf{A}$.

The monad and the comonad thus obtained are called the *induced monad* and the *induced comonad* of the adjunction.

PROOF. We obtain the monadic laws

$$U\varepsilon F \circ UF\eta = \mathrm{id}_{UF} = U\varepsilon F \circ \eta UF$$

for the monad $UF, \eta, U\varepsilon F$ by composing the identities

$$\varepsilon F \circ F\eta = \mathrm{id}_F \qquad \text{and} \qquad U\varepsilon \circ \eta U = \mathrm{id}_U$$

of 4.4 with U from the left and F from the right, using the formal laws of 3.4. The third monadic identity follows in the same way from the special case

$$\varepsilon \circ FU\varepsilon = \varepsilon \circ \varepsilon FU$$

of the last formal law of 3.4.

The second part of the Theorem is dual to the first part.

8.4. Algebras and coalgebras. An *algebra* for a monad $\mathcal{T} = (T, \eta, \mu)$ on a category $\mathbf{C}$ is defined as a pair (A, α), consisting of an object A of $\mathbf{C}$ and a morphism $\alpha : TA \to A$ which satisfies the two formal laws

$$\alpha \circ \eta_A = \mathrm{id}_A \qquad \text{and} \qquad \alpha \circ \mu_A = \alpha \circ TA.$$

The morphism α is often called the *algebra structure* of (A, α). A *homomorphism of algebras* (A, α) and (B, β) for $\mathcal{T}$ is defined as a morphism $f : A \to B$ of $\mathbf{C}$ which satisfies the law

$$f \circ \alpha = Tf \circ \beta.$$

Dually, a *coalgebra* (C, γ) for a comonad $\mathcal{G} = (G, \varepsilon, \psi)$ on $\mathbf{C}$ consists of an object C of $\mathbf{C}$ and a morphism $\psi : C \to GC$ which satisfies

$$\varepsilon_C \circ \gamma = \mathrm{id}_C \qquad \text{and} \qquad \psi_C \circ \gamma = G\gamma \circ \gamma.$$

A *homomorphism of coalgebras* $f : (B, \beta) \to (C, \gamma)$ is a morphism $f : B \to C$ of $\mathbf{C}$ such that

$$\gamma \circ f = Gf \circ \beta.$$

8.5. Examples. For the monad of 8.2, an algebra (A, α) is interpreted as an action $\alpha : (m, x) \mapsto m \cdot x : M \times A \to A$ of the monoid M on the set A. The formal laws of 8.4 state in this case that

$$e \cdot x = x \qquad \text{and} \qquad a \cdot (b \cdot x) = (ab) \cdot x,$$

for $x \in A$, the neutral element e of M, and a, b in M. An algebra of this kind is called an *M-set*. A homomorphism $f : A \to B$ of M-sets must satisfy $a \cdot f(x) = f(a \cdot x)$, for $x \in A$ and $a \in M$.

Free groups (see 4.1) define a left adjoint of the forgetful functor from groups to sets. Algebras for the monad on sets induced by this adjunction are groups, with TA the underlying set of the free group over A for a set A. If A is the underlying set of a group, then the algebra structure $\alpha : TA \to A$ maps a formal composition in TA into the actual composition in A.

For the monad induced by a forgetful functor from categories to graphs, and its left adjoint free category functor, algebras are categories, and their homomorphisms are functors. Morphisms of a free category are paths, and the algebra structure of a category **C** assigns to a path in **C** the composition of the morphisms in the path.

8.6. Free monadic algebras. Let $\mathcal{T} = (T, \eta, \mu)$ be a monad on a category **C**. We denote by $\mathbf{C}^\mathcal{T}$ the category of $\mathcal{T}$-algebras, defined in 8.4. Putting $U^\mathcal{T}(A, \alpha) = A$ for a $\mathcal{T}$-algebra (A, α) defines a forgetful functor $U^\mathcal{T}$ from $\mathcal{T}$-algebras to **C**.

Two of the formal laws of 8.1 state that (TC, μ_C) is a $\mathcal{T}$-algebra for every object C of **C**. We denote this algebra by $F^\mathcal{T} C$ and call it the *free $\mathcal{T}$-algebra* over C. We obtain a free algebra functor $F^\mathcal{T} : \mathbf{C} \to \mathbf{C}^\mathcal{T}$ by putting $F^\mathcal{T} f = Tf :$ $(TC, \mu_C) \to (T, \mu_{C'})$ for $f : C \to C'$ in **C**. This is a homomorphism of $\mathcal{T}$-algebras by naturality of μ.

Theorem. *For a monad* $\mathcal{T} = (\mathcal{T}, \eta, \mu)$, *the functor* $F^\mathcal{T}$ *is left adjoint to the forgetful functor* $U^\mathcal{T}$, *and the monad induced by this adjunction is* $\mathcal{T}$. *The unit of the adjunction is* η, *and the counit at an algebra* (B, β) *the homomorphism* $\beta : F^\mathcal{T} B \to (B, \beta)$.

PROOF. The adjunction is obtained by

$$f = \beta \circ Tg \quad \Longleftrightarrow \quad g = f \circ \eta_A,$$

for $f : (TA, \mu_A) \to (B, \beta)$ in $\mathbf{C}^\mathcal{T}$ and $g : A \to B$ in **C**, if A is an object of **C** and (B, β) a $\mathcal{T}$-algebra; we omit the easy proof that this works. The unit is obtained by putting $f = \mathrm{id}_{(B,\beta)}$, the counit by $g = \mathrm{id}_B$, and the induced monad clearly is $\mathcal{T}$.

8.7. Theorem. *If* $F \dashv U : \mathbf{A} \to \mathbf{B}$ *is an adjunction, with unit* η *and counit* ε, *and if* $\mathcal{T}$ *is the induced monad on* **B**, *then putting*

$$KA = (UA, U\varepsilon_A),$$

for an object A *of* **A**, *defines a functor* $K : \mathbf{A} \to \mathbf{B}^\mathcal{T}$ *such that* $U^\mathcal{T} K = U$ *and* $KF = F^\mathcal{T}$. *If every pair of morphisms* $\varepsilon_{FB}, F\beta$, *for a* $\mathcal{T}$-algebra (B, β), *has a coequalizer* $FB \to L(B, \beta)$ *in* **B**, *then* K *has a left adjoint, and these coequalizers are adjoint to the unit for an adjunction* $L \dashv K$.

This functor K is called the *comparison functor* for the given data.

PROOF. It is easily verified that KA is a $\mathcal{T}$-algebra for an object A of **A**, and that $Uf : KA \to KA'$ is a homomorphism of $\mathcal{T}$-algebras for $f : A \to A'$ in **A**; we omit the details. Thus K is a functor with $U^\mathcal{T} K = U$. If $A = FB$, then clearly $KA = F^\mathcal{T} B$; thus $KF = F^\mathcal{T}$.

For an algebra (B, β), a morphism $f : B \to GA$ with adjoint $\hat{f} : FB \to A$ satisfies $U\varepsilon_A \cdot UFf = U\hat{f}$. It follows that f is a homomorphism $f : (B, \beta) \to KA$ iff $\hat{f} \cdot F\beta = \hat{f} \cdot \varepsilon_{FB}$, and the second part of the Theorem follows.

8.8. Creation. We say that a functor $F : \mathbf{A} \to \mathbf{B}$ *creates* limits of diagrams with domain Δ if for a diagram D in $\mathbf{A}$ with domain Δ, and for a limit cone $\lambda : B \to FD$ in $\mathbf{B}$, there is exactly one cone $\lambda' : A \to D$ in $\mathbf{A}$ such that $\lambda = F\lambda'$, and this cone is a limit cone in $\mathbf{A}$. Other kinds of creation are defined similarly.

Theorem. *For a monad* $\mathcal{T} = (T, \eta, \mu)$, *the forgetful functor* $U^{\mathcal{T}}$ *for* $\mathcal{T}$-*algebras reflects isomorphisms and creates limits, and* $U^{\mathcal{T}}$ *creates all colimits which* T *preserves.*

PROOF. We prove the last part; the other parts are proved similarly.

Let D be a diagram in $\mathbf{C}^{\mathcal{T}}$, and let $\lambda : U^{\mathcal{T}}D \to C$ be a colimit cone with $T\lambda$ and $TT\lambda$ colimit cones. The algebra structures of the vertices of D provide a morphism of diagrams in $\mathbf{C}$ which we denote by αD. To lift λ to a cone in $\mathbf{C}^{\mathcal{T}}$, we need an algebra structure γ of C such that $\gamma \circ T\lambda = \lambda \circ \alpha D$. Since $T\lambda$ is a colimit cone, this determines γ uniquely. Using colimit cones λ and $TT\lambda$, it is easily seen that γ is an algebra structure of C, and that the unique cone in $\mathbf{C}^{\mathcal{T}}$ which lifts λ is a colimit cone.

8.9. Remarks. The second example in 8.5 is typical for categories of algebras. Thus if $F \dashv U : \mathbf{A} \to \mathbf{B}$ is an adjunction, then we say that $\mathbf{A}$ is *algebraic* over $\mathbf{B}$, or that U is a *monadic functor*, if the comparison functor $K : \mathbf{A} \to \mathbf{B}^{\mathcal{T}}$ for the induced monad is an isomorphism, or at least an equivalence, of categories. A *comonadic functor* is defined dually.

The results of 8.6 and 8.7 can of course be dualized for comonads. A comonad $\mathcal{G} = (G, \varepsilon, \psi)$ has a forgetful functor $U_{\mathcal{G}} : (C, \gamma) \mapsto C$ for coalgebras, and it has *cofree coalgebras* $F_{\mathcal{G}}C = (GC, \psi_C)$ which define an adjunction $U_{\mathcal{G}} \dashv F_{\mathcal{G}}$. If $\mathcal{G}$ is induced by an adjunction $F \dashv U : \mathbf{A} \to \mathbf{B}$, then the comparison functor is given by $KB = (FB, F\eta_B)$, for an object B of $\mathbf{B}$.

8.10. Left adjoint triangles. We consider adjunctions $F \dashv U : \mathbf{C} \to \mathbf{A}$ and $F' \dashv U' : \mathbf{C} \to \mathbf{B}$, with units η and η' and counits ε and ε', and a functor $L : \mathbf{B} \to \mathbf{A}$ such that $FL = F'$.

We say that a pair $A \underset{g}{\overset{f}{\rightrightarrows}} B$ of morphisms of a category $\mathbf{C}$ is *coreflexive* if $tf = \mathrm{id}_A = tg$ in $\mathbf{C}$ for some morphism t.

The first hypothesis of the following result is satisfied in particular if the functor F is comonadic.

Proposition (DUBUC [26]). *If η_a is an equalizer of $UF\eta_a$ and η_{UFa} for every object a of* **A**, *and if* **B** *has equalizers of coreflexive pairs, then the functor L has a right adjoint.*

PROOF. We refer to 3.4 for formal laws used in this proof. Putting $f_1 = \eta_A \cdot f$ defines a bijection between morphisms $f : LB \to A$ of **A**, and morphisms $f_1 : LB \to UFA$ with $UF\eta_A \cdot f_1 = \eta_{UFA} \cdot f_1$. If f_1 is adjoint to $f_2 : FLB \to FA$ for the adjunction $F \dashv U$, then $UF\eta_A \cdot f_1$ and $\eta_{UFA} \cdot f_1$ are adjoint to $F\eta_A \cdot f_2$ and $Ff_1 = FUf_2 \cdot F\eta_{LB}$. If f_2 is adjoint to $f_3 : B \to U'FA$ for the adjunction $F' \dashv U'$, then $F\eta_A \cdot f_2$ and $FUf_2 \cdot F\eta_{LB}$ are adjoint to $U'F\eta_A \cdot f_3$ and to $U'FUf_2 \cdot U'F\eta_{LB} \cdot \eta'_B$. Using $f_2 = \epsilon'_{FA} \cdot F'f_3$, and naturality of units and counits, we have

$$U'FUf_2 \circ U'F\eta_{LB} \circ \eta'_B = U'FU\epsilon'_{FA} \circ U'FUF'f_3 \circ U'F\eta_{LB} \circ \eta'_B$$
$$= U'FU\epsilon'_{FA} \circ U'F\eta_{LU'FA} \circ U'F'f_3 \circ \eta'_B$$
$$= U'FU\epsilon'_{FA} \circ UF\eta_{LU'FA} \circ \eta'_{U'FA} \circ f_3 .$$

We note that $U'F\eta_A$ and $U'FU\epsilon'_{FA} \cdot U'F\eta_{LU'FA} \cdot \eta'U'FA$ have an equalizer $\rho_A : RA \to B$ in **B** since

$$U'\epsilon_{FA} \circ U'FU\epsilon'_{FA} \circ U'F\eta_{LU'FA} \circ \eta'U'FA$$
$$= U'\epsilon'_{FA} \circ U'\epsilon FLU'FA \circ U'F\eta_{LU'FA} \circ \eta'U'FA$$
$$= U'\epsilon'_{FA} \circ \eta'_{U'FA} = \mathrm{id}_{U'FA} = U'\epsilon_{FA} \circ U'F\eta_A .$$

Putting $f_3 = \rho_A \cdot g$ now provides a bijection between morphisms $f : LB \to A$ of **A** and $g : RA \to B$ in **B**.

The bijections used in this construction are clearly natural in B; thus they define the desired right adjoint functor R of the functor L.

9. Cartesian Closed Categories

9.1. Definitions. A category **E** with finite products is called *cartesian closed* if for every pair of objects A and B of **E**, there is an *exponential object* $[A, B]$ and an *evaluation morphism* $\mathrm{ev}_{A,B} : [A, B] \times A \to B$, with the couniversal property that for every morphism $f : X \times A \to B$ of **E**, there is a unique morphism $g : X \to [A, B]$ of **E** such that $f = \mathrm{ev}_{A,B} \circ (g \times \mathrm{id}_A)$ in **E**.

This means that every functor $- \times A$ on **E** has a right adjoint $[A, -]$, with the counit defined by evaluation morphisms $\mathrm{ev}_{A,B}$, and with

$$g \circ \mathrm{ev}_{A,B} = \mathrm{ev}_{A,C} \circ ([A, g] \times \mathrm{id}_A)$$

for $g : B \to C$ in **E**. It follows easily that objects $[A, B]$ produce a functor of two variables, contravariant in A and covariant in B, with $[f, B] : [C, B] \to [A, B]$ determined by

$$\mathrm{ev}_{A,B} \circ ([f, B] \times \mathrm{id}_A) = \mathrm{ev}_{C,B} \circ (\mathrm{id}_{[C,B]} \times f),$$

for $f : A \to C$ in **E**. The resulting contravariant functor $[-, B]$, for an object B of **E**, is easily seen to be self-adjoint on the right.

9.2. Examples. The categories of sets and of finite sets are cartesian closed. In these categories, a set $[A, B] = B^A$ is the set of all functions $f : A \to B$, and $\mathrm{ev}_{A,B}(f, x) = f(x)$ for $f : A \to B$ and $x \in A$. Because of these examples, objects $[A, B]$ are often denoted B^A in cartesian closed categories, and the natural bijection between morphisms $f : X \times A \to B$ and $g : X \to B^A$ is called *exponential adjunction*.

Topoi and quasitopoi are cartesian closed; thus we shall encounter many other examples.

9.3. The cartesian closed structure of CAT. For categories **A, B** and **C**, there is a natural bijection between functors $F : \mathbf{C} \times \mathbf{A} \to \mathbf{B}$ and functors $\Phi : \mathbf{C} \to [\mathbf{A}, \mathbf{B}]$, obtained by letting Φ correspond to F if always

$$(\Phi C)f = F(\mathrm{id}_C \times f),$$

for an object C of **C** and a morphism f of **A**, and

$$(\Phi g)_A = F(g \times \mathrm{id}_A),$$

for a morphism g of **C** and an object A of **A**. This bijection between objects of the categories $[\mathbf{C} \times \mathbf{A}, \mathbf{B}]$ and $[\mathbf{C}, [\mathbf{A}, \mathbf{B}]]$ is in fact part of an isomorphism of the two categories. We omit proofs and further details.

The bijection between functors $F : \mathbf{C} \times \mathbf{A} \to \mathbf{B}$ and functors $\Phi : \mathbf{C} \to [\mathbf{A}, \mathbf{B}]$ is clearly natural in **C**. Thus it defines a cartesian closed structure for categories.

If we restrict ourselves to ordered sets A, B, C, then $C \times A$ and $[A, B]$ also are ordered sets, the latter by 3.5. Thus the cartesian closed structure of CAT induces a cartesian closed structure of the category of ordered sets.

9.4. In the remainder of this Section, we prove an important special case of a result of B.DAY [22].

Theorem. *For a full reflective subcategory* **C** *of a cartesian closed category* **B**, *with reflections* $\rho_B : B \to RB$, *the following are logically equivalent.*

(1) *Every reflection* $\rho_{[B,C]} : [B,C] \to R[B,C]$, *with* C *an object of* **C**, *is an isomorphism.*

(2) *Every morphism* $R(\rho_A \times \mathrm{id}_B)$ *is an isomorphism.*

(3) *Every morphism* $R(\rho_A \times \rho_B)$ *is an isomorphism.*

(4) *The reflector* R *preserves products.*

If these statements are valid, then **C** *is cartesian closed. Conversely, if the full embedding* **C** $\to$ **B** *is dense* (7.8) *and* **C** *cartesion closed, then* (1)–(4) *are valid.*

We call **C** *closed-reflective* in **B** if (1)–(4) are valid.

PROOF. We show in 9.5 that (3) $\iff$ (4), in 9.6 that (1) $\implies$ (2) $\implies$ (3), in 9.7 that (3) $\implies$ (1), and the last part of the Theorem in 9.8.

9.5. We note first that products in **C** are products in **B**. Thus the functor R preserves products iff Rp and Rq always are the projections of a product in **B** for a product $A \xleftarrow{p} A \times B \xrightarrow{q} B$ in **B**. We consider the following commutative diagram in **B**, with p' and q' the projections of the product $RA \times RB$.

$$
\begin{array}{ccccc}
RA & \xleftarrow{\ Rp\ } & R(A \times B) & \xrightarrow{\ Rq\ } & RB \\
\Big\downarrow{\scriptstyle R\rho_A} & & \Big\downarrow{\scriptstyle R(\rho_A \times \rho_B)} & & \Big\downarrow{\scriptstyle R\rho_B} \\
RRA & \xleftarrow{\ Rp'\ } & R(RA \times RB) & \xrightarrow{\ Rq'\ } & RRB
\end{array}
$$

Since R, restricted to **C**, is an equivalence, Rp' and Rq' are the projections of a product, and $R\rho_A = \rho_{RA}$ and $R\rho_B$ are isomorphisms. It follows that Rp and Rq are the projections of a product iff $R(\rho_A \times \rho_B)$ is an isomorphism.

9.6. If (1) is valid, then we have bijections between morphsims

$$
\begin{aligned}
f_1 &: R(A \times B) \to C, & \qquad f_4 &: RA \to [B,C], \\
f_2 &: A \times B \to C, & \qquad f_5 &: RA \times B \to C, \\
f_3 &: A \to [B,C], & \qquad f_6 &: R(RA \times B) \to C,
\end{aligned}
$$

with corresponding morphisms related by

$$
f_1 \circ \rho_{A \times B} = f_2 = \mathrm{ev}_{B,C} \circ (f_3 \times \mathrm{id}_B),
$$

$$
f_3 = f_4 \circ \rho_A, \qquad \mathrm{ev}_{B,C} \circ (f_4 \times \mathrm{id}_B) = f_5 = f_6 \circ \rho_{RA \times B}.
$$

It follows that
$$
f_2 = f_5 \circ (\rho_A \times \mathrm{id}_B),
$$
and
$$
f_1 \circ \rho_{A \times B} = f_6 \circ \rho_{RA \times B} \circ (\rho_A \times \mathrm{id}_B) = f_6 \circ R(\rho_A \times \rho_B) \circ \rho_{A \times B},
$$

by naturality of ρ. Thus we have

$$f_1 = f_6 \circ R(\rho_A \times \mathrm{id}_B).$$

Since this defines a bijection between morphisms f_1 and f_6, we must have (2). If (2) is valid, then $R(\mathrm{id}_A \times \rho_B)$ is an isomorphism by symmetry. Thus

$$R(\rho_A \times \rho_B) = R(\rho_A \times \mathrm{id}_{RB}) \circ R(\mathrm{id}_A \times \rho_B)$$

is an isomorphism, and (3) is valid.

9.7. Put $r = \rho_{[B,C]}$. We only need $\hat{s} : R[B,C] \to [B,C]$ such that $\hat{s} \circ r = \mathrm{id}_{[B,C]}$, for then also $r \circ \hat{s} = \mathrm{id}_{R[B,C]}$, by the universal property of r. Exponential adjunction replaces $\hat{s}$ by $s : R[B,C] \times B \to C$ which must satisfy $s \circ (r \times \mathrm{id}_B) = \mathrm{ev}_{B,C}$.

There is $e : R([B,C] \times B) \to C$ such that

$$\mathrm{ev}_{B,C} = e \circ \rho_{[B,C] \times B},$$

and we have

$$R(r \times \rho_B) \circ \rho_{[B,C] \times B} = \rho_{R[B,C] \times RB} \circ (r \times \rho_B)$$

by naturality of ρ. It follows that

$$s = e \circ (R(r \times \rho_B))^{-1} \circ \rho_{R[B,C] \times RB} \circ (\mathrm{id}_{R[B,C]} \times \rho_B)$$

satisfies $s \circ (r \times \mathrm{id}_B) = \mathrm{ev}_{B,C}$ if (3) is valid.

9.8. If (1) is valid, then objects $R[A,C]$, for objects A and C of $\mathbf{C}$, clearly provide a cartesian closed structure of $\mathbf{C}$.

Conversely, if $\mathbf{C}$ is cartesian closed with exponential objects $H(A,C)$, then for objects X and C of $\mathbf{C}$ and B of $\mathbf{B}$, we have bijections, natural in X, between morphisms

$$u_1 : X \to H(RB,C), \qquad u_4 : B \to [X,C],$$
$$u_2 : X \times RB \to C, \qquad u_5 : X \times B \to C,$$
$$u_3 : RB \to [X,C], \qquad u_6 : X \to [B,C].$$

If the embedding $\mathbf{C} \to \mathbf{B}$ is dense, then corresponding morphisms u_1 and u_6 satisfy $u_6 = h u_1$ and $u_1 = k u_6$ for unique morphisms h and k. By unicity, h and k are inverse isomorphisms. Now (1) is valid since $[B,C]$ is isomorphic to an object of $\mathbf{C}$.

10. Diagonal Polarity

10.1. Diagonally polar pairs. Let $\mathcal{E}$ and $\mathcal{M}$ be classes of morphisms in a category $\mathbf{C}$. We say that $(\mathcal{E}, \mathcal{M})$ is a *diagonally polar pair* in $\mathbf{C}$ if for every commutative square $mf = ge$ in $\mathbf{C}$ with $e \in \mathcal{E}$ and $m \in \mathcal{M}$, there is a unique "diagonal" morphism t in $\mathbf{C}$ such that $f = te$ and $g = mt$ in $\mathbf{C}$.

We note that diagonal polarity is self-dual; a pair $(\mathcal{E}, \mathcal{M})$ is diagonally polar in $\mathbf{C}$ if and only if the pair $(\mathcal{M}, \mathcal{E})$ is diagonally polar in $\mathbf{C}^{\mathrm{op}}$.

10.2. For classes $\mathcal{E}$ and $\mathcal{M}$ of morphisms of $\mathbf{C}$, we denote by $D_* \mathcal{E}$ the class of all morphisms m of $\mathbf{C}$ such that $(\mathcal{E}, \{m\})$ is a diagonally polar pair in $\mathbf{C}$, and by $D^* \mathcal{M}$ the class of all morphisms e of $\mathbf{C}$ such that $(\{e\}, \mathcal{M})$ is a diagonally polar pair. It follows that

$$(1) \qquad \mathcal{M} \subset D_* \mathcal{E} \iff \mathcal{E} \subset D^* \mathcal{M};$$

these inclusions hold iff $(\mathcal{E}, \mathcal{M})$ is a diagonally polar pair.

The assignments D^* and D_* clearly reverse inclusions; thus it follows from (1) that

$$(2) \qquad \begin{aligned} \mathcal{M} &\subset D_* D^* \mathcal{M} & \text{and} \quad D^* \mathcal{M} &= D^* D_* D^* \mathcal{M}, \\ \mathcal{E} &\subset D^* D_* \mathcal{E} & \text{and} \quad D_* \mathcal{E} &= D_* D^* D_* \mathcal{E}, \end{aligned}$$

for all classes $\mathcal{M}$ and $\mathcal{E}$ of morphisms of $\mathbf{C}$. It follows that if $(\mathcal{E}, \mathcal{M})$ is a diagonally polar pair, then the larger classes $D^* D_* \mathcal{E}$ and $D_* D^* \mathcal{M}$ also form a diagonally polar pair.

Classes $D_* \mathcal{E}$ and $D^* \mathcal{M}$ have dual properties; we shall obtain properties of classes $D_* \mathcal{E}$.

10.3. Proposition. *For a class $\mathcal{E}$ of morphisms of $\mathbf{C}$, the class $D_* \mathcal{E}$ has the following properties.*

(1) *$D_* \mathcal{E}$ contains all isomorphisms of $\mathbf{C}$, and $\mathcal{E} \cap D_* \mathcal{E}$ consists of isomorphisms of $\mathbf{C}$.*

(2) *If $m_1 m$ is defined in $\mathbf{C}$ with $m_1 \in D_* \mathcal{E}$, then $m_1 m \in D_* \mathcal{E}$ if and only if $m \in D_* \mathcal{E}$.*

(3) *If*

$$\begin{array}{ccc} \cdot & \xrightarrow{\;v\;} & \cdot \\ {\scriptstyle m'}\downarrow & & \downarrow{\scriptstyle m} \\ \cdot & \xrightarrow{\;u\;} & \cdot \end{array}$$

is a pullback square in $\mathbf{C}$ with $m \in D_* \mathcal{E}$, then $m' \in D_* \mathcal{E}$.

(4) *If a morphism* $\mu : D \to D'$ *of diagrams in* **C** *has all components in* $D_*\mathcal{E}$, *and if* $\kappa : K \to D$ *and* $\lambda : L \to D'$ *are limit cones, then the morphism* $m : K \to L$ *for which* $\mu \circ \kappa = \lambda \circ m$ *is in* $D_*\mathcal{E}$.

PROOF. The first part of (1) follows immediately from the definitions. For $u : A \to B$ in $\mathcal{E} \cap D_*\mathcal{E}$, there is $t : B \to A$ so that $tu = \mathrm{id}_A$ and $ut = \mathrm{id}_B$, using the square $u \cdot \mathrm{id}_A = \mathrm{id}_B \cdot u$; thus u is an isomorphism.

We note for (2) that if $m_1 \in D_*\mathcal{E}$, then $m_1 mf = ge$, with $e \in \mathcal{E}$, iff $mf = he$ with $g = m_1 h$, and then g and h determine each other. In this situation, $f = te$ and $g = m_1 mt$ iff $f = te$ and $h = mt$. It follows that $m_1 m \in D_*\mathcal{E}$ iff $m \in D_*\mathcal{E}$.

For the pullback square in (3), t with $m't = g$ and s with $ms = ug$ determine each other by $vt = s$. If $m'f = ge$ in this situation, then $te = f$ iff $se = vf$. If $e \in \mathcal{E}$, this determines s and hence t uniquely; thus $m' \in D_*\mathcal{E}$.

For (4), let $mf = ge$ with $e \in \mathcal{E}$. Then $\mu_i \kappa_i f = \lambda_i ge$ for each vertex i of the common domain of D and D', and it follows that $\tau_i e = \kappa_i f$ and $\mu_i \tau_i = \lambda_i g$ for a unique τ_i in **C**. By their unicity, the morphisms τ_i form a cone; thus $\tau e = \kappa f$ and $\mu \tau = \lambda g$ for a unique cone $\tau : L \to D$. We have $te = f$ and $mt = g$ for a morphism t iff

$$\kappa t e = \kappa f = \tau e \qquad \text{and} \qquad \mu \kappa t = \lambda mt = \lambda g = \mu \tau ,$$

and hence iff $\kappa t = \tau$. Since κ is a limit cone, this determines t uniquely, and thus $m \in D_*\mathcal{E}$.

10.4. Examples. If $\mathcal{E}$ consists of isomorphisms, then $D_*\mathcal{E}$ is the class of all morphisms of **C**, and $D^*D_*\mathcal{E}$ the class of all isomorphisms of **C**.

If $\mathcal{E}$ is the class of all epimorphisms of **C**, then a monomorphism in $D_*\mathcal{E}$ is called a *strong monomorphism* of **C**. It follows from 7.3.(1) and 7.4 that the class of strong monomorphisms has the four properties of 10.3. It is easily seen that $D_*\mathcal{E}$ is the class of all strong monomorphisms if **C** has coequalizers.

Dually, a *strong epimorphism* is an epimorphism in $D^*\mathcal{M}$, for $\mathcal{M}$ the class of all monomorphisms. We note that $D^*\mathcal{M}$ always consists of epimorphisms if **C** has equalizers and $\mathcal{M}$ contains all equalizers.

We refer to C.M. RINGEL [90] for more examples and a fuller treatment of diagonal polarity.

10.5. Proposition. *Right adjoint functors preserve strong monomorphisms, and left adjoint functors preserve strong epimorphisms.*

We omit the easy proof.

11. Concrete and Topological Categories

11.1. Concrete categories generalized. We define a *concrete category* over a category $\mathbf{E}$ as a pair $(\mathbf{A}, p)$ consisting of a category $\mathbf{A}$ and a faithful and amnestic functor $p : \mathbf{A} \to \mathbf{E}$. The functor p is called *amnestic* if every isomorphism u in a fibre of p, i.e. with $pu = \mathrm{id}_A$ for an object A of $\mathbf{E}$, is an identity morphism in $\mathbf{A}$. We say that $(\mathbf{A}, p)$ and p are *transportable* if every isomorphism $h : pA \to E$ of $\mathbf{E}$, for an object A of $\mathbf{A}$, lifts to a (unique) isomorphism $h : A \to B$ of $\mathbf{A}$, with $pB = E$.

We shall follow the usual *abus de langage* of using the same notation for a morphism f of $\mathbf{A}$ and the underlying morphism pf of $\mathbf{E}$, and of calling $\mathbf{A}$ concrete over $\mathbf{E}$ if the functor p is understood from the context. If $\mathbf{A}$ is concrete over $\mathbf{E}$, then the objects of $\mathbf{A}$ can be considered as objects of $\mathbf{E}$ provided with a structure of some kind, and the morphisms $f : X \to Y$ of $\mathbf{A}$ as morphisms $f : pX \to pY$ of $\mathbf{E}$ which "preserve structure".

Examples abound. Thus the category of rings is concrete over commutative groups and over monoids as well as over sets, and topological groups are concrete over groups and over topological spaces. A concrete category over the terminal category $\mathbf{1}$ (7.6) is an ordered class. Concreteness is now self-dual: $(\mathbf{A}^{\mathrm{op}}, p^{\mathrm{op}})$ is concrete over $\mathbf{E}^{\mathrm{op}}$ if $(\mathbf{A}, p)$ is concrete over $\mathbf{E}$.

11.2. Sources and sinks. We define a *source* for a faithful functor $p : \mathbf{A} \to \mathbf{E}$, at an object E of $\mathbf{E}$, as a collection of pairs (A_i, f_i), consisting of an object A_i an object of $\mathbf{A}$ and a morphism $f_i : E \to pA_i$ of $\mathbf{E}$. The collection may be large; we do not require that the pairs (A_i, f_i) can be indexed by a set. By *abus de langage*, we usually just speak of a source of morphisms $f_i : E \to pA_i$, with p, E, and the objects A_i, understood from the context.

An *initial lift* (A, h) for a source of morphisms $f_i : E \to pA_i$ consists of an object A of $\mathbf{A}$ and an isomorphism $h : pA \to E$ of $\mathbf{E}$, with the property that a morphism $g : pC \to pA$ in $\mathbf{E}$, for an object C of $\mathbf{A}$, is the underlying morphism of, or lifts to, a morphism $g : C \to A$ in $\mathbf{A}$ if and only if each morphism $f_i hg$ lifts to $f_i hg : C \to A_i$ in $\mathbf{A}$. It follows that $f_i h : A \to A_i$ in $\mathbf{A}$ for every f_i; take $g = \mathrm{id}_{pA}$.

An initial lift (A, h) for a source is determined by the source up to isomorphism. If p is transportable, then we can and always shall assume that $h = \mathrm{id}_E$. With this convention, the initial lift becomes unique if p is amnestic.

We define sinks and their final lifts dually. A *sink* for p is dual to a source for the functor p^{op}, and a *final lift* for a sink of morphisms $f_i : pB_i \to E$ consists of an object B of $\mathbf{A}$ and an isomorphism $h : E \to pB$ of $\mathbf{E}$, so that every morphisms $g : pB \to pC$ of $\mathbf{E}$ with $ghf_i : B_i \to C$ in $\mathbf{A}$ for every f_i lifts to $g : B \to C$ in $\mathbf{A}$.

11.3. Topological categories. We define a *topological category* (T, p) over a category E as a transportable concrete category with the property that all sources for p have initial lifts. We note that this concept is self-dual.

Proposition. *If a concrete category* (T, p) *over a category* E *is topological over* E, *then the dual concrete category* (T^{op}, p^{op}) *is topological over* E^{op}, *i.e.* p *admits final lifts for all sinks.*

PROOF. For a sink of morphisms $f_i : pB_i \rightarrow E$, we construct the source of all pairs (A, g), with $g : E \rightarrow pA$ in E such that $gf_i : B_i \rightarrow A$ in T for every f_i. It is easily seen that an initial lift B of this source, with $pB = E$, is a final lift for the given sink.

11.4. Examples and Discussion. Topological categories over a category E were defined in [101, 102] under the name *top categories*, as categories of objects of E with structure, with morphisms characterized by 11.8.(1). Sources and sinks, and initial lifts for sources and final lifts for sinks, were introduced in the thesis of R.-E. HOFFMANN [51]. If (T, p) is a topological category, then the functor p is often called a *topological functor*. Topological functors are due to R.-E. HOFFMANN [52] and H. HERRLICH [44, 43].

Every category is a topological category over itself. Non-trivial examples of topological categories over sets are the categories TOP ot topological spaces, UNIF of uniform spaces, PROX of proximity spaces, and many other categories considered in general topology. Categories of topological algebras, such as topological groups or topological vector spaces, are topological over the corresponding set-based categories of algebras. A topological category over the terminal category **1** is a complete lattice. Objects of a topological category over sets are often called *spaces*, and morphisms *continuous functions*.

We note that the categories of Hausdorff spaces and of separated uniform spaces, and other full subcategories of TOP or UNIF characterized by separation or completeness axioms, are not topological categories as defined above.

11.5. Trivial and discrete objects. For a concrete category (T, p) over a category E, an initial lift (tE, h) for the empty source at an object E of E is called an *indiscrete* or *trivial object* of T at E. This object is characterized by the condition that every morphism $f : pX \rightarrow E$ of E, with X an object of T, lifts to a morphism $h^{-1}f : X \rightarrow tE$ of T. A *discrete object* of T is defined dually, as a final lift for the empty sink at an object E of E.

If T is topological, then discrete objects define a left inverse left adjoint functor of the forgetful functor p, and trivial objects define a left inverse right adjoint functor of p.

11.6. Constant maps. A mapping $f : E \to F$ between sets is called *constant* if its range $f^{\to}(E)$ is empty or a singleton. We say that a concrete category $(\mathbf{T}, p)$ over sets has *constant maps* if every constant mapping $f : pX \to pY$, for objects X and Y of $\mathbf{T}$, lifts to $f : X \to Y$ in $\mathbf{E}$.

A topological category $(\mathbf{T}, \mathbf{p})$ over sets has constant maps if and only if $\mathbf{E}$ has exactly one object A with pA empty, and for a singleton $\{x\}$ exactly one object A with $pA = \{x\}$. These objects are then both discrete and trivial.

The categories used in Topology usually have constant maps.

11.7. Limits and colimits. By 7.2 and 11.5, the forgetful functor p of a topological category $(\mathbf{T}, p)$ preserves all limits and colimits. Conversely, $\mathbf{T}$ has *lifted limits* and *lifted colimits* in the following sense. If D is a diagram in $\mathbf{T}$ such that the diagram pD in $\mathbf{E}$ has a limit cone $\lambda : A \to pD$ in $\mathbf{E}$, consisting of morphisms $\lambda_i : A \to pX_i$ in $\mathbf{E}$, then it is easily verified that the initial lift X for the source of all morphism λ_i provides a limit cone $\lambda : X \to D$ in $\mathbf{T}$, conisting of all morphisms $\lambda_i : X \to X_i$ of $\mathbf{T}$.

Colimit cones are lifted dually.

11.8. Fine and coarse morphisms. For a concrete category $(\mathbf{T}, p)$ over $\mathbf{E}$ and an object E of $\mathbf{E}$, we denote by $p^{\leftarrow}E$ the *fibre* of p over E, the subcategory of $\mathbf{T}$ defined by the morphisms u of $\mathbf{T}$ with $pu = \mathrm{id}_E$. Every fibre is an ordered class. We put $A \leqslant B$ for objects A, B of $p^{\leftarrow}E$, and we say that A is *finer* than B and B *coarser* than A, if $\mathrm{id}_E : A \to B$ in $\mathbf{T}$. If $\mathbf{T}$ is topological, then $p^{\leftarrow}E$ is a complete lattice, with initial lifts as infima and final lifts as suprema. Topologists sometimes use the dual order for fibres.

We say that a morphism $f : A \to B$ of $\mathbf{T}$ is a *coarse morphism* if A is an initial lift for the source given by B and $f : pA \to pB$, i.e. if every morphism $h : pC \to pA$ of $\mathbf{E}$ with $fh : C \to B$ in $\mathbf{T}$ lifts to $h : C \to A$ in $\mathbf{T}$. Dually, we say that f is a *fine morphism* of $\mathbf{T}$ if B is the final lift of the sink given by A and $f : pA \to pB$.

If $\mathbf{T}$ is topological, then a morphism $f : E \to F$ in $\mathbf{E}$ induces functors $f^* : p^{\leftarrow}F \to p^{\leftarrow}E$ and $f_* : p^{\leftarrow}E \to p^{\leftarrow}F$, with f^*Y the initial lift for $f : E \to pY$, and f_*X the final lift for $f : pX \to F$. The following result, which follows immediately from the definitions, shows that f_* and f^* define a Galois connection, *i.e.* $f_* \dashv f^*$.

Proposition. *If* $(\mathbf{T}, p)$ *is a topological category over* $\mathbf{E}$ *and* $f : E \to F$ *in* $\mathbf{E}$, *then*

$$(1) \qquad X \leqslant f^*Y \quad \Longleftrightarrow \quad f : X \to Y \quad \Longleftrightarrow \quad f_*X \leqslant Y,$$

for objects X *and* Y *of* $\mathbf{T}$ *with* $pX = E$ *and* $pY = F$.

11.9. Embeddings and quotient morphisms. Strong monomorphisms of a topological category $(\mathbf{T}, p)$, over a category $\mathbf{E}$, are also called *embeddings*, and strong epimorphisms *quotient morphisms*.

Since p is faithful, with a left adjoint and a right adjoint, a morphism $f : X \to Y$ of $\mathbf{T}$ is monomorphic or epimorphic if and only if the underlying morphism $f : pX \to pY$ is monomorphic or epimorphic in $\mathbf{E}$.

Proposition. *A monomorphism m in a topological category $(\mathbf{T}, p)$ over a category $\mathbf{E}$ is an embedding if and only if m is coarse, and a strong monomorphism in $\mathbf{E}$. Dually, an epimorphism e of $\mathbf{T}$ is a quotient morphism if and only if f is fine, and a strong epimorphism in $\mathbf{E}$.*

PROOF. We factor $m : X \to Y$ as $m \circ \mathrm{id}_{pX}$, with $\mathrm{id}_{pX} : X \to m^*Y$ monomorphic and epimorphic. If m is a strong monomorphism of $\mathbf{T}$, then m is strong in $\mathbf{E}$ by 10.5, and $\mathrm{id}_{pX} : X \to m^*Y$ is isomorphic, so that $X = m^*Y$. Conversely, consider a commutative square $ve = mu$ in $\mathbf{T}$, with e epimorphic. If m is strongly monomorphic in $\mathbf{E}$, then $f = te$ and $g = mt$ for a unique morphism t of $\mathbf{E}$. If m is coarse, then t lifts to a morphism of $\mathbf{T}$; thus m is a strong monomorphism of $\mathbf{T}$. The proof for strong epimorphisms is dual.

11.10. The Taut Lift Theorem. Let $(\mathbf{T}, p)$ and $(\mathbf{S}, q)$ be topological categories, over categories $\mathbf{E}$ and $\mathbf{F}$. We say that a functor $\Phi : \mathbf{T} \to \mathbf{S}$ *lifts* a functor $F : \mathbf{E} \to \mathbf{F}$ if $q\Phi = Fp$. We say that the lift of F by Φ is *taut* if it preserves initial lifts of sources, i.e. ΦX is an initial lift for the source of morphisms $F f_i : FA \to q\Phi X_i$ whenever X is an initial lift for a source of morphisms $f_i : A \to pX_i$ of $\mathbf{E}$. *Cotaut lifts* are defined dually.

If Φ lifts F, then the Taut Lift Theorem [101, 4] states that Φ has a left adjoint which lifts a functor $G : \mathbf{F} \to \mathbf{E}$ if and only if Φ lifts F tautly and G is left adjoint to F. In this situation, every adjunction $G \dashv F$ has a unique lift to an adjunction $\Psi \dashv \Phi$ with Ψ lifting G, and $g : \Psi Y \to X$ adjoint to $f : Y \to \Phi X$ if and only if $g : GqY \to pX$ is adjoint to $f : qY \to FpX$. The lifting of G to Ψ then is cotaut.

We say that a concrete full embedding $\Phi : \mathbf{T} \to \mathbf{S}$, with $q\Phi = p$ for topological categories $(\mathbf{T}, p)$ and $(\mathbf{S}, q)$ over the same base category $\mathbf{E}$, is a *taut embedding* if it preserves initial lifts ofsources. By the Taut Lift Theorem, this is the case if and only if $\mathbf{T}$ is a reflective subcategory of $\mathbf{S}$, with reflections $\mathrm{id}_{qY} : Y \to \Psi Y$ for objects Y of $\mathbf{S}$.

A *cotaut embedding* is defined dually; a cotaut embedding $\mathbf{T} \to \mathbf{S}$ embeds $\mathbf{T}$ into $\mathbf{S}$ as a coreflective full subcategory, with coreflections $\mathrm{id}_{qY} : \Psi Y \to Y$.

Chapter 1

BASIC PROPERTIES

A topos or quasitopos is defined to be a category **E** with finite limits which has the following seven properties.

(1) **E** has finite colimits.

(2) **E** is cartesian closed.

(3) Partial morphisms in **E** are represented.

(4) Pullback functors in **E** have right adjoints.

(5) **E** is locally cartesian closed.

(6) Subobjects in **E** are classified by an object Ω of truth values.

(7) Relations in **E** are represented by powerset objects.

The concepts used in (3) – (7) will be defined in this chapter.

Fortunately, the properties listed above are far from being independent. Properties (4) and (5) are logically equivalent, (2) is a special case of (5), and (6) a special case of (3) and of (7). Other dependencies are less obvious, but we shall see that either (1), (2) and (3), or (1), (5) and (6) suffice to obtain all seven properties.

The basic difference between topoi and quasitopoi is the definition of a subobject, on which the definitions of relations and of partial morphisms are based. For an object A of **E**, every monomorphism of **E** with codomain A represents a subobject of A if **E** is a topos, but only strong monomorphisms with codomain A represent subobjects of A if **E** is a quasitopos. This simple difference has far-reaching consequences.

It does not seem possible to reduce the list of defining properties for quasitopoi further than either (1), (2) and (3), or (1), (5) and (6) above, but the situation for topoi is very different. If **E** is cartesian closed and has a subobject classifier, for the class of all monomorphisms of **E**, then **E** is a topos. It even suffices to demand that relations in **E**, again for the class of all monomorphisms of **E**, are represented.

We begin the chapter with a discussion of subobjects. Sections 13 – 18 introduce the concepts used in (3) – (7) above, and their properties. Topoi are formally defined in Section 13, quasitopoi in Section 19. Sections 20 and 21 introduce concepts which do not strictly belong to the present chapter, but which are needed for the proof, given in Section 22, that the dual category of a topos **E** is algebraic over **E**. It follows that finite colimits can be constructed from finite limits if $\mathbf{E}$ is a topos; this is not the case for a quasitopos. The last section discusses some basic properties of colimits in a topos or quasitopos.

We give almost no examples in this chapter. Some basic examples of topoi and quasitopoi will be given in Chapter 2; other examples will be given in later chapters as applications of the constructions discussed in these chapters.

Most of the results of this Chapter are well known, but some of them are new, notably the splitting of 18.6 into 18.4 and 18.5.

12. Subobjects

12.1. Subobjects. Throughout these Notes, $\mathbf{E}$ shall be a category with finite limits. Thus $\mathbf{E}$ has a terminal object, which we often denote by 1, products $A \times B$ of objects, and all possible pullbacks and equalizers.

For monomorphisms m_1 and m_2 of $\mathbf{E}$, we put $m_1 \leqslant m_2$ if $m_1 = m_2 u$ in $\mathbf{E}$, for a (unique) morphism u of $\mathbf{E}$. This relation is clearly a preorder, *i.e.* reflexive and transitive. We call m_1 and m_2 *equivalent*, in symbols $m_1 \simeq m_2$, if $m_1 \leqslant m_2$ and $m_2 \leqslant m_1$. This means that $m_1 = m_2 u$ for an isomorphism u of $\mathbf{E}$. This is clearly an equivalence relation, and equivalent monomorphisms have the same codomain. An equivalence class of monomorphisms of $\mathbf{E}$ with codomain an object A of $\mathbf{E}$ is called a *subobject* of A.

Equivalence classes for the relation $\simeq$ usually are proper classes. Thus we always replace subobjects by representatives, using a monomorphism in the equivalence class instead of the equivalence class.

In the category of sets, monomorphisms are injective mappings. Every equivalence class of monomorphisms contains exactly one subset inclusion; thus we can identify subobjects with subsets. This is also the case for algebraic categories over sets, with subobjects identified with subalgebras.

12.2. Inverse images. For a morphism $f : A \to B$ and a monomorphism $m : B' \to B$ of $\mathbf{E}$, we define the *inverse image* $f^{\frown}m$ of m by f by putting $f^{\frown}m \simeq m'$ in a pullback square

$$
\begin{array}{ccc}
\cdot & \longrightarrow & B' \\
{\scriptstyle m'}\downarrow & & \downarrow{\scriptstyle m} \\
A & \xrightarrow{\ f\ } & B
\end{array}
$$

in $\mathbf{E}$. The inverse image $f^{\frown}m$ is determined by f and m up to equivalence, and clearly $f^{\frown}m_1 \simeq f^{\frown}m$ if $m_1 \simeq m$. Thus the inverse image of a subobject of B is a uniquely determined subobject of A. If there is a *set* of subobjects of A, for every object A of **E**, then subobjects and inverse images define a contravariant functor from **E** to sets.

For sets, if $f : A \to B$ is a mapping and T a subset of B, with inverse image $f^\leftarrow(T)$ by f in the usual sense, then the subset inclusion $f^\leftarrow(T) \to A$ is an inverse image by f, as defined in the preceding paragraph, of the subset inclusion $T \to B$.

12.3. Restricting subobjects. For sets and categories of algebras over sets, every monomorphism is equivalent to a subset or subalgebra inclusion. Thus every monomorphism represents a subobject in these categories. In topological categories and categories of topological algebras, this is no longer the case; only topological embeddings represent subspaces or subalgebras.

For the general situation, we denote by $\mathcal{M}$ the class of monomorphisms of **E** which represent subobjects. This class should satisfy the following three conditions.

(1) $\mathcal{M}$ is stable for pullbacks, *i.e.* if $m_1 \simeq f^\leftarrow m$ in **E**, for $m \in \mathcal{M}$ and a morphism f with the same codomain as m, then $m_1 \in \mathcal{M}$.

(2) $\mathcal{M}$ contains all identity morphisms and is closed under composition.

(3) $\mathcal{M}$ contains all equalizers in **E**.

We assume from now on that a class $\mathcal{M}$ of monomorphisms of **E** is given which satisfies these conditions.

12.4. Discussion. It follows from (1) that if $m \in \mathcal{M}$ and $m' \simeq m$, then also $m' \in \mathcal{M}$. Thus subobjects and their inverse images by morphisms can be defined in the usual way if $\mathcal{M}$ satisfies (1). If we regard subobjects as generalized objects, then (2) states that the subobject relation is reflexive and transitive, and thus an order relation. Condition (3) will be used when we consider relations.

The class of all monomorphisms of **E** satisfies the three conditions of 12.3. If $\mathcal{M}$ is the class of all equalizers in **E**, then conditions (1) and (3), and the first half of (2), are satisfied, but a composition of equalizers is not necessarily an equalizer.

For topoi and quasitopoi, $\mathcal{M}$ will always be the class of strong monomorphisms, defined in 10.4. A monomorphism m of **E** is strong iff every commutative square $mu = ve$ in **E** with e epimorphic has a diagonal, *i.e.* there is a morphism t of **E** such that $u = te$ and $v = mt$ in **E**. This determines t uniquely, and if t satisfies one of the equations, then t also satisfies the other one. It is easily seen that the class of strong monomorphisms satisfies the three conditions of 12.3. Strong monomorphisms need not be equalizers, but we note the following result.

12.5. Proposition. *If f in $\mathbf{E}$ has a cokernel pair, and a composition of equalizers in $\mathbf{E}$ is always an equalizer, then f factors $f = me$ with e epimorphic and m an equalizer. Conversely, if every monomorphism of $\mathbf{E}$ factors in this way, then every strong monomorphism of $\mathbf{E}$ is an equalizer, and equalizers are closed under composition.*

PROOF. We recall that (u, v) is a *cokernel pair* of f in $\mathbf{E}$ if $uf = vf$ in $\mathbf{E}$, and the square $uf = vf$ is a pushout square. Let (u, v) be a cokernel pair of f, and let m be an equalizer of u and v. Then f can be factored $f = me$. If $ae = be$, let m_1 be an equalizer of a and b. Then e factors $e = m_1 e_1$. If mm_1 is an equalizer of x and y, then $xf = yf$ follows, and thus $x = tu$ and $y = tv$ for a morphism t. But then $xm = ym$, and thus $m = mm_1 s$ for a morphism s. Since m and m_1 are monomorphic, it follows that m_1 and s are inverse isomorphisms. But then $a = b$, and e is epimorphic.

If a strong monomorphism $f : A \to B$ factors $f = me$ with e epimorphic and m an equalizer, then there is a morphism t in $\mathbf{E}$ such that $te = \mathrm{id}_A$ and $ft = m$ in $\mathbf{E}$. Since e is epimorphic, it follows that t and e are inverse isomorphisms. But then $f \simeq m$, and f is an equalizer. If this works for all strong monomorphisms, then equalizers are closed under composition since a composition of equalizers always is a strong monomorphism.

13. Relations and Powerset Objects

13.1. Relations. We consider a category $\mathbf{E}$ with finite limits, and with a class $\mathcal{M}$ of monomorphisms representing subobjects in $\mathbf{E}$, satisfying the conditions of 12.3.

For objects A and B of $\mathbf{E}$, a relation from A to B is usually defined as a subobject of $A \times B$. Since subobjects in large categories are proper classes, we modify this definition and define relations as spans which represent subobjects. We recall that a *span* $(u, v) : A \to B$ in $\mathbf{E}$ is a pair $A \xleftarrow{u} X \xrightarrow{v} B$ of morphisms of $\mathbf{E}$, with codomains A and B and a common domain. The span (u, v) is called a *relation* in $\mathbf{E}$ for $\mathcal{M}$ if the morphism $\langle u, v \rangle : X \to A \times B$ is in $\mathcal{M}$. Relations $(u, v) : A \to B$ are preordered by putting $(x, y) \leqslant (u, v)$ if $x = ut$ and $y = vt$ in $\mathbf{E}$ for a (unique) morphism t. We call relations ρ and σ *equivalent*, and write $\rho \simeq \sigma$, if $\rho \leqslant \sigma$ and $\sigma \leqslant \rho$. Thus $(x, y) \simeq (u, v)$ iff $x = ut$ and $y = vt$ for an isomorphism t of $\mathbf{E}$.

Every morphism $f : A \to B$ of $\mathbf{E}$ induces a relation $(\mathrm{id}_A, f) : A \to B$, since $\langle \mathrm{id}_A, f \rangle$ is a coretraction, and thus an equalizer.

13.2. Composition. We define a composition $(u,v) \circ f \simeq (x,y)$: $X \to B$, for a morphism $f : X \to A$ and a relation $(u,v) : A \to B$, by a pullback square

$$
(1) \quad
\begin{array}{ccc}
\cdot \xrightarrow{\ f'\ } \cdot \xrightarrow{\ v\ } B \\
\downarrow x \qquad \downarrow u \\
X \xrightarrow{\ f\ } A
\end{array}
\quad \text{or} \quad
\begin{array}{ccc}
\cdot \xrightarrow{\quad f'\quad} \cdot \\
\downarrow \langle x,y\rangle \qquad \downarrow \langle u,v\rangle, \\
X \times B \xrightarrow{\ f \times \mathrm{id}\ } A \times B
\end{array}
$$

with $vf' = y$. This defines the composition up to equivalence; we note that the lefthand square in (1) is a pullback iff the righthand square is one. Since $\mathcal{M}$ is closed under pullbacks, $(u,v) \circ f$ is a relation if (u,v) is one.

It is easily seen that in particular

$$(\mathrm{id}_A, g) \circ f \simeq (\mathrm{id}_X, gf),$$

for $X \xrightarrow{f} A \xrightarrow{g} B$ in $\mathbf{E}$.

13.3. Proposition. *If* $X \xrightarrow{f} A \xrightarrow{g} B$ *in* $\mathbf{E}$, *then*

$$(u,v) \circ (g \circ f) \simeq ((u,v) \circ g) \circ f$$

for a relation $(u,v) : B \to C$ *in* $\mathbf{E}$.

PROOF. In this situation, we have pullback squares

$$
\begin{array}{ccccc}
\cdot & \xrightarrow{\ f'\ } & \cdot & \xrightarrow{\ g'\ } & \cdot & \xrightarrow{\ v\ } & C \\
\downarrow x & & \downarrow & & \downarrow u \\
X & \xrightarrow{\ f\ } & A & \xrightarrow{\ g\ } & B
\end{array}
$$

in $\mathbf{E}$, with both compositions equivalent to $(x, vg'f')$.

13.4. Powerset objects and topoi. For an object B of $\mathbf{E}$, we say that relations with codomain B are *represented* by an object PB and a relation $\ni_B : PB \to B$ if for every relation $(u,v) : A \to B$ with codomain B there is exactly one morphism $f : A \to PB$ such that $(u,v) \simeq \ni_B \circ f$. We call this morphism f the *characteristic morphism* of the relation (u,v) and denote it by $\chi(u,v)$.

If relations with codomain B are represented in $\mathbf{E}$, then we call PB a *powerset object* of B in $\mathbf{E}$, and we note that two relations with codomain B have the same characteristic morphism if and only if they are equivalent.

We say that *relations in* $\mathbf{E}$ *are represented* if relations with codomain B are represented for every object B of $\mathbf{E}$.

We define a *TOPOS* as a category $\mathbf{E}$ with finite limits in which relations for the class $\mathcal{M}$ of *all* monomorphisms of $\mathbf{E}$ are represented. Quasitopoi will be defined in Section 19, after a full discussion of their defining properties.

For sets and finite sets, with $\mathcal{M}$ the class of all injective mappings, relations in the usual sense are represented, with PB the powerset of a set B, and $S \ni_B x$, for $x \in B$ and $S \subset B$, iff $x \in S$. For a relation $\rho : A \to B$ and for $(x, y) \in A \times B$, we have $y \in (\chi\rho)(x)$ iff $x\rho y$. Thus the categories of sets and of finite sets are topoi.

In a lattice L regarded as a category, with $\mathcal{M}$ the class of identity morphisms, the only relations are spans $a \leftarrow a \wedge b \to b$. Thus relations in L are trivially represented, with $Pb = 1$ for every $b \in L$.

13.5. Proposition. *If relations in* $\mathbf{E}$ *are represented, then* $\chi \ni_B = \mathrm{id}_{PB}$, *and*

$$(1) \qquad \chi((u, v) \circ f) = \chi(u, v) \circ f,$$

for a morphism $f : X \to A$ *and a relation* $(u, v) : A \to B$ *in* $\mathbf{E}$.

PROOF. The first part follows immediately from the definitions, and (1) from a special case of 13.3 since $(u, v) \simeq \ni_B \circ \chi(u, v)$.

13.6. Dual relations. We define the *converse* or *dual relation* $(u, v)^{\mathrm{op}}$ of a relation $(u, v) : A \to B$ by putting

$$(u, v)^{\mathrm{op}} = (v, u) : B \to A.$$

If $(u, v) : A \to B$ and $g : Y \to B$ in $\mathbf{E}$, then we put

$$g^{\mathrm{op}} \circ (u, v) \simeq ((v, u) \circ g)^{\mathrm{op}}.$$

By 13.2, this composition is obtained from pullback squares

$$
\begin{array}{ccc}
A \xleftarrow{\ u\ } \cdot \xleftarrow{\ g'\ } \cdot & & \cdot \xrightarrow{\ \ g'\ \ } \cdot \\
\Big\downarrow{v} \qquad \Big\downarrow{y} \quad \text{and} & & \Big\downarrow{\langle x, y\rangle} \qquad \Big\downarrow{\langle u, v\rangle} \\
B \xleftarrow{\ g\ } Y & & A \times Y \xrightarrow{\ \mathrm{id} \times g\ } A \times B
\end{array}
$$

in $\mathbf{E}$, with $x = ug'$.

If relations in $\mathbf{E}$ are represented, then dual relations generate a bijection between morphisms

$$(1) \qquad \varphi = \chi(u, v) : A \to PB \qquad \text{and} \qquad \psi = \chi(v, u) : B \to PA$$

of $\mathbf{E}$, for relations $(u, v) : A \to B$ in $\mathbf{E}$.

13.7. Proposition. *If* $X \xrightarrow{f} A \xrightarrow{(u,v)} B \xleftarrow{g} Y$ *in* **E**, *with a relation* (u,v) *in the middle, then*

$$g^{\mathrm{op}} \circ ((u,v) \circ f) \simeq (g^{\mathrm{op}} \circ (u,v)) \circ f.$$

If also $h : Z \to Y$ *in* **E**, *then*

$$(g \circ h)^{\mathrm{op}} \circ (u,v) \simeq h^{\mathrm{op}} \circ (g^{\mathrm{op}} \circ (u,v)).$$

PROOF. Using the righthand pullback squares in 13.2 and 13.6, both compositions in the first part are equivalent to (x, y) in a single pullback square

$$
\begin{array}{ccc}
\cdot & \xrightarrow{\;h\;} & \cdot \\
{\scriptstyle\langle x, y\rangle}\Big\downarrow & & \Big\downarrow{\scriptstyle\langle u, v\rangle} \\
X \times Y & \xrightarrow{f \times g} & A \times B
\end{array}
$$

in **E**. The two compositions in the second part are dual to the equivalent compositions $(v, u) \circ (g \circ h)$ and $((v, u) \circ g) \circ h$.

13.8. Theorem. *If relations in* **E** *are represented, then powerset objects in* **E** *define a contravariant functor* P *on* **E**, *with*

(1) $$\mathsf{P}g \circ \chi(u, v) = \chi(g^{\mathrm{op}} \circ (u, v))$$

for a relation $(u, v) : A \to B$ *and a morphism* $g : Y \to B$ *in* **E**. *The functor* $\mathsf{P} : \mathbf{E}^{\mathrm{op}} \to \mathbf{E}$ *is self-adjoint on the right.*

The functor P *is called the* powerset functor *on* **E**.

PROOF. The bijection 13.6.(1) between morphisms $\varphi : A \to \mathsf{P}B$ and $\psi : B \to \mathsf{P}A$ is clearly symmetric; we must show that it is natural in A, and then also in B.

We must put $\mathsf{P}g = \chi(g^{\mathrm{op}} \circ \ni_B)$. Then (1) in the Theorem follows from 13.5 and the first part of 13.7, since $(u, v) \simeq \ni_B \circ \chi(u, v)$. This proves the desired naturality, and it follows that the morphisms $\mathsf{P}g$ define a contravariant functor P. This functor is self-adjoint on the right since the bijections given by 13.6.(1) are symmetric.

14. Subobject Classifiers

14.1. Definition. We use again the assumptions of 13.1. A *subobject classifier* for $\mathcal{M}$ in $\mathbf{E}$ is a monomorphism $\top : T \to \Omega$ such that for every morphism $m : X \to A$ in $\mathcal{M}$, there is exactly one pullback square

(1)
$$
\begin{array}{ccc}
X & \xrightarrow{\ h\ } & T \\
{\scriptstyle m}\downarrow & & \downarrow{\scriptstyle \top} \\
A & \xrightarrow{\ f\ } & \Omega
\end{array}
$$

in $\mathbf{E}$. The morphism $f : A \to \Omega$ in this pullback square is then denoted by $\operatorname{ch} m$ and called the *characteristic morphism* of m.

For sets and finite sets, with $\mathcal{M}$ the class of all injective mappings, we have $\Omega = \{0,1\}$, and $\top : \{*\} \to \Omega$ is given by $\top(*) = 1$. The characteristic morphism of an injective mapping $m : X \to A$ is then the characteristic function in the usual sense of the range $m^{\to}(X)$ of m.

For a lattice L regarded as a category, with $\mathcal{M}$ the class of identity morphisms, $\top = \operatorname{id}_1$, and $\operatorname{ch} \operatorname{id}_a$ is the unique morphism $a \to 1$ for $a \in L$.

14.2. Proposition. *If $\mathbf{E}$ has a subobject classifier $\top : T \to \Omega$ for $\mathcal{M}$, then T is a terminal object of $\mathbf{E}$, and two morphisms m, m' in $\mathcal{M}$ have the same characteristic morphism if and only if $m \simeq m'$, i.e. $m' = mu$ for an isomorphism u of $\mathbf{E}$.*

We use this to replace T by a specified terminal object 1 of $\mathbf{E}$, and the subobject classifier $\top$ by the equivalent monomorphism $\top : 1 \to \Omega$. We note that a morphism with domain 1 always is a coretraction, hence an equalizer.

PROOF. If we apply 14.1.(1) to morphisms $m = \operatorname{id}_A$ in $\mathcal{M}$, we get $f = \top h$. It follows that for every object A of $\mathbf{E}$, there is exactly one morphism $h : A \to T$ in $\mathbf{E}$. Thus T is a terminal object of $\mathbf{E}$.

For the second part, we simply note that a pullback 14.1.(1) of f and $\top$ is determined up to an isomorphism at the vertex X.

14.3. Proposition. *If $\top : 1 \to \Omega$ is a subobject classifier for $\mathcal{M}$, then $\mathcal{M}$ is the class of all equalizers in $\mathbf{E}$, with $m : X \to A$ in $\mathcal{M}$ an equalizer of $\operatorname{ch} m$ and the morphism $\top \tau_A$ of $\mathbf{E}$, with $\tau_A : A \to 1$ in $\mathbf{E}$.*

PROOF. We have $(\operatorname{ch} m)m = \top \tau_X = \top \tau_A m$, and $u = mt$ for a morphism t of $\mathbf{E}$ iff $(\operatorname{ch} m)u = \top v$ for $v = \tau_X t = \tau_A u$, hence iff $(\operatorname{ch} m)u = \top \tau_A u$. Thus m is an equalizer as claimed.

14.4. Proposition. *If* **E** *has a subobject classifier, then*

$$\mathrm{ch}(f^{\leftarrow}m) = \mathrm{ch}\, m \circ f$$

for morphisms f *and* m *of* **E***, with a common codomain and with* $m \in \mathcal{M}$.

PROOF. By the definitions of $f^{\leftarrow}m$ (see 12.2) and $\mathrm{ch}\,m$, we have two pullback squares

$$
\begin{array}{ccccc}
Y & \longrightarrow & X & \longrightarrow & 1 \\
\downarrow{\scriptstyle f^{\leftarrow}m} & & \downarrow{\scriptstyle m} & & \downarrow{\scriptstyle \top} \\
C & \xrightarrow{\ f\ } & A & \xrightarrow{\ \mathrm{ch}\,m\ } & \Omega
\end{array}
$$

in **E**, with the outer rectangle also a pullback.

14.5. Proposition. *If relations in* **E** *are represented, then* **E** *has a subobject classifier, and* $\mathcal{M}$ *is the class of all equalizers in* **E**.

Corollary. *A topos* **E** *has a subobject classifier, and every monomorphism of* **E** *is an equalizer in* **E**.

PROOF. We denote by $\tau_A : A \to 1$ the unique morphism from an object A of **E** to the terminal object 1. As the projection $A \times 1 \to A$ is an isomorphism, with inverse $\langle \mathrm{id}_A, \tau_A \rangle$, a span $(u, \tau_X) : A \to 1$ is a relation iff u is a monomorphism in $\mathcal{M}$. Thus if $\ni_1 = (e, \tau_T) : \mathrm{P}1 \to 1$, there is for every monomorphism m in $\mathcal{M}$ a unique pullback square

$$
\begin{array}{ccc}
\cdot & \xrightarrow{\ f'\ } & T \\
\downarrow{\scriptstyle m} & & \downarrow{\scriptstyle e} \\
\cdot & \xrightarrow{\ f\ } & \mathrm{P}1
\end{array}
$$

in **E**, so that $\top = e$ and $\Omega = \mathrm{P}1$ define a subobject classifier.

14.6. If relations in **E** are represented, then a relation $(u, v) : A \to B$ in **E** has by 14.5 two characteristic morphisms,

$$\mathrm{ch}\langle u, v\rangle : A \times B \to \Omega \qquad \text{and} \qquad \chi(u, v) : A \to \mathrm{P}B .$$

These characteristic morphisms define a bijection between morphisms $\varphi : A \times B \to \Omega$ and morphisms $\hat{\varphi} : A \to \mathrm{P}B$ in **E**.

Proposition. *The bijection between morphisms* $\varphi : A \times B \to \Omega$ *and morphisms* $\hat{\varphi} : A \to \mathrm{P}B$ *in* **E** *is natural in* A.

PROOF. If $(x, y) \simeq (u, v) \circ f$, then $\langle x, y \rangle \simeq (f \times \mathrm{id}_B)^{\leftarrow} \langle u, v \rangle$ by 13.2.(1). Thus

$$\chi(x, y) = \chi(u, v) \circ f \quad \text{and} \quad \mathrm{ch}\langle x, y \rangle = \mathrm{ch}\langle u, v \rangle \circ (f \times \mathrm{id}_B),$$

by 13.5 and 14.4.

14.7. Discussion. If relations in $\mathbf{E}$ are represented and $\Omega = \mathrm{P}1$, then the adjunction of 13.8 allows us to represent a monomorphism $m : X \to A$ in $\mathcal{M}$ by a morphism

$$m^{\#} = \chi(\tau_X, m) : 1 \to \mathrm{P}A,$$

adjoint to $\mathrm{ch}\, m$ by 13.6.(1). This provides a bijection between subobjects of A and *global elements* of $\mathrm{P}A$, *i.e.* morphisms $1 \to \mathrm{P}A$. By 13.8 and 14.4, we have

$$(f^{\leftarrow} m)^{\#} = \mathrm{P}f \circ m^{\#}$$

if the inverse image $f^{\leftarrow} m$ is defined. Thus we can regard $\mathrm{P}A$ as an object of subobjects of A, and $\mathrm{P}f$ as an internal version of the inverse image mapping $f^{\leftarrow}$.

15. Topoi are Cartesian Closed

15.1. External relations. A relation $\rho = A \xleftarrow{u} Z \xrightarrow{v} B$ in $\mathbf{E}$ defines for every object X of $\mathbf{E}$ an *external relation*

$$\rho_X = (u, v)_X : \mathbf{E}(X, A) \to \mathbf{E}(X, B),$$

by putting $x \, \rho_X \, y$ iff $x = ut$ and $y = vt$ for a (unique) morphism $t : X \to Z$ in $\mathbf{E}$. It is easily seen that two relations define the same external relation if and only if they are equivalent.

We usually omit subscripts for external relations, using the letter ρ, without subscript, for all external relations ρ_X. If $x \, \rho \, y$, then $xt \, \rho \, yt$ for all morphisms t of $\mathbf{E}$ for which xt and hence yt is defined.

Proposition. If $X \xrightarrow{f} A \xrightarrow{\rho} B$ in $\mathbf{E}$, for a morphism f and a relation ρ, then $s(\rho \circ f)t$ for morphisms s and t of $\mathbf{E}$ iff $(fs) \rho t$.

PROOF. We put $\rho = (u, v)$ and use the lefthand diagram in 13.2.(1). If $(fs = ur$ and $t = vr$, then $s = xr'$ and $r = f'r'$ for a unique r', with $t = yr'$. Conversely, if $s = xr'$ and $t = yr'$, then $fx = ur$ and $t = vr$ for $r = f'r'$.

15.2. Equivalence relations. External relations are often used to carry over concepts from relations over sets to relations in a category $\mathbf{E}$. Thus we say that a relation $(u,v) : A \to B$ in $\mathbf{E}$ is an *equivalence relation* in $\mathbf{E}$ if every external relation $(u,v)_X$ is an equivalence relation.

Every pullback square

$$
\begin{array}{ccc}
X & \xrightarrow{\;v\;} & B \\
\downarrow u & & \downarrow g \\
A & \xrightarrow{\;f\;} & \cdot
\end{array}
$$

in $\mathbf{E}$ defines a relation $(u,v) : A \to B$ in $\mathbf{E}$, with $x\,(u,v)\,y$ for morphisms x and y of $\mathbf{E}$, iff $fx = gy$ in $\mathbf{E}$. If $g = f$, then we say that (u,v) is a *kernel pair* of f. Kernel pairs clearly are equivalence relations; we note a converse.

15.3. Proposition. *If relations in $\mathbf{E}$ are represented, then every equivalence relation (u,v) in $\mathbf{E}$ is a kernel pair of its characteristic morphism $\chi(u,v)$.*

PROOF. For $\rho = (u,v)$, we must show that $\chi(u,v) \circ x = \chi(u,v) \circ y$ in $\mathbf{E}$, for morphisms x and y of $\mathbf{E}$, iff $x\,\rho\,y$. This is equivalent to saying that $\rho \circ x \simeq \rho \circ y$ iff $x\,\rho\,y$.

If $\rho : A \to A$ is an equivalence relation, then $y\,\rho\,y$ for a morphism $y : X \to A$, and thus $\mathrm{id}_X\,(\rho \circ y)\,y$ by 15.1. If $\rho \circ x \simeq \rho \circ y$, it follows that $\mathrm{id}_X\,(\rho \circ x)\,y$, and hence $x\,\rho\,y$. Conversely, if $x\,\rho\,y$, then always $xa\,\rho\,ya$, and

$$
a\,(\rho \circ x)\,b \iff xa\,\rho\,b \iff ya\,\rho\,b \iff a\,(\rho \circ y)\,b
$$

if ρ is an equivalence relation, so that $\rho \circ x \simeq \rho \circ y$.

15.4. We have an identity relation $(\mathrm{id}_B, \mathrm{id}_B) : B \to B$ for every object B of $\mathbf{E}$. If relations in $\mathbf{E}$ are represented, then we put

$$
s_B = \chi(\mathrm{id}_B, \mathrm{id}_B) : B \to \mathrm{P}B .
$$

For sets, we have $s_B(x) = \{x\}$, for a set B and $x \in B$.

Lemma. s_B *is a monomorphism.*

PROOF. An identity relation clearly is an equivalence relation. Thus $(\mathrm{id}_B, \mathrm{id}_B)$ is a kernel pair of s_B by 15.3; it follows immediately that s_B is monomorphic.

15.5. Theorem (KOCK–WRAITH [65]). *Every topos is cartesian closed.*

PROOF. By 15.2, we have a monomorphism $s_B : B \to PB$ for every object B of a topos **E**. By 14.3, this is an equalizer of morphisms $\varphi = \operatorname{ch} s_B$ and $\psi = \top \circ \tau_{PB}$ with codomain Ω. Changing $f : X \to A \times B$ to $u = s_B \circ f$, we have a bijection between morphisms $f : X \times A \to B$, and morphisms $u : X \times A \to PB$ such that $\varphi \circ u = \psi \circ u$. Characteristic morphisms of relations define a bijection between morphisms

$$u = \chi(\langle x, a \rangle, b) : X \times A \to PB \qquad \text{and} \qquad v = \chi(x, \langle a, b \rangle) : X \to P(A \times B),$$

for morphisms x, a, b with a common domain for which $\langle x, a, b \rangle$ is monomorphic with codomain $X \times A \times B$. This correspondence clearly is natural in X. Thus if $u_0 = e_{A,B}$ corresponds to $v_0 = \operatorname{id} P(A \times B)$, then we always have $u = e_{A,B} \circ (v \times \operatorname{id}_B)$ for corresponding u and v.

Now let the morphisms

$$\Phi : P(A \times B) \to PB \qquad \text{and} \qquad \varphi \circ e_{A,B} : P(A \times B) \times B \to \Omega,$$

and the morphisms

$$\Psi : P(A \times B) \to PB \qquad \text{and} \qquad \psi \circ e_{A,B} : P(A \times B) \times B \to \Omega,$$

correspond by the natural bijections of 14.6. Then $\Phi \circ v$ corresponds to

$$\varphi \circ e_{A,B} \circ (v \times \operatorname{id}_B) = \varphi \circ u,$$

and $\Psi \circ v$ to $\psi \circ u$, if u corresponds to v by the preceding paragraph. Thus

$$\Phi \circ v = \Psi \circ v \iff \varphi \circ u = \psi \circ u$$

in this situation.

Now if $\sigma_{A,B} : [A, B] \to P(A \times B)$ is an equalizer of Φ and Ψ, it follows that $v : X \to P(A \times B)$ factors $v = \sigma_{A,B} \circ g$ iff u factors $u = s_B \circ f$. This produces the desired bijection between morphisms $f : X \times A \to B$ and $g : X \to [A, B]$ of **E**. The bijection is clearly natural in X; thus it provides the desired cartesian closed structure of **E**.

15.6. The theorem just proved has the following partial converse.

Proposition. *If a category* **E** *is cartesian closed and has a subobject classifier, then relations in* **E** *are represented, with* $PB = [B, \Omega] = \Omega^B$ *for an object B of* **E**.

PROOF. For a relation $(u, v) : A \to B$, we have a morphism $\chi(u, v) : A \to \Omega^B$, exponentially adjoint to $\operatorname{ch}\langle u, v \rangle : A \times B \to \Omega$, with $\chi(u, v) \circ f$ exponentially adjoint to

$$\operatorname{ch}\langle u, v \rangle \circ (f \times \operatorname{id}_B) = \operatorname{ch}((f \times \operatorname{id}_B)^{\frown}\langle u, v \rangle).$$

Thus $\chi(u, v) \circ f = \chi((u, v) \circ f)$, and it follows that $(u, v) \simeq \ni_B \circ \chi(u, v)$ for the relation $\ni_B : PB \to B$ with $\chi \ni_B = \operatorname{id}_{PB}$.

16. Partial Morphisms

16.1. Definition. We return to a category $\mathbf{E}$ which satisfies the assumptions of 13.1. A relation $(m, f) : A \to B$ in $\mathbf{E}$ is called a *partial morphism* in $\mathbf{E}$ if $m \in \mathcal{M}$.

Every span $(m, f) : A \to B$ with $m \in \mathcal{M}$ is a partial morphism since $\langle m, f \rangle = (m \times \mathrm{id}_B) \circ \langle \mathrm{id}, f \rangle$, with $\langle \mathrm{id}, f \rangle$ a coretraction, hence an equalizer, and $m \times \mathrm{id}_B$ a pullback of m by the projection $A \times B \to A$.

Partial morphisms with codomain 1 are the same as relations with codomain 1.

If $(m, f) : A \to B$ is a partial morphism in $\mathbf{E}$ and $g : X \to A$ in $\mathbf{E}$, then $(m, f) \circ g \simeq (m', fg')$ for a pullback

$$
\begin{array}{ccccc}
\cdot & \xrightarrow{g'} & \cdot & \xrightarrow{f} & B \\
\downarrow{\scriptstyle m'} & & \downarrow{\scriptstyle m} & & \\
X & \xrightarrow{g} & A & &
\end{array}
,
$$

with $m' \in \mathcal{M}$. Thus $(m, f) \circ g$ is a partial morphism.

16.2. Representing partial morphisms. We say that partial morphisms in $\mathbf{E}$ with codomain B are *represented* by a partial morphism $(t, u) : \tilde{B} \to B$ if for every object A of $\mathbf{E}$ and every partial morphism $(m, f) : A \to B$ in $\mathbf{E}$, there is exactly one morphism $\bar{f} : A \to \tilde{B}$ in $\mathbf{E}$ such that $(m, f) \simeq (t, u) \circ \bar{f}$. We may denote this morphism $\bar{f}$ by $\mathrm{ch}_p(m, f)$.

We say that *partial morphisms* in $\mathbf{E}$ *are represented* if partial morphisms with codomain B are represented for every object B of $\mathbf{E}$.

If partial morphisms in $\mathbf{E}$ are represented, then $\mathbf{E}$ has a subobject classifier, representing partial morphisms with codomain 1.

16.3. Proposition. *If* $(t, u) : \tilde{B} \to B$ *represents partial morphisms in* $\mathbf{E}$ *with codomain* B, *then* u *is an isomorphism of* $\mathbf{E}$.

PROOF. If (t, u) represents partial morphisms with codomain B, then we have for $(\mathrm{id}_A, f) : A \to B$ in $\mathbf{E}$ a unique pullback square

$$
\begin{array}{ccccc}
A & \xrightarrow{h} & \cdot & \xrightarrow{u} & B \\
\downarrow{\scriptstyle \mathrm{id}_A} & & \downarrow{\scriptstyle t} & & \\
A & \longrightarrow & \tilde{B} & &
\end{array}
$$

with $uh = f$, and thus a unique h in $\mathbf{E}$ with $uh = f$. Applying this to $f = \mathrm{id}_B$ and then to $f = u$, we see that u must be an isomorphism.

16.4. We apply 16.3 to replace (t, u) by an equivalent partial morphism $(\vartheta_B, \mathrm{id}_B)$ which also represents partial morphisms with codomain B, with $\vartheta_B = tu^{-1}$, and we say that the morphism $\vartheta_B : B \to \tilde{B}$ in $\mathcal{M}$ *represents partial morphisms with codomain B.*

For sets and for finite sets, $\tilde{B}$ is a one point extension

$$\tilde{B} = \{\emptyset\} \cup \{\,\{x\} \mid x \in B\,\}$$

of B, with $\vartheta_B(x) = \{x\}$ for $x \in B$. For a lattice L, considered as a category, with $\mathcal{M}$ the set of identity morphisms, $\vartheta_b = \mathrm{id}_b$ represents partial morphisms with codomain b for $b \in L$.

Proposition. *A morphism $\vartheta_B : B \to \tilde{B}$ in $\mathcal{M}$ represents partial morphisms in $\mathbf{E}$ with codomain B if and only if for every partial morphism $(m, f) : A \to B$ in $\mathbf{E}$ with codomain B, there is a unique morphism $\bar{f} : A \to \tilde{B}$ in $\mathbf{E}$, denoted by $\mathrm{ch}_p(m, f)$, for which the square*

$$
\begin{array}{ccc}
\cdot & \xrightarrow{\ f\ } & B \\
{\scriptstyle m}\downarrow & & \downarrow{\scriptstyle \vartheta_B} \\
A & \xrightarrow[\ \bar{f}\ ]{} & \tilde{B}
\end{array}
$$

(1)

is a pullback square.

PROOF. We note that $(m, f) \simeq (\vartheta_B, \mathrm{id}_B) \circ \bar{f}$ iff (1) is a pullback square, and then the result follows immediately from the definitions.

16.5. We consider from now on relations $(m, f) : A \to B$ with m monomorphic. We assume that relations in $\mathbf{E}$ with codomain B are represented, and we put

$$d_B = \chi(s_B, \mathrm{id}_B) : PB \to PB,$$

with $s_B = \chi(\mathrm{id}_B, \mathrm{id}_B)$. We recall from 15.4 that s_B is a monomorphism.

Lemma. *A commutative square*

$$
\begin{array}{ccc}
\cdot & \xrightarrow{\ f\ } & B \\
{\scriptstyle m}\downarrow & & \downarrow{\scriptstyle s_B} \\
A & \xrightarrow[\ \varphi\ ]{} & PB
\end{array}
$$

(1)

in **E** *is a pullback square in* **E** *iff* m *is monomorphic in* **E** *with* $\langle m, f \rangle \in \mathcal{M}$, *and* $\chi(m, f) = d_B\,\varphi$.

PROOF. We put $\ni_B = (a, b)$, and we consider a diagram

$$
\begin{array}{ccccccc}
\cdot & \xrightarrow{\;f\;} & B & \xrightarrow{\;t\;} & \cdot & \xrightarrow{\;b\;} & B \\
\downarrow{\scriptstyle m} & & \downarrow{\scriptstyle s_B} & & \downarrow{\scriptstyle a} & & \\
A & \xrightarrow{\;\varphi\;} & PB & \xrightarrow{\;d_B\;} & PB & &
\end{array}
\quad,
$$

with a pullback square in the middle and $b\,t = \mathrm{id}_B$. This middle square exists by the definition of d_B. Now the lefthand square is a pullback iff the outer rectangle is a pullback. Since $b\,t\,f = f$, this is the case iff $\langle m, f \rangle \in \mathcal{M}$ and $d_B\,\varphi = \chi(m, f)$. In this situation, m is monomorphic since s_B is.

16.6. Lemma. *If* $(m, f) : A \to B$ *is a relation with* m *monomorphic, then* 16.5.(1) *is a pullback square for* $\varphi = \chi(m, f)$.

PROOF. Using the notation of 15.1 and 15.2, we must show that $u\,(m, f)\,v$ in **E** iff $\varphi \circ u = s_B \circ v$, hence iff $(m, f) \circ u \simeq (\mathrm{id}_B, \mathrm{id}_B) \circ v$. Thus we must show that $u\,(m, f)\,v$ iff always

(1) $$u x\,(m, f)\,y \iff v x\,(\mathrm{id}_B, \mathrm{id}_B)\,y \iff v x = y,$$

for morphisms x and y of **E**.

If (1) is valid, then $u\,(m, f)\,v$; put $x = \mathrm{id}$ and $y = v$. Conversely, if $u\,(m, f)\,v$, then $u x\,(m, f)\,y$ if $y = v\,x$. On the other hand, if $u = m t, v = f t$, and $u x = m s$, $y = f s$, then $s = t x$ since m is monomorphic, and hence $y = v\,x$.

16.7. Proposition. *If relations in* **E** *with codomain* B *are represented, then relations* (m, f) *in* **E** *with codomain* B *and* m *monomorphic are represented.*

PROOF. By 16.5 and 16.6, we have $d_B\,\chi(m, f) = \chi(m, f)$ for (m, f) with codomain B, with m monomorphic. In particular, $d_B\,d_B = d_B$. If $e_B : \bar{B} \to PB$ is an equalizer of d_B and id_{PB}, it follows that $\chi(m, f)$ factors through e_B for a relation (m, f) with codomain B and m monomorphic. In particular, s_B factors $s_B = e_B\,t_B$, with $t_B : B \to \bar{B}$ monomorphic. Now consider a commutative diagram

(1)

$$
\begin{array}{ccccc}
\cdot & \xrightarrow{\ f\ } & B & \xrightarrow{\ \mathrm{id}_B\ } & B \\
\downarrow{m} & & \downarrow{t_B} & & \downarrow{s_B} \\
A & \xrightarrow{\ \bar{f}\ } & \bar{B} & \xrightarrow{\ e_B\ } & PB
\end{array}
$$

If $e_B\,\bar{f} = \chi(m,f)$, then the outer rectangle is a pullback by 16.6; thus the lefthand square is a pullback. Conversely, if the lefthand square is a pullback, then $d_B\,e_B\,\bar{f} = \chi(m,f)$ by 16.5, and $e_B\,\bar{f} = \chi(m,f)$ follows since $d_B e_B = e_B$. Thus 16.2.(1) is a pullback for exactly one morphism $\bar{f} : A \to \bar{B}$ of $\mathbf{E}$.

16.8. Corollary. *If $s_B \in \mathcal{M}$, then partial morphisms with codomain B are represented. In particular, partial morphisms in a topos are represented.*

PROOF. If $s_B \in \mathcal{M}$, then the morphisms t_B and m in 16.7.(1) are in $\mathcal{M}$. Thus every relation $(m,f) : A \to B$ with m monomorphic is a partial morphism, and t_B represents partial morphisms with codomain B.

17. Slice Categories

17.1. Definition. For an object A of an arbitrary category $\mathbf{E}$, a *slice category* $\mathbf{E}/A$ and a *domain functor* $D_A : \mathbf{E}/A \to \mathbf{E}$ are defined as follows. Objects of $\mathbf{E}/A$ are all morphisms of $\mathbf{E}$ with codomain A, and $D_A\,u = X$ if $u : X \to A$ in $\mathbf{E}$. For objects u and v of $\mathbf{E}/A$, morphisms $h : u \to v$ of $\mathbf{E}/A$ are morphisms $h = D_A\,h : D_A\,u \to D_A\,v$ of $\mathbf{E}$ such that $u\,h = v$ in $\mathbf{E}$. Composition in $\mathbf{E}/A$ is lifted from composition in $\mathbf{E}$, and $\mathrm{id}_u = \mathrm{id}_X$ in $\mathbf{E}/A$ if $u : X \to A$ in $\mathbf{E}$.

The notation $(\mathbf{E},A)$ or $\mathbf{E}\downarrow A$ is sometimes used for a slice category $\mathbf{E}/A$.

We note that $D_1 : \mathbf{E}/1 \to \mathbf{E}$ is an isomorphism of categories for a terminal object 1 of $\mathbf{E}$, and that $(\mathbf{E}/A)/u$ is isomorphic to $\mathbf{E}/D_A\,u$ for an object u of $\mathbf{E}/A$.

The slice category $\mathbf{E}/A$ always has a terminal object id_A, with $u : u \to \mathrm{id}_A$ in $\mathbf{E}/A$ for an object u of $\mathbf{E}/A$.

17.2. Limits. A diagram $\Delta : \mathcal{C} \to \mathbf{E}/A$ in $\mathbf{E}/A$ consists of a diagram $D_A\,\Delta : \mathcal{C} \to \mathbf{E}$ and a cone $\delta : D_A\,\Delta \to A$ in $\mathbf{E}$, with $\delta_i = \Delta\,i$ in $\mathbf{E}$ for a vertex i of $\mathcal{C}$. These data can be combined into a diagram $\bar{\Delta} : \mathcal{C}_\star \to \mathbf{E}$ as follows. We get $\mathcal{C}_\star$ by adding a vertex $\star$ to $\mathcal{C}$, and for every vertex i of $\mathcal{C}$ an arrow $i_\star : i \to \star$, with $j_\star a = i_\star$ for an arrow $a : i \to j$ of $\mathcal{C}$. We put $\bar{\Delta} \star = A$, and $\bar{\Delta} i_\star = \delta_i$ for a vertex i of $\mathcal{C}$.

For an object u of $\mathbf{E}/A$, there is a bijection, natural in u, between cones $\lambda : u \to \Delta$ in $\mathbf{E}/A$ and cones $\mu : D_A u \to \bar{\Delta}$ in $\mathbf{E}$ for which $\mu \star = u$, with μ extending $D_A \lambda$ to $\bar{\Delta}$. From this observation, we get the following result.

Theorem. *A diagram $\Delta : \mathcal{C} \to \mathbf{E}/A$ in a slice category $\mathbf{E}/A$ has a limit in $\mathbf{E}/A$ if and only if the extended diagram $\bar{\Delta} : \mathcal{C}_\star \to \mathbf{E}$ has a limit in $\mathbf{E}$, and then a limit cone for $\bar{\Delta}$ in $\mathbf{E}$ lifts to a unique limit cone for Δ in $\mathbf{E}/A$.*

17.3. Remarks. If $\mathbf{E}$ has finite limits, then every slice category $\mathbf{E}/A$ has finite limits by 17.2. Products $u \times_A v$ in $\mathbf{E}/A$ are obtained as pullbacks in $\mathbf{E}$; these products are sometimes called *fibred products*. It is easily seen that the functor D_A preserves and creates pullbacks and equalizers.

If $\mathbf{E}$ has finite products, then the functor D_A has a right adjoint, obtained by natural bijections between morphisms $h : D_A u \to Y$ in $\mathbf{E}$ and morphisms $\langle h, u \rangle : u \to q_Y$, where $q_Y : Y \times A \to A$ is the second projection of the product $Y \times A$.

The functor D_A preserves and reflects monomorphisms; it also preserves and reflects epimorphisms if $\mathbf{E}$ has finite products. If an object m of $\mathbf{E}/A$ is monomorphic in $\mathbf{E}$, then m is also a monomorphism $m : m \to \mathrm{id}_A$ in $\mathbf{E}/A$. An equalizer in $\mathbf{E}/A$ is clearly an equalizer in $\mathbf{E}$. Conversely, if an object m of $\mathbf{E}/A$ is an equalizer of g and h in $\mathbf{E}$, with common codomain B, then $m : m \to \mathrm{id}_A$ is an equalizer of $\langle \mathrm{id}_A, g \rangle$ and $\langle \mathrm{id}_A, h \rangle$ in $\mathbf{E}/A$, with the projection $A \times B \to A$ as common codomain.

If $\mathbf{E}$ satisfies the assumptions of 13.1, then so does $\mathbf{E}/A$, with $\mathcal{M}$ for $\mathbf{E}/A$ the class of all morphisms h with $D_A h$ in $\mathcal{M}$ for $\mathbf{E}$.

17.4. Colimits. If the category $\mathbf{E}$ has finite products, then the functor $D_A : \mathbf{E}/A \to \mathbf{E}$ has a right adjoint, and D_A preserves colimits. Even without this assumption, we have the following result.

Theorem. *If $\Delta : \mathcal{C} \to \mathbf{E}/A$ is a diagram in a slice category $\mathbf{E}/A$ such that the diagram $D_A \Delta$ has a colimit in $\mathbf{E}$, then Δ has a colimit in $\mathbf{E}/A$.*

PROOF. As in 17.2, Δ consists of $D_A \Delta$ and a cone $\delta : D_A \Delta \to A$ in $\mathbf{E}$. A cone $\lambda : \Delta \to v$ in $\mathbf{E}/A$ is determined by the cone $D_A \lambda$ in $\mathbf{E}$; this cone must satisfy $v \circ D_A \lambda = \delta$.

If $D_A \Delta$ has a colimit cone $\sigma : D_A \Delta \to S$ in $\mathbf{E}$, then $\delta = s \circ \sigma$ in $\mathbf{E}$ for a unique morphism $s : S \to A$ in $\mathbf{E}$; thus $\sigma = D_A \rho$ for a unique cone $\rho : \Delta \to s$ in $\mathbf{E}/A$. It is easily verified that ρ is the desired colimit cone for Δ.

17.5. Pullback and composition functors. For an arbitrary category $\mathbf{E}$, a morphism $f : A \to B$ of $\mathbf{E}$ determines a *composition functor* which we denote by $f_> : \mathbf{E}/A \to \mathbf{E}/B$, with $f_> u = fu$ for an object u of $\mathbf{E}/A$, and with $f_> h = h : fu \to fu'$ for a morphism $h : u \to u'$ of $\mathbf{E}/A$. We note that $f_>$ is clearly faithful, and that $D_B\, f_> = D_A$ for the domain functors. The functor $f_>$ is sometimes denoted by Σ_f.

The *abus de langage* implicit in our notations should be noted; a morphism in a slice category $\mathbf{E}/A$ is not distinguished from the underlying morphism of $\mathbf{E}$.

If $\mathbf{E}$ has pullbacks, then we also have a *pullback functor* $f^* : \mathbf{E}/B \to \mathbf{E}/A$, described in the following result.

Theorem. *If $f : A \to B$ in a category $\mathbf{E}$ with pullbacks, then pullback squares*

(1)
$$\begin{array}{ccc} \cdot & \xrightarrow{\ \varepsilon_v\ } & \cdot \\ {\scriptstyle f^*v}\downarrow & & \downarrow{\scriptstyle v} \\ A & \xrightarrow{\ f\ } & B \end{array}$$

in $\mathbf{E}$ determine a functor $f^* : \mathbf{E}/B \to \mathbf{E}/A$. This functor is right adjoint to the functor $f_> : \mathbf{E}/A \to \mathbf{E}/B$, with the counit given by the morphisms $\varepsilon_v : f_> f^* v \to v$ in pullback squares (1).

PROOF. We have $h : f_> u \to v$ in $\mathbf{E}/B$ for a morphism h of $\mathbf{E}$ iff $fu = vh$ in $\mathbf{E}$ For the pullback (1), this means that $h = \varepsilon_v\, k$ in $\mathbf{E}$, and hence $h = \varepsilon_v \circ f_> k$ in $\mathbf{E}/B$, for a unique morphism $k : u \to f^* v$ of $\mathbf{E}/A$. Thus ε_v has the desired couniversal property.

17.6. We recall that an adjunction $F \dashv U : \mathbf{A} \to \mathbf{B}$ induces a monad on $\mathbf{B}$ and a comonad on $\mathbf{A}$, with comparison functors for algebras and coalgebras. The functor U is called *monadic* if the comparison functor from $\mathbf{A}$ to the category of algebras for the induced monad is an equivalence of categories. Dually, F is called *comonadic* if the comparison functor from $\mathbf{B}$ to the category of coalgebras for the induced comonad is an equivalence.

Theorem. *If $\mathbf{E}$ has pullbacks and $f : A \to B$ in $\mathbf{E}$, then the comparison functor for coalgebras, from $\mathbf{E}/A$ to the category of coalgebras for the induced comonad of $f_> \dashv f^*$, is an isomorphism of categories.*

PROOF. The comparison functor K assigns to an object u of $\mathbf{E}/A$ the coalgebra $Ku = (fu, f_> \eta_u)$, where η is the unit of the adjunction. The functor

K is faithful since $f_>$ is faithful. If $s : Ku \to Ku'$, then $fu's = fu$ and $\eta_{u'} \circ s = f^*s \circ \eta_u$ in $\mathbf{E}$. As $\eta_u : u \to f^*f_>u$ in $\mathbf{E}/A$, and similarly for $\eta_{u'}$, we have

$$u's = f^*f_>u' \circ \eta_{u'} \circ s = f^*f_>u' \circ f^*f_>s \circ \eta_u = f^*f_>u \circ \eta_u = u$$

in $\mathbf{E}$, and $s : u \to u'$ in $\mathbf{E}/A$. Thus K is full as well as faithful.

The functor K is injective on objects, because $f^*v \circ t = u$ in $\mathbf{E}$ for a coalgebra $(v,t) = Ku = (fu, \eta_u)$. On the other hand, if (v,t) is a coalgebra, with $t : v \to f_>f^*v$ in $\mathbf{E}/B$, then $v = fu$ in $\mathbf{E}$ if we put $u = f^*v \circ t$. We have

$$\varepsilon_v \circ t = \mathrm{id}_v \quad \text{and} \quad f_>f^*t \circ t = f_>\eta(f^*v) \circ t$$

in $\mathbf{E}/B$. Since $f_>$ does not change morphisms at the $\mathbf{E}$ level, and $t : u \to f^*v$ in $\mathbf{E}/A$, we also have

$$f^*t \circ t = \eta(f^*v) \circ t = f^*f_>t \circ \eta_u = f^*t \circ \eta_u$$

in $\mathbf{E}/A$. Compose this with $f^*\varepsilon_v$ on the left, and we get $t = \eta_u$ since $\varepsilon_v \circ t = \mathrm{id}_v$ in $\mathbf{E}/B$. Thus $(v,t) = Ku$, and K is bijective on objects.

17.7. Proposition. *If $\mathbf{E}$ has pullbacks and*

$$
\begin{array}{ccc}
\cdot & \xrightarrow{\;t\;} & C \\
{\scriptstyle s}\downarrow & & \downarrow{\scriptstyle g} \\
A & \xrightarrow{\;f\;} & B
\end{array}
$$

is a pullback square in $\mathbf{E}$, then the functors $t_> s^*$ and $g^* f_>$ are equivalent.

Proof. For an object u of $\mathbf{E}/A$, consider two pullback squares

$$
\begin{array}{ccccc}
\cdot & \xrightarrow{\;v\;} & \cdot & \xrightarrow{\;t\;} & C \\
\downarrow & & {\scriptstyle s}\downarrow & & \downarrow{\scriptstyle g} \\
\cdot & \xrightarrow{\;u\;} & A & \xrightarrow{\;f\;} & B
\end{array}
$$

in $\mathbf{E}$, with $v = s^*u$. Then $t_> s^*u = tv$, and the outer rectangle is a pullback. Thus tv and $g^* f_> u$ are isomorphic in $\mathbf{E}/C$.

18. Locally Cartesian Closed Categories

18.1. Definition. We consider in this section a category $\mathbf{E}$ with finite limits. We say that $\mathbf{E}$ is *locally cartesian closed* if every slice category $\mathbf{E}/A$, for an object A of $\mathbf{E}$, is cartesian closed.

If $\mathbf{E}$ is locally cartesian closed, then $\mathbf{E}$ is cartesian closed since $\mathbf{E}$ is isomorphic to $\mathbf{E}/1$ for a terminal object 1 of $\mathbf{E}$.

18.2. Theorem (B. DAY [23]). *A category $\mathbf{E}$ with finite limits is locally cartesian closed if and only if every pullback functor $f^* : \mathbf{E}/B \to \mathbf{E}/A$, for a morphism $f : A \to B$ of $\mathbf{E}$, has a right adjoint.*

We denote this adjoint by $f_* : \mathbf{E}/A \to \mathbf{E}/B$; it is also denoted by Π_f.

PROOF. For an object u of a category $\mathbf{E}/A$, the product functor $- \times_A u$ is equivalent to $u_> u^*$. If u^* always has a right adjoint u_*, then $- \times_A u$ has a right adjoint $u_* u^*$, and $\mathbf{E}/A$ is cartesian closed.

The converse follows immediately from 17.6 and 8.10, using 8.10 with $u_> \dashv u^*$ for $F \dashv U$, and $L = u^*$. The functors F' and U' are then $- \times_A u$ and its right adjoint.

18.3. Proposition. *If $\mathbf{E}$ is locally cartesian closed and*

$$
\begin{array}{ccc}
\cdot & \xrightarrow{\ t\ } & C \\
{\scriptstyle s}\downarrow & & \downarrow{\scriptstyle g} \\
A & \xrightarrow{\ f\ } & B
\end{array}
$$

is a pullback square in $\mathbf{E}$, then the functors $s_* t^*$ and $f^* g_*$ are equivalent.

PROOF. The two functors are right adjoint to the equivalent functors $t_> s^*$ and $g^* f_>$ of 17.7.

18.3.1. Corollary. *If m is monomorphic in $\mathbf{E}$ and the pullback functor m^* has a right adjoint m_*, then $m^* m_*$ is an equivalence.*

PROOF. We apply the Proposition to $f = g = m$, with $s = t = \mathrm{id}$.

18.4. We assume now that $\mathbf{E}$ satisfies the assumptions of 13.1.

Proposition. *If partial morphisms in $\mathbf{E}$ are represented, then every pullback functor m^* for $m \in \mathfrak{M}$ has a right adjoint, and $\mathbf{E}$ has a subobject classifier. Conversely, if $\mathbf{E}$ has a subobject classifier $\top : 1 \to \Omega$ and the functor $\top^*$ has a right adjoint, then partial morphisms in $\mathbf{E}$ are represented.*

PROOF. Partial morphisms with codomain 1 correspond bijectively to monomorphisms in $\mathcal{M}$; thus a morphism ϑ_1 representing these partial morphisms classifies subobjects. For $m : A \to B$ in $\mathcal{M}$ and $u : X \to A$ in $\mathbf{E}$, we construct $m_* u$ by three pullback squares

$$
\begin{array}{ccc}
A \xrightarrow{\ \mathrm{id}_A\ } A & X \xrightarrow{\ u\ } A & \cdot \longrightarrow \tilde{X} \\[2pt]
\downarrow m \quad \downarrow \vartheta_A & \downarrow \vartheta_X \quad \downarrow \vartheta_A & \downarrow m_* u \quad \downarrow \tilde{u} \\[2pt]
B \xrightarrow{\ \varphi\ } \tilde{A} & \tilde{X} \xrightarrow{\ \tilde{u}\ } \tilde{A} & B \xrightarrow{\ \varphi\ } \tilde{A}
\end{array}
$$

in $\mathbf{E}$, with ϑ_A and ϑ_X representing partial morphisms. For $v : Y \to B$, we have a bijection between morphism $f : v \to m_* u$ in $\mathbf{E}/B$, and $g : Y \to \tilde{X}$ in $\mathbf{E}$ with $\tilde{u}g = \varphi v$. Now pullback squares

$$
\begin{array}{ccc}
\cdot \xrightarrow{\ m^* v\ } A \xrightarrow{\ \mathrm{id}_A\ } A & & \cdot \xrightarrow{\ h\ } X \xrightarrow{\ u\ } A \\[2pt]
\downarrow z \quad \downarrow m \quad \downarrow \vartheta_A & \text{and} & \downarrow z \quad \downarrow \vartheta_X \quad \downarrow \vartheta_A \\[2pt]
Y \xrightarrow{\ v\ } B \xrightarrow{\ \varphi\ } A & & Y \xrightarrow{\ g\ } \tilde{X} \xrightarrow{\ \tilde{u}\ } \tilde{A}
\end{array}
$$

with the same outer rectangle provide a bijection between morphisms g as above and morphisms $h : m^* v \to u$ in $\mathbf{E}/A$. These bijections are natural in v; thus they define the desired right adjoint m_* of m^*.

Conversely, if the functor T^*, for $\mathsf{T} : 1 \to \Omega$ representing subobjects, has a right adjoint T_*, then $\mathsf{T}_* \tau_A$, for $\tau_A : A \to 1$ in $\mathbf{E}$, is the characteristic function of a strong monomorphism ϑ_A which represents partial morphisms with codomain A. This follows immediately from the more general result 19.2.

18.5. Proposition. *For a cartesian closed category $\mathbf{E}$, every pullback functor q^* for a projection $q : A \times B \to B$ of a product has a right adjoint. Conversely, if every pullback functor τ^* for a projection $\tau : B \to 1$ has a right adjoint, then $\mathbf{E}$ is cartesian closed.*

PROOF. We claim that a pullback square

$$
\begin{array}{ccc}
\cdot & \xrightarrow{\ h'\ } & X^B \\[2pt]
\downarrow{\scriptstyle q_* u} & & \downarrow{\scriptstyle u^B} \\[2pt]
A & \xrightarrow{\ h\ } & (A \times B)^B
\end{array}
$$

(1)

for $u : X \to A \times B$ in $\mathbf{E}$, with h exponentially adjoint to $\mathrm{id}_{A \times B}$, defines the desired right adjoint. For $v : Y \to A$, we have $q^* v = v \times \mathrm{id}_B$ up to isomorphism, and if $g : q^*v \to u$, then $ug = v \times \mathrm{id}_B$ is exponentially adjoint to $u^B \hat{g} = hv$, for $\hat{g}$ exponentially adjoint to g. Then $\hat{g} = h'f$ for a unique $f : v \to q_*u$ in (1), and this provides the adjunction $q^* \dashv q_*$.

For the converse, we observe first that τ is equivalent to the projection $1 \times B \to 1$. If we replace $\mathbf{E}$ by the isomorphic category $\mathbf{E}/1$, then $- \times B$ becomes $\tau_> \tau^*$, and this has a right adjoint $\tau_* \tau^*$ if $\tau^* \dashv \tau_*$.

18.6.　The following result has been called the Main Theorem of Topos Theory.

Theorem (LAWVERE–TIERNEY). *If $\mathbf{E}$ is cartesian closed and partial morphisms in $\mathbf{E}$ are represented, then every pullback functor $f^* : \mathbf{E}/B \to \mathbf{E}/A$, for $f : A \to B$ in $\mathbf{E}$, has a right adjoint $f_* : \mathbf{E}/A \to \mathbf{E}/B$.*

Corollary. *If $\mathbf{E}$ is a topos, then every functor $f^* : \mathbf{E}/B \to \mathbf{E}/A$, for $f : A \to B$ in $\mathbf{E}$, has a right adjoint $f_* : \mathbf{E}/A \to \mathbf{E}/B$.*

PROOF.　This follows immediately from 18.4 and 18.5 since f factors $f = q \cdot \langle \mathrm{id}_A, f \rangle$, with $\langle \mathrm{id}_A, f \rangle$ a section, hence a strong monomorphism, and $q : A \times B \to B$ the projection.

18.7. Example. For a set A, the slice category SET/A is equivalent to the category SET^A of families $(X_i)_{i \in A}$ of sets, indexed by A, with families $(f_i : X_i \to Y_i)_{i \in A}$ of mappings, indexed by A, as morphisms. The equivalence replaces $u : X \to A$ by the family of sets $(f^\frown(\{i\}))_{i \in A}$, and $(X_i)_{i \in A}$ by $u : X \to A$, with X the set of all pairs (x, i) with $i \in A$ and $x \in X_i$, and $u(x, i) = i$.

For $f : A \to B$, it is easily seen that f^* then sends a family $(Y_j)_{j \in B}$ to the family $(Y_{f(i)})_{i \in A}$ obtained by relabeling the sets Y_j, and $f_>$ and f_* send a family $(X_i)_i \in A$ to the families $(Y_j)_{j \in B}$ with

$$Y_j = \coprod_{f(i)=j} X_i \quad \text{and} \quad Y_j = \prod_{f(i)=j} X_i$$

respectively.

19. Two Definitions of Quasitopoi

19.1.　We consider in this Section a category $\mathbf{E}$ which satisfies the assumptions of 13.1. In Section 21, we shall also need the following additional condition.

19.1.1.　If f is a monomorphism of $\mathbf{E}$, then f can be factored $f = me$, with e monomorphic and epimorphic, and with $m \in \mathcal{M}$.

This condition is satisfied if **E** is a topos, with e an identity morphism. It is also satisfied, by 14.3 and 12.5, if **E** has cokernel pairs and a subobject classifier. By 14.3, $\mathcal{M}$ is the class of all equalizers in **E** if **E** has a subobject classifier. With 12.5, it follows from this and 19.1.1 that $\mathcal{M}$ is also the class of all strong monomorphisms of **E**.

19.2. Our next result extends a construction of J. PENON [83].

Proposition. *Let* $t : A \to B$ *in* **E**. *If* $\beta : B \to \tilde{B}$ *represents partial morphisms with codomain* B *and the pullback functor* β^* *has a right adjoint* β_*, *then there is a pullback square*

(1)
$$
\begin{array}{ccc}
A & \xrightarrow{\ t\ } & B \\
\Big\downarrow{\alpha} & & \Big\downarrow{\beta} \\
\tilde{A} & \xrightarrow{\beta_* t} & \tilde{B}
\end{array}
$$

in **E** *with* $\alpha : A \to \tilde{A}$ *representing partial morphisms with codomain* A.

PROOF. The pullback square (1) exists by 18.3.1, with $\alpha \in \mathcal{M}$ since $\beta \in \mathcal{M}$. For a partial morphism $X \xleftarrow{\ m\ } \cdot \xrightarrow{\ u\ } A$, consider two diagrams

(2)
$$
\begin{array}{ccccc}
\cdot & \xrightarrow{\ u\ } & A & \xrightarrow{\ t\ } & B \\
\Big\downarrow{m} & & \Big\downarrow{\alpha} & & \Big\downarrow{\beta} \\
X & \xrightarrow{\ \bar{u}\ } & \tilde{A} & \xrightarrow{\beta_* t} & \tilde{A}
\end{array}
\quad \text{and} \quad
\begin{array}{ccc}
\cdot & \xrightarrow{\ tu\ } & B \\
\Big\downarrow{m} & & \Big\downarrow{\beta} \\
X & \xrightarrow{\ v\ } & \tilde{B}
\end{array}
$$

of pullback squares. This determines v in the square at right uniquely, and we must have $\bar{u} : v \to \beta_* t$ in the squares at left. There is a pullback square $\beta \cdot \beta^* v = v \cdot m_1$, and then $\beta^* v = tuz$ and $m_1 = mz$ for a unique isomorphism z. If $\bar{u}$ is adjoint to $u_1 : \beta^* v \to t$ for the adjunction $\beta^* \dashv \beta_*$, then $u_1 = \varepsilon_t \cdot \beta^* \bar{u}$ for the counit ε_t of the adjunction, with $\beta^* \bar{u} : \beta^* v \to \beta^* \beta_* t$ a pullback of $\bar{u}$. Since ε_t is an isomorphism by 18.3.1, it follows that $u_1 : \beta^* v \to t$ and $\bar{u} : v \to \beta_* t$ are adjoint for $\beta^* \dashv \beta_*$ iff $\bar{u} m_1 = \alpha u_1$ is a pullback square. But then the lefthand square in (2) is a pullback square iff $\bar{u}$ is adjoint to uz for $\beta^* \dashv \beta_*$.

19.3. Definition. By 18.6 and 19.2, applied to $\beta = \top$, the following conditions are equivalent for a category **E** which satisfies the assumptions of 19.1.

(1) **E** is cartesian closed, and partial morphisms in **E** are represented.

(2) **E** is locally cartesian closed and has a subobject classifier.

We say that **E** is a *QUASITOPOS* if **E** satisfies these conditions, and **E** has finite colimits.

Heyting algebras (see Section 24) are quasitopoi, but not topoi, except for one-element lattices. Other examples will be given in Chapter 2 and in later chapters.

19.4. Theorem. *If* **E** *is a topos or a quasitopos, then every slice category* **E**/A *is a topos or a quasitopos.*

PROOF. Since $(\mathbf{E}/A)/u$ is isomorphic to $\mathbf{E}/D_A u$ for an object u of $\mathbf{E}/A$, the category $\mathbf{E}/A$ is locally cartesian closed. For the class of morphisms $m : u \to v$ of $\mathbf{E}/A$ with $m \in \mathcal{M}$, the projection $\omega : A \times \Omega \to A$ and the morphism $\langle \mathrm{id}_A, \top \tau_A \rangle : \mathrm{id}_A \to \omega$ of $\mathbf{E}/A$ define a subobject classifier, since it is easily seen that the righthand square of

$$
\begin{array}{ccc}
D_A u & \longrightarrow & 1 \\
\downarrow{\scriptstyle m} & & \downarrow{\scriptstyle \top} \\
D_A v & \xrightarrow{\ f\ } & \Omega
\end{array}
\qquad \text{and} \qquad
\begin{array}{ccc}
u & \xrightarrow{\ u\ } & \mathrm{id}_A \\
\downarrow{\scriptstyle m} & & \downarrow{\scriptstyle \langle \mathrm{id}_A, \top \tau_A \rangle} \\
v & \xrightarrow{\langle v, f \rangle} & \omega
\end{array}
$$

is a pullback in $\mathbf{E}/A$ iff the lefthand square is a pullback in **E**.

19.5. Discussion. Every topos is a quasitopos, but we cannot prove this until Section 22, where we show that topoi have finite colimits. Until this is done, we shall only use 19.1.1, and no colimits. This condition is trivial for topoi; it follows from 12.5 for quasitopoi.

For quasitopoi, the existence of finite colimits cannot be proved from conditions (1) and (2) of 19.1. An example for this is the half-open real interval $(0, 1]$ with the usual order, considered as a category. This category is locally cartesian closed and has a trivial subobject classifier, but it does not have an initial object. If we add elements α and β with $\frac{1}{2} < \alpha < t$ and $\frac{1}{2} < \beta < t$ for every $t > \frac{1}{2}$ in $(0, 1]$, then we still have a locally cartesian closed category, but now another finite colimit, the join $\alpha \vee \beta$, does not exist.

As noted in 19.1, $\mathcal{M}$ is the class of strong monomorphisms of **E** for every quasitopos **E**. Thus we have the following result.

19.6. Proposition. *A quasitopos* **E** *is a topos if and only if every monomorphism in* **E** *is strong.*

20. Universal Quantifiers

20.1. Notations. For an object A of a topos or quasitopos $\mathbf{E}$, we denote by $\operatorname{Mon} A$ and $\mathcal{M} A$ the full subcategories of $\mathbf{E}/A$ with monomorphisms and strong monomorphisms, with codomain A, as their objects. Both $\operatorname{Mon} A$ and $\mathcal{M} A$ are preordered classes, regarded as categories, and if $t : u \to v$ in $\operatorname{Mon} A$ or $\mathcal{M} A$, then t is also a monomorphism or strong monomorphism of $\mathbf{E}$.

If $f : A \to B$ in $\mathbf{E}$, then the pullback functor $f^* : \mathbf{E}/B \to \mathbf{E}/A$ maps $\operatorname{Mon} B$ into $\operatorname{Mon} A$, and $\mathcal{M} B$ into $\mathcal{M} A$. We shall denote by $f^\leftarrow : \mathcal{M} B \to \mathcal{M} A$ or $f^\leftarrow : \operatorname{Mon} B \to \operatorname{Mon} A$ the resulting restriction of f^*. We shall deal only with classes $\mathcal{M} A$ from now on, but the results which do not use relations or characteristic morphisms are also valid for classes $\operatorname{Mon} A$.

For objects A and B of $\mathbf{E}$, the preordered class $\operatorname{Rel}(A, B)$ of relations from A to B in $\mathbf{E}$ is isomorphic to $\mathcal{M}(A \times B)$, by the bijection $(u, v) \mapsto \langle u, v \rangle$. For $f : X \to A$ and $g : B \to C$ in $\mathbf{E}$, the compositions $- \circ f$ and $g^{\mathrm{op}} \circ -$ of 13.2 and 13.6 become functors $(f \times \operatorname{id}_B)^\leftarrow$ and $(\operatorname{id}_A \times g)^\leftarrow$ by these isomorphisms. By 15.1, the functors $(f \times \operatorname{id}_B)^\leftarrow$ and $(\operatorname{id}_A \times g)^\leftarrow$ can be regarded as substitutions. For topoi and quasitopoi, these substitution functors have left and right adjoints which can be regarded as defining existential and universal quantifiers. In this section, we consider right adjoints; existential quantifiers will be obtained in Section 33.

20.2. External universal quantifiers. For $f : A \to B$ in $\mathbf{E}$, the right adjoint f_* of the pullback functor f^* maps a strong monomorphism $m : m \to \operatorname{id}_A$ in $\mathbf{E}/A$ into a strong monomorphism $f_* m : f_* m \to \operatorname{id}_B$, up to an isomorphism. Thus f_* maps $\mathcal{M} A$ into $\mathcal{M} B$. We denote by $\forall_f : \mathcal{M} A \to \mathcal{M} B$ the resulting restriction of f_*, and we call this functor an *external universal quantifier* in $\mathbf{E}$.

Proposition. *For $f : A \to B$ in $\mathbf{E}$, the functor $\forall_f : \mathcal{M} A \to \mathcal{M} B$ is right adjoint to $f^\leftarrow : \mathcal{M} B \to \mathcal{M} A$.*

Corollary. $\forall_g \forall_f \simeq \forall_{gf}$ *if gf is defined in $\mathbf{E}$, and $\forall_f \simeq \operatorname{Id}_{\mathcal{M} A}$ for $f = \operatorname{id}_A$.*

PROOF. The Proposition follows immediately from the definitions, and then the Corollary follow immediately from the corresponding properties of pullback functors $f^\leftarrow$.

20.3. Example. For sets, subobjects can be replaced by subsets, and thus a functor $f^\leftarrow : \mathcal{M} B \to \mathcal{M} A$, for a mapping $f : A \to B$, becomes a mapping $f^\leftarrow : \mathrm{P} B \to \mathrm{P} A$. This is the usual inverse image mapping, with

$$x \in f^\leftarrow(T) \iff f(x) \in T$$

for $x \in A$ and $T \subset B$. The right adjoint $\forall_f$ then becomes a mapping $\forall_f : PA \to PB$, with

$$f^{\leftarrow}(T) \subset S \iff T \subset \forall_f(S)$$

for $S \subset A$ and $T \subset B$. It follows that

$$y \in \forall_f(S) \iff (\forall x \in A)(y = f(x) \implies x \in S),$$

for $S \subset A$ and $y \in B$. In particular, if f is a projection $A \times B \to B$ of a product, then

$$y \in \forall_f(S) \iff (\forall x \in A)((x, y) \in S)$$

for $S \subset A \times B$ and $y \in B$. Thus operators $\forall_f$ generalize universal quantifiers.

20.4. For $g : B \to C$ in $\mathbf{E}$, we can consider $\forall_{\mathrm{id} \times g}$ as a functor from $\mathrm{Rel}(A, B)$ to $\mathrm{Rel}(A, C)$, using the isomorphisms of 20.1. We then have the following result.

Proposition. *If $f : X \to A$ and $g : B \to C$ in $\mathbf{E}$, then*

$$\forall_{\mathrm{id} \times g}((u, v) \circ f) \simeq (\forall_{\mathrm{id} \times g}(u, v)) \circ f$$

for a relation $(u, v) : A \to B$ in $\mathbf{E}$.

PROOF. We note that we have a pullback square

$$
\begin{array}{ccc}
X \times B & \xrightarrow{\mathrm{id}_X \times g} & X \times C \\
{\scriptstyle f \times \mathrm{id}_B} \downarrow & & \downarrow {\scriptstyle f \times \mathrm{id}_C} \\
A \times B & \xrightarrow{\mathrm{id}_A \times g} & A \times C
\end{array}
$$

in $\mathbf{E}$, and thus

$$(\mathrm{id}_X \times g)_* (f \times \mathrm{id}_B)^* \simeq (f \times \mathrm{id}_C)^* (\mathrm{id}_A \times g)_*$$

by 18.3.1. Applying restrictions of the two equivalent functors to $\mathcal{M}(A \times B)$ and using the definitions, we get the desired result.

20.5. Internal universal quantifiers. For $g : B \to C$ in $\mathbf{E}$, we put

$$\forall g = \chi(\forall_{\mathrm{id} \times g} \ni_B) : PB \to PC.$$

This defines an *internal universal quantifier* in $\mathbf{E}$. Using this with 20.4 and $(u, v) \simeq \ni_B \circ \chi(u, v)$, we have the following result.

Proposition. *If $(u, v) : A \to B$ is a relation in $\mathbf{E}$, then*

$$\chi(\forall_{\mathrm{id} \times g}(u, v)) = \forall g \circ \chi(u, v)$$

for a morphism $g : B \to C$ of $\mathbf{E}$.

20.6. Proposition. *Internal universal quantifiers define a functor* $\forall$:
$\mathbf{E} \to \mathbf{E}$.

PROOF. This follows immediately from 20.2, 20.5, and the definitions.

20.7. Theorem. *If*

$$
\begin{array}{ccc}
\cdot & \xrightarrow{\ t\ } & C \\
{\scriptstyle s}\big\downarrow & & \big\downarrow{\scriptstyle g} \\
A & \xrightarrow{\ f\ } & B
\end{array}
$$

is a pullback square in a quasitopos $\mathbf{E}$, *then* $\forall_s\, t^{\leftarrow} \simeq f^{\leftarrow}\, \forall_g$, *and*

$$\forall s \circ \mathsf{P}t = \mathsf{P}f \circ \forall g$$

in $\mathbf{E}$.

PROOF. The first part follows immediately from 18.3 and the definitions.
For the second part, we observe that

$$
\begin{array}{ccc}
\cdot & \xrightarrow{\ \mathrm{id}_X \times t\ } & X \times C \\
{\scriptstyle \mathrm{id}_X \times s}\big\downarrow & & \big\downarrow{\scriptstyle \mathrm{id}_X \times g} \\
X \times A & \xrightarrow{\ \mathrm{id}_X \times f\ } & X \times B
\end{array}
$$

is also a pullback square in $\mathbf{E}$. Thus

$$\forall_{\mathrm{id}\times s}(t^{\mathrm{op}} \circ (u, v)) \simeq f^{\mathrm{op}} \circ (\forall_{\mathrm{id}\times g}(u, v))$$

for a relation $(u, v) : X \to C$ in $\mathbf{E}$, by the first part and the definitions. Now
the result follows immediately from 20.5 and 13.8.

20.8. Corollary. *If* $f : A \to B$ *is monomorphic in* $\mathbf{E}$, *then* $\mathsf{P}f$ *is a
retraction in* $\mathbf{E}$, *with* $\mathsf{P}f \circ \forall f = \mathrm{id}_{\mathsf{P}A}$.

PROOF. Use 20.7 with $g = f$, and $s = t = \mathrm{id}_A$.

21. Coarse Objects of a Quasitopos

21.1. Definition. Throughout this section, $\mathbf{E}$ will be a quasitopos, with
$\mathcal{M}$ the class of all strong monomorphisms of $\mathbf{E}$, although some results are valid
with weaker assumptions. We say that an object A of $\mathbf{E}$ is *coarse* if for every
pair consisting of a morphism $f : X \to A$ and a monomorphic and epimorphic

morphism $u : X \to Y$ of $\mathbf{E}$, there is a (unique) morphism $g : Y \to A$ in $\mathbf{E}$ such that $f = g u$ in $\mathbf{E}$. We denote by $\mathbf{E}^{\mathrm{gr}}$ the full subcategory of $\mathbf{E}$ with coarse (in French: grossier) objects of $\mathbf{E}$ as its objects.

For a Heyting algebra L considered as a category, every morphism is monomorphic and epimorphic; thus only the terminal object of L, its greatest element 1, is coarse.

21.2. We show that the theory of this section is trivial for topoi.

Proposition. *If $\mathbf{E}$ is a topos, then every object of $\mathbf{E}$ is coarse.*

PROOF. If $\mathbf{E}$ is a topos, then every monomorphism of $\mathbf{E}$ is an equalizer. If u is an equalizer and epimorphic, then u is an isomorphism. Thus the defining property of coarse objects is trivial if $\mathbf{E}$ is a topos.

21.3. Proposition. *The contravariant functor P on a quasitopos $\mathbf{E}$ is faithful, and a morphism u of $\mathbf{E}$ is monomorphic and epimorphic if and only if $\mathsf{P}u$ is an isomorphism.*

PROOF. For $f : A \to B$ in $\mathbf{E}$, the morphisms

$$\mathsf{P}f \circ s_B = \chi(f^{\mathrm{op}} \circ (\mathrm{id}_B, \mathrm{id}_B)) = \chi(f, \mathrm{id}_A) : B \to \mathsf{P}A$$

and $\qquad s_B \circ f = \chi((\mathrm{id}_B, \mathrm{id}_B) \circ f) = \chi(\mathrm{id}_A, f) : A \to \mathsf{P}B$

are exponentially adjoint. If $\mathsf{P}f = \mathsf{P}g$ for $g : A \to B$, it follows that $f = g$ since s_B is monomorphic by 15.4.

Now the faithful functor P reflects monomorphisms and epimorphisms; thus u is monomorphic and epimorphic if $\mathsf{P}u$ is isomorphic. On the other hand, $\mathsf{P}u$ is monomorphic if u is epimorphic, since the functor P is self-adjoint on the right. If $u : X \to Y$ is also monomorphic, then $\mathsf{P}u \circ \forall u = \mathrm{id}_{\mathsf{P}X}$ by 20.8; thus $\mathsf{P}u$ and $\forall u$ are inverse isomorphisms.

21.4. Proposition. *Every object $\mathsf{P}A$ of $\mathbf{E}$ is coarse. In particular, Ω is coarse.*

PROOF. If $f : X \to \mathsf{P}A$ and $g : Y \to \mathsf{P}A$, then $f = g u$ for $u : X \to Y$ iff $f^{\#} = \mathsf{P}u \circ g^{\#}$ for the exponentially adjoint morphisms $f^{\#} : A \to \mathsf{P}X$ of f and $g^{\#} : A \to \mathsf{P}Y$ of g. If u is monomorphic and epimorphic, then $\mathsf{P}u$ is isomorphic by 21.3; thus $g^{\#}$ and hence g always exist if f and u are given. We note for the last part that Ω is isomorphic to $\mathsf{P}1$ by 15.6.

21.5. Proposition. *If $m : A \to B$ is in $\mathcal{M}$ and B is coarse, then A is coarse.*

PROOF. If $f : X \to A$ and $u : X \to Y$, with u monomorphic and epimorphic, then $m f = g u$ for a morphism $g : Y \to B$. Since m is a strong monomorphism and u an epimorphism, we have $f = t u$ and $g = m t$ for a morphism $t : Y \to A$; this shows that A is coarse.

21.6. Theorem. *For an object A of a quasitopos* **E**, *the following statements are logically equivalent.*

(1) *The object A of* **E** *is coarse.*

(2) *Every monomorphic and epimorphic morphism of* **E** *with domain A is an isomorphism of* **E**.

(3) *Every monomorphism of* **E** *with domain A is strong.*

(4) *The monomorphism $s_A : A \to \mathsf{P}A$ of* **E** *is strong.*

PROOF. If (1) is valid, then for $u : A \to X$ monomorphic and epimorphic in **E**, there is $v : X \to A$ in **E** such that $v u = \mathrm{id}_A$. This is a two-sided inverse since u is epimorphic; thus (2) is valid.

If a monomorphism $f : A \to B$ in **E** factors $f = m u$ by 19.1.1, with m a strong monomorphism and u monomorphic and epimorphis, then u is an isomorphism, and hence f a strong monomorphism, if (2) holds. Thus $(2) \Longrightarrow (3)$.

$(3) \Longrightarrow (4)$ is trivial, and $(4) \Longrightarrow (1)$ follows immediately from 21.4 and 21.5.

21.7. Proposition. *The full subcategory* **E**^{gr} *of* **E** *with coarse objects of* **E** *as its objects is a reflective subcategory of* **E**, *with monomorphic and epimorphic reflections.*

PROOF. For an object A of **E**, factor $s_A = A \xrightarrow{u} X \xrightarrow{m} \mathsf{P}A$, with u monomorphic and epimorphic, and m a strong monomorphism, by 19.1.1. Then X is coarse by 21.4 and 21.5, and if $f : A \to B$ in **E** with B coarse, then f factors $f = g u$ in **E** for a unique morphism $g : X \to A$. Thus u is a reflection for **E**^{gr}.

21.8. Remarks. The proof of 21.7 shows that every monomorphic and epimorphic morphism u of **E** with coarse codomain is a reflection for **E**^{gr} in **E**. We also note that if $u : A \to X$ is a reflection for **E**^{gr}, then $\mathsf{P}u : \mathsf{P}X \to \mathsf{P}A$ is an isomorphism by 21.2.

21.9. Theorem (PENON [82]). *If* **E** *is a quasitopos, then the full subcategory* **E**^{gr} *of* **E** *with coarse objects of* **E** *as its objects is a topos.*

PROOF. All monomorphisms of **E**^{gr} are equalizers in **E** by 21.6 and 14.5. If m is an equalizer of morphisms f and g with codomain X, and if $u : X \to Y$

is a reflection for $\mathbf{E}^{\mathrm{gr}}$, then m is also an equalizer of uf and ug since u is monomorphic. Thus every monomorphism of $\mathbf{E}^{\mathrm{gr}}$ is an equalizer in $\mathbf{E}^{\mathrm{gr}}$.

Products in $\mathbf{E}^{\mathrm{gr}}$ are products in $\mathbf{E}$ since $\mathbf{E}^{\mathrm{gr}}$ is reflective in $\mathbf{E}$. For objects A and B of $\mathbf{E}^{\mathrm{gr}}$, it follows with 21.5 that relations $(u,v) : A \to B$ are the same in $\mathbf{E}^{\mathrm{gr}}$ as in $\mathbf{E}$. Since PB is coarse by 21.4, and pullbacks in $\mathbf{E}^{\mathrm{gr}}$ are pullbacks in $\mathbf{E}$, relations $(u,v) : A \to B$ in $\mathbf{E}^{\mathrm{gr}}$ have characteristic morphisms $\chi(u,v) : A \to PB$ in $\mathbf{E}^{\mathrm{gr}}$ which are inherited from $\mathbf{E}$. Thus $\mathbf{E}^{\mathrm{gr}}$ is indeed a topos.

22. The Dual Category of a Topos is Algebraic

22.1. The double powerset monad. We saw in 13.8 that the contravariant powerset functor $P : \mathbf{E}^{\mathrm{op}} \to \mathbf{E}$ on a quasitopos $\mathbf{E}$ is self-adjoint on the right. We denote by $h : \mathrm{Id}\,\mathbf{E} \to P\,P^{\mathrm{op}}$ the unit/counit of this adjunction, and by

$$\mathcal{P} = (P\,P^{\mathrm{op}}, h, P\,h^{\mathrm{op}}\,P^{\mathrm{op}})$$

the *double powerset monad* on $\mathbf{E}$ obtained from this adjunction. An algebra for this monad is called a *double powerset algebra* or $\mathcal{P}$-*algebra*. We denote by $\mathbf{E}^{\mathcal{P}}$ the category of double powerset algebras and their homomorphisms.

22.2. We recall that an object of $\mathbf{E}^{\mathcal{P}}$ is a pair (A, α), consisting of an object A and a morphism $\alpha : PPA \to A$ of $\mathbf{E}$ such that

$$\alpha \circ h_A = \mathrm{id}_A \quad \text{and} \quad \alpha \circ P\,P\alpha = \alpha \circ P\,h_{PA}$$

in $\mathbf{E}$. A homomorphism $f : (A, \alpha) \to (B, \beta)$ of $\mathcal{P}$-algebras is a morphism $f : A \to B$ of $\mathbf{E}$ such that

$$f \circ \alpha = \beta \circ PPf$$

in $\mathbf{E}$. By the general theory, the functor P lifts to a comparison functor $K : \mathbf{E}^{\mathrm{op}} \to \mathbf{E}^{\mathcal{P}}$, given by

$$K\,X = (PX, Ph_X)$$

for an object X of $\mathbf{E}$. We shall study this functor.

22.3. Proposition. *The comparison functor* $K : \mathbf{E}^{\mathrm{op}} \to \mathbf{E}^{\mathcal{P}}$ *has an adjoint on the right* $H : (\mathbf{E}^{\mathcal{P}})^{\mathrm{op}} \to \mathbf{E}$, *given by equalizer forks*

$$(1) \qquad H(A, \alpha) \xrightarrow{\rho_{A,\alpha}} PA \underset{h_{PA}}{\overset{P\alpha}{\rightrightarrows}} PPPA$$

in $\mathbf{E}$.

PROOF. We have $f : (A, \alpha) \to K X$ in $\mathbf{E}^{\mathcal{P}}$ iff the morphism $f : A \to \mathsf{P}X$ of $\mathbf{E}$ satisfies

$$f \circ \alpha = \mathsf{P}h_X \circ \mathsf{PP}f = \mathsf{P}f^\# \,,$$

with $f^\# = \mathsf{P}f \circ h_X : X \to \mathsf{P}A$ exponentially adjoint to f. By the adjunction $\mathsf{P}^{\mathrm{op}} \dashv \mathsf{P}$, this is the case iff

$$\mathsf{P}\alpha \circ f^\# = h_{\mathsf{P}A} \circ f^\# \,,$$

and thus iff $f^\# = \rho_{A,\alpha} \circ g$ for a unique morphism $g : X \to H(A, \alpha)$. This construction is clearly natural in X and thus defines the desired adjunction.

22.4. We note that the contravariant adjunction $H^{\mathrm{op}} \dashv K$ has two units, $\mathrm{Id} \to K H^{\mathrm{op}}$ and $\mathrm{Id} \to H K^{\mathrm{op}}$, instead of a unit and a counit.

Proposition. *The unit* $\mathrm{Id}\,\mathbf{E}^{\mathcal{P}} \to K H^{\mathrm{op}}$ *is a natural isomorphism, given by isomorphisms* $f : (A, \alpha) \to K H(A, \alpha)$ *in* $\mathbf{E}^{\mathcal{P}}$, *with* $f^\# : H(A, \alpha) \to \mathsf{P}A$ *an equalizer of* $\mathsf{P}\alpha$ *and* $h_{\mathsf{P}A}$ *in* $\mathbf{E}$, *and with* $K f^\# = f \circ \alpha$.

PROOF. For an algebra (A, α) and $X = H(A, \alpha)$, we obtain this unit at (A, α) by putting $g = \mathrm{id}_X$ in the proof of 22.3, with $f^\# = \rho_{A,\alpha} : X \to \mathsf{P}A$ the equalizer in 22.2.(1). We have

$$\mathsf{P}h_A \circ h_{\mathsf{P}A} = \mathrm{id}_{\mathsf{P}A} = \mathsf{P}h_A \circ \mathsf{P}\alpha \,,$$

since h is the unit of a self-adjunction, and $\alpha h_A = \mathrm{id}_A$. Thus $x = y$ if $h_{\mathsf{P}A} \circ x = \mathsf{P}\alpha \circ y$, and

$$
\begin{array}{ccc}
X & \xrightarrow{\;f^\#\;} & \mathsf{P}A \\[4pt]
\downarrow{\scriptstyle f^\#} & & \downarrow{\scriptstyle h_{\mathsf{P}A}} \\[4pt]
\mathsf{P}A & \xrightarrow{\;\mathsf{P}\alpha\;} & \mathsf{PPP}A
\end{array}
$$

is a pullback square. Now $\forall f^\# \circ \mathsf{P}f^\# = \mathsf{PP}\alpha \circ \forall h_{\mathsf{P}A}$ and $\mathsf{P}h_{\mathsf{P}A} \circ \forall h_{\mathsf{P}A} = \mathrm{id}_{\mathsf{P}A}$ by 20.7 and 20.8, and thus

$$\alpha \circ \forall f^\# \circ \mathsf{P}f^\# = \alpha \circ \mathsf{PP}\alpha \circ \forall h_{\mathsf{P}A} = \alpha \circ \mathsf{P}h_{\mathsf{P}A} \circ \forall h_{\mathsf{P}A} = \alpha \,.$$

Since also $f \circ \alpha = \mathsf{P}f^\#$ by 22.3, and α and $\mathsf{P}f^\#$ are epimorphisms, it follows that f and $\alpha \circ \forall f^\#$ are inverse isomorphisms.

22.5. Proposition. *Units* $X \to H K X$ *are monomorphic and epimorphic, and isomorphisms for coarse objects* X.

PROOF. This unit is g in the proof of 22.3, for $f = \mathrm{id}_{KX}$ and $f^\# = h_X$. Thus $h_X = m g$ for an equalizer fork $X \xrightarrow{\;m\;} \mathsf{P}X \underset{\mathsf{PP}h_X}{\overset{h_{\mathsf{PP}X}}{\rightrightarrows}} \mathsf{PPP}X$ in $\mathbf{E}$. In this situation, $m^\# : KX \to KHKX$ is a unit for $H^{op} \dashv K$, by 22.4. Thus $m^\#$ is an isomorphism, with inverse Pg since $Pg \circ m^\# = \mathrm{id}_{PX}$ by exponential adjunction. Now g is monomorphic and epimorphic by 21.3, and an isomorphism by 21.6.(3) if X is coarse.

22.6. In [81], R. PARÉ proved that the dual category $\mathbf{E}^{op}$ of a topos $\mathbf{E}$ is equivalent to the algebraic category $\mathbf{E}^{\mathcal{P}}$ over $\mathbf{E}$. We generalize this result to quasitopoi.

Theorem. *The category* $\mathbf{E}^{\mathcal{P}}$ *of double powerset algebras for a quasitopos* $\mathbf{E}$ *is equivalent to the dual category of the topos* $\mathbf{E}^{\mathrm{gr}}$.

PROOF. Every object $H(A, \alpha)$ of $\mathbf{E}$ is coarse by 21.4 and 21.5, as the domain of an equalizer with codomain PA. Thus if we restrict the domain of K and the codomain of H to coarse objects, we obtain functors $K' : (\mathbf{E}^{\mathrm{gr}})^{op} \to \mathbf{E}^{\mathcal{P}}$ and $H' : (\mathbf{E}^{\mathcal{P}})^{op} \to \mathbf{E}^{\mathrm{gr}}$ which are adjoint on the right, with both units natural isomorphisms.

22.7. Corollary. *Every topos has finite colimits.*

PROOF. If $\mathbf{E}$ is a topos, then $\mathbf{E}^{\mathcal{P}}$ is equivalent to $\mathbf{E}^{op}$, by 22.6 and 21.2. Since the category $\mathbf{E}^{\mathcal{P}}$ of $\mathcal{P}$-algebras inherits finite limits from $\mathbf{E}$, it follows that $\mathbf{E}^{op}$ has finite limits, *i.e.* $\mathbf{E}$ has finite colimits.

22.8. Corollary. *Every topos is a quasitopos.*

PROOF. By 16.8 and 18.6, a topos satisfies conditions (1) and (2) of 19.3. and we just proved that a topos has finite colimits. Thus a topos meets all conditions of 19.3.

23. Exactness Properties of Quasitopoi

23.1. In this section, we consider again a topos or quasitopos $\mathbf{E}$. We recall from 10.4 that an epimorphism e of $\mathbf{E}$ is called *strong* if every commutative square $mu = ve$ in $\mathbf{E}$ with m monomorphic admits a (unique) morphism t such that $u = te$ and $v = mt$. Thus strong epimorphisms are dual to strong monomorphisms.

Theorem. *Pullbacks in a quasitopos* $\mathbf{E}$ *preserve epimorphisms and strong epimorphisms.*

PROOF. We have noted that a morphism e with codomain B is an epimorphism or strong epimorphism in $\mathbf{E}$ iff $e : e \to \mathrm{id}_B$ is an epimorphism or strong epimorphism in $\mathbf{E}/B$. If $f : A \to B$ in $\mathbf{E}$, then $f^* \, \mathrm{id}_B \simeq \mathrm{id}_A$; thus f^* changes $e : e \to \mathrm{id}_B$ to $f^*e : f^*e \to \mathrm{id}_A$, up to an isomorphism. Since f^* has a right adjoint, it also preserves epimorphisms and strong epimorphisms, in $\mathbf{E}/B$ and hence in $\mathbf{E}$.

23.2. Proposition. *If e_1 and e_2 are epimorphisms or strong epimorphisms in $\mathbf{E}$, then $e_1 \times e_2$ is also an epimorphism or strong epimorphism.*

PROOF. We note that $e_1 \times e_2$ is a composition of morphisms $e_1 \times \mathrm{id}$ and $\mathrm{id} \times e_2$. These morphisms are pullbacks of e_1 and e_2, by projections of products, and thus epimorphisms or strong epimorphisms by 23.1.

23.3. Theorem. *Every morphism f in a quasitopos $\mathbf{E}$ can be factored $f = m_1 \, e_1$ with e_1 epimorphic and m_1 an equalizer in $\mathbf{E}$ of a cokernel pair of f, and also $f = m_2 \, e_2$ with e_2 a coequalizer of a kernel pair of f and m_2 monomorphic in $\mathbf{E}$.*

Corollary. *Every strong monomorphism of $\mathbf{E}$ is an equalizer, and every strong epimorphism a coequalizer. If $\mathbf{E}$ is a topos, then every epimorphism of $\mathbf{E}$ is a coequalizer.*

PROOF. We have proved the first part of the Theorem in 12.5. For the second part, let (a, b) be a kernel pair of f, and e a coequalizer of a and b. Then f factors $f = m e$; we must show that m is monomorphic.

If $mu = mv$, construct a pullback

$$\begin{array}{ccc} \cdot & \xrightarrow{\langle x, y \rangle} & \cdot \\ {\scriptstyle t}\big\downarrow & & \big\downarrow {\scriptstyle e \times e} \\ \cdot & \xrightarrow[\langle u, v \rangle]{} & \cdot \end{array}$$

in $\mathbf{E}$, with t epimorphic by 23.1 and 23.2. By this construction,

$$fx = mex = mut = mvt = mey = fy$$

in $\mathbf{E}$; thus $x = as$ and $y = bs$ for a morphism s. But then

$$ut = eas = ebs = vt \,,$$

and $u = v$ since t is epimorphic. Thus m is monomorphic.

The first part of the Corollary was proved in 12.5; the second part is proved dually. For the last part, factor an epimorphism f as $f = m_2 e_2$ with e_2 a coequalizer and m_2 monomorphic. Then m_2 is also epimorphic, hence an isomorphism by 14.5.

23.4. Let $f : A \to B$ in $\mathbf{E}$. If $\lambda : \Delta \to B$ is a cone in $\mathbf{E}$, then we can regard Δ as a diagram in $\mathbf{E}/B$, with vertices given by λ, and λ as a cone $\lambda : \Delta \to \mathrm{id}_B$ in $\mathbf{E}/B$. The pullback functor f^* sends this cone to a cone $f^*\lambda : f^*\Delta \to \mathrm{id}_A$ in $\mathbf{E}/A$, or equivalently to a cone $f^*\lambda : f^*\Delta \to A$ in $\mathbf{E}$. We say that *colimits in $\mathbf{E}$ are universal* if $f^*\lambda$ is always a colimit cone for a colimit cone λ.

Theorem. *Colimits in a quasitopos are universal.*

PROOF. If $\lambda : \Delta \to B$ is a colimit cone in $\mathbf{E}$, then $\lambda : \Delta \to \mathrm{id}_B$ is a colimit cone in $\mathbf{E}/B$ by 17.4. Since the functor f^* has a right adjoint, the cone $f^*\lambda : f^*\Delta \to \mathrm{id}_A$ is a colimit cone in $\mathbf{E}/A$.

For an object X of $\mathbf{E}$, let $A \xleftarrow{p} A \times X \xrightarrow{q} X$ be the projections of the product. For a cone $\rho : f^*\Delta \to X$ in $\mathbf{E}$, there is a unique cone $\sigma : \Delta \to A \times X$ with $p\sigma = f^*\lambda$ and $q\sigma = \rho$, and $\sigma : f^*\Delta \to p$ in $\mathbf{E}/A$. Thus $\sigma = s \circ f^*\lambda$ for a unique morphism $s : \mathrm{id}_A \to p$. If $t = qs$ in $\mathbf{E}$, then $s = \langle \mathrm{id}_A, t \rangle$, and $\rho = t \circ f^*\lambda$ follows.

Conversely, if $\rho = t \circ f^*\lambda$ in $\mathbf{E}$, then $\sigma = s \circ f^*\lambda$ for $s = \langle \mathrm{id}_A, t \rangle : \mathrm{id}_A \to p$ in $\mathbf{E}/A$, and this determines s and hence t uniquely. Thus $f^*\lambda$ is a colimit cone in $\mathbf{E}$ as claimed.

23.5. Theorem. *If $f : P \to 0$ in a quasitopos $\mathbf{E}$ with 0 an initial object of $\mathbf{E}$, then f is an isomorphism and P an initial object of $\mathbf{E}$.*

Corollary. *If 0 is an initial object of $\mathbf{E}$, then every morphism $h : 0 \to A$ of $\mathbf{E}$ is monomorphic in $\mathbf{E}$.*

PROOF. The initial object 0 is a colimit of an empty cone. Pulling this back by f, we get P as colimit of an empty cone, by 23.4. Thus P is also an initial object, and f an isomorphism.

Now there is at most one morphism $f : X \to 0$ for an object X of $\mathbf{E}$; thus a morphism h with domain 0 is trivially monomorphic.

23.6. Theorem. *If m is a strong monomorphism and*

(1)
$$
\begin{array}{ccc}
\cdot & \xrightarrow{f} & \cdot \\
{\scriptstyle m}\Big\downarrow & & \Big\downarrow{\scriptstyle m'} \\
\cdot & \xrightarrow{f'} & \cdot
\end{array}
$$

a pushout square in a quasitopos $\mathbf{E}$, then m' is a strong monomorphism of $\mathbf{E}$, and (1) is also a pullback square.

PROOF. By 19.3.(1) and the definitions, we have a pullback square

$$
\begin{array}{ccc}
\cdot & \xrightarrow{\ f\ } & \cdot \\
\downarrow m & & \downarrow s \\
\cdot & \xrightarrow{\ \varphi\ } & \cdot
\end{array}
\ ,
$$

with s a strong monomorphism. Then $\varphi = tf'$ and $s = tm'$ for a morphism t, and it follows easily that (1) is also a pullback square, and m' strongly monomorphic.

23.7. Corollary. *If* **E** *is a topos, then the injections* $A \xrightarrow{\ h\ } A \amalg B \xleftarrow{\ k\ } B$ *of a coproduct are monomorphisms, and the pushout square*

$$
\begin{array}{ccc}
0 & \longrightarrow & B \\
\downarrow & & \downarrow k \\
A & \xrightarrow{\ h\ } & A \amalg B
\end{array}
$$

is also a pullback square.

PROOF. This follows immediately from 23.6 and the Corollary of 23.5.

23.8. Proposition. *For a pullback square*

(1)
$$
\begin{array}{ccc}
X \cap Y & \xrightarrow{\ r\ } & X \\
\downarrow s & & \downarrow a \\
Y & \xrightarrow{\ b\ } & A
\end{array}
$$

of monomorphisms in **E**, *there is a pushout square*

(2)
$$
\begin{array}{ccc}
X \cap Y & \xrightarrow{\ r\ } & X \\
\downarrow s & & \downarrow u \\
Y & \xrightarrow{\ v\ } & X \cup Y
\end{array}
$$

in **E**, *with a monomorphism* $m : X \cup Y \to A$ *such that* $a = mu$ *and* $b = mv$ *in* **E**.

PROOF. We form a coproduct $X\xrightarrow{p}X \amalg Y\xleftarrow{q}Y$ in $\mathbf{E}$. If $a = fp$ and $b = fq$, then $a\xrightarrow{p}f\xleftarrow{q}b$ is a coproduct in $\mathbf{E}/A$. Now let (g,h) be a kernel pair of f in $\mathbf{E}$; then $fg = fh$ is a product

$$f \times_A f = (a \amalg b) \times_A (a \amalg b)$$

in $\mathbf{E}/A$. Since $\mathbf{E}/A$ is cartesian closed, then this is also a coproduct of products

$$a \times_A a, \quad a \times_A b, \quad b \times_A a, \quad b \times_A b$$

in $\mathbf{E}/A$, with $a \times_A a \simeq a$ and $b \times_A b \simeq b$ since a and b are monomorphic in $\mathbf{E}/A$, with codomain the terminal object id_A. This is also a coproduct in $\mathbf{E}$. The injections

$$x : a \to f, \quad y : a \times_A b \to f, \quad z : b \times_A a \to f, \quad w : b \to f$$

of this coproduct are given by

$$gx = p = hx; \qquad gy = pr, \quad hy = qs;$$
$$gz = qs, \quad hz = pr; \qquad gw = q = hw.$$

If we factor $f = me$ with e a coequalizer of g and h, and m monomorphic in $\mathbf{E}$ by 23.3, then we claim that (2) is a pushout square, in $\mathbf{E}$ and in $\mathbf{E}/A$, for $u = ep$ and $v = eq$, and hence $a = mu$ and $b = mv$.

Indeed, if $\alpha r = \beta s$ in $\mathbf{E}$, then $\alpha = \gamma p$ and $\beta = \gamma q$ for a unique morphism γ. We have

$$\gamma gx = \gamma hx \qquad \text{and} \qquad \gamma gw = \gamma hw$$

trivially, and also

$$\gamma gy = \gamma pr = \alpha r = \beta s = \gamma qs = \gamma hy,$$

and

$$\gamma gz = \gamma qs = \beta s = \alpha r = \gamma pr = \gamma hz,$$

by our construction. Since x, y, z, w are the injections of a coproduct, $\gamma g = \gamma h$ follows. But then γ factors $\gamma = te$ for a unique morphism t, with $\alpha = tu$ and $\beta = tv$.

Chapter 2

EXAMPLES OF TOPOI AND QUASITOPOI

As we already noted in Chapter 1, Heyting algebras are quasitopoi. We begin this chapter with a look at Heyting algebras, and in particular complete Heyting algebras. This includes a brief discussion, in Section 25, of the duality between complete Heyting algebra and topological spaces. As an application, we obtain a construction of all finite Heyting algebras, and hence up to equivalence of all finite quasitopoi, from finite ordered sets.

Categories $[\mathcal{C}^{op}, \mathrm{SET}]$ of set-valued functors, and categories of set-valued sheaves over a topological space, are standard and important examples of topoi. We obtain the basic constructions for topoi $[\mathcal{C}^{op}, \mathrm{SET}]$ in Section 26, and we give examples and complements in Section 27. In Sections 28 and 29, we construct the topos $\mathrm{Sh}\,H$ of set-valued sheaves, and the quasitopos $\mathrm{Sep}\,H$ of set-valued separated presheaves, over a complete Heyting algebra H. We specialize to sheaves over a topological space in Section 30, with emphasis on the representation of sheaves on a space X by local homeomorphisms $p : Y \to X$, which generalize covering spaces of X.

The examples in Sections $26-29$ will be generalized in Chapters 4 and 5. Except for results in Section 30 which involve local homeomorphisms, most results of Sections $26-30$ are special cases of general properties of topoi and quasitopoi, but these special cases are of special and often independent interest.

Quasitopoi also occur in topology. Topological categories typically have morphisms which are monomorphic and epimorphic, but not isomorphic. Thus non-trivial topological categories are not topoi. We discuss topological quasitopoi over sets, with constant maps, in Section 31, with some examples. It follows from the results of Chapter 7 that every topological category over sets, with constant maps, can be embedded into a topological quasitopos over sets, with constant maps.

24. Heyting Algebras

24.1. Definitions. A lattice L is called a *Heyting algebra* if L, considered as a category, is cartesian closed. This means that L has a binary operation $\to$ such that always

$$a \cap b \leqslant c \iff a \leqslant b \to c,$$

for elements a, b, c of L. We say that L is a *Heyting prealgebra* if L is a preordered class with finite meets and finite joins, *i.e.* a prelattice with 0 and 1, and L is cartesian closed when considered as a category. If the lattice operations are interpreted as logical connectives, then $\rightarrow$ becomes implication.

Since partial morphisms in a prelattice are trivially represented, every Heyting algebra or prealgebra is a quasitopos.

The algebraic theory of Heyting algebras has the nullary and binary operations $\top, \bot, \cap, \cup$ of lattice theory, denoting greatest and least element, meet and join, and an additional binary operation $\rightarrow$. It has the usual formal laws of lattice theory,

$$a \cap b = b \cap a, \qquad\qquad a \cup b = b \cup a,$$
$$a \cap (b \cap c) = (a \cap b) \cap c, \qquad a \cup (b \cup c) = (a \cup b) \cup c,$$
$$\top \cap a = a, \qquad\qquad \bot \cup a = a,$$
$$a \cap (a \cup b) = a, \qquad\qquad a \cup (a \cap b) = a,$$

and four additional formal laws as follows:

$$a \cap (a \rightarrow b) = a \cap b, \qquad\qquad a \rightarrow a = \top,$$
$$b \cap (a \rightarrow b) = b, \qquad\qquad a \rightarrow (b \cap c) = (a \rightarrow b) \cap (a \rightarrow c).$$

Order is then defined by

$$a \cap b = a \iff a \leqslant b \iff a \cup b = b;$$

and this relation has the expected properties.

We refer to RASIOWA and SIKORSKI [89] for a detailed discussion of Heyting algebras, there called *pseudocomplemented lattices*, from a logician's viewpoint. The Compendium of Continuous Lattices [35] also contains some material on Heyting algebras and complete Heyting algebras.

24.2. Theorem. *A category* **E** *is a Heyting prealgebra if and only if* **E** *is a quasitopos, and the only strong monomorphisms of* **E** *are the isomorphisms.*

PROOF. If **E** has finite limits and the only strong monomorphisms are the isomorphisms, then two parallel morphisms $f : X \rightarrow Y$ and $g : X \rightarrow Y$ of **E** have an equalizer id_X in **E**, and thus $f = g$. But then **E** is a preordered class with finite meets, with at most one morphism $f : X \rightarrow Y$ for objects X and Y. Conversely, every morphism of a preordered class is monomorphic and epimorphic; thus only isomorphisms are strong monomorphisms. In this situation, partial morphisms in **E** are trivially represented by identity morphisms, and **E** is a quasitopos iff **E** has finite joins and is cartesian closed, *i.e.* iff **E** is a Heyting prealgebra.

24.3. We recall from 20.1 that $\operatorname{Mon} A$ and $\mathcal{M} A$ denote the full subcategories of $\mathbf{E}/A$ with monomorphisms and strong monomorphisms of $\mathbf{E}$, with codomain A, as objects.

Proposition. *If $\mathbf{E}$ is a quasitopos, then the preordered classes $\operatorname{Mon} A$ and $\mathcal{M} A$ are Heyting prealgebras for every object A of $\mathbf{E}$.*

PROOF. Meets in $\operatorname{Mon} A$ and in $\mathcal{M} A$ are pullbacks. Joins in $\operatorname{Mon} A$ are obtained as pushouts by 23.8. Both classes have a greatest element id_A, and $o_A : 0 \to A$ is a least element of $\operatorname{Mon} A$ by 23.5. If m is a finite join of strong monomorphisms in $\operatorname{Mon} A$, and we factor $m = m_1 u$ by 11.5 or 23.3, with u monomorphic and epimorphic and m_1 a strong monomorphism, then m_1 clearly is the corresponding join in $\mathcal{M} A$. Thus $\operatorname{Mon} A$ and $\mathcal{M} A$ are prelattices.

A functor $a \cap -$, on $\operatorname{Mon} A$ or $\mathcal{M} A$, is a restriction of the functor $a_> a^*$ on $\mathbf{E}/A$. The functor $a_> a^*$ has a right adjoint $a_* a^*$ which preserves monomorphisms and strong monomorphisms. The restriction of this right adjoint, to $\operatorname{Mon} A$ or $\mathcal{M} A$, defines a right adjoint $a \to -$ of $a \cap -$. Thus $\operatorname{Mon} A$ and $\mathcal{M} A$ are Heyting prealgebras.

24.4. Proposition. *A linearly ordered set with a least element 0 and a greatest element 1 is a Heyting algebra, with $a \to b = 1$ if $a \leqslant b$, and $a \to b = b$ otherwise.*

PROOF. In any lattice, we have $a \to b = 1$ if $a \leqslant b$. If $b < a$ in a linearly ordered set, then $a \cap x \leqslant b \iff x \leqslant b$.

24.5. Proposition. *Every Boolean algebra is a Heyting algebra, with $a \to b = a' \cup b$ for the complement a' of a.*

PROOF. If $x \cap a \leqslant b$, then $x = x \cap (a' \cup a) \leqslant a' \cup b$. Conversely, $(a' \cup b) \cap a = b \cap a \leqslant b$.

24.6. Complete Heyting algebras. We define a *complete Heyting algebra* as a complete lattice which is also a Heyting algebra. A *homomorphism* $f : L \to M$ of complete Heyting algebras is a mapping of the underlying sets which preserves finite meets and arbitrary joins. We denote by **CHA** the category of complete Heyting algebras and their homomorphisms.

Theorem. *A complete lattice L is a complete Heyting algebra if and only if L satisfies the infinite distributive law*

$$(1) \qquad a \cap \left(\bigcup_{i \in I} x_i \right) = \bigcup_{i \in I} (a \cap x_i),$$

for $a \in L$ and every family $(x_i)_{i \in I}$ of elements of L.

PROOF. If L is a complete Heyting algebra, then the functor $a \cap -$ on L has a right adjoint. Thus it preserves arbitrary joins, which are colimits. Conversely,

$$a \to b = \bigcup_{a \cap x \leqslant b} x$$

satisfies $t \leqslant a \to b \iff a \cap t \leqslant b$ if L satisfies (1).

24.7. Proposition. *Open sets of a topological space X form a complete Heyting algebra $\mathcal{O}X$, and if $f : X \to Y$ is a continuous map of topological spaces, then inverse images by f induce a homomorphism $\mathcal{O}f : \mathcal{O}Y \to \mathcal{O}X$ of complete Heyting algebras.*

Thus lattices of open sets, and inverse images by continuous maps, define a contravariant functor $\mathcal{O} : \mathrm{TOP}^{\mathrm{op}} \to \mathrm{CHA}$ from topological spaces to complete Heyting algebras.

PROOF. Open sets of X form a complete lattice, with set unions as joins and finite set intersections as finite meets. The general distributive law

$$A \cap \left(\bigcup B_i \right) = \bigcup (A \cap B_i)$$

is satisfied for set intersections and set unions, and hence in $\mathcal{O}_X$. The map $f^\leftarrow$ always preserves set unions and set intersections, and $f^\leftarrow$ preserves open sets if f is continuous. Thus the restriction $\mathcal{O}f : \mathcal{O}Y \to \mathcal{O}X$ of $f^\leftarrow$ to $\mathcal{O}Y$ and $\mathcal{O}X$ is a morphism of CHA.

24.8. Prime elements and filters. We recall that a *filter* in a lattice L is a subset φ of L with the following three properties. (i) The greatest element $\top$ of L is in φ. (ii) For elements a and b of φ, the element $a \cap b$ of L always is in φ. (iii) If $a \leqslant b$ in L and $a \in \varphi$, then also $b \in \varphi$. A filter *on* a set A is a filter in the lattice $\mathsf{P}A$ of subsets of A, ordered by set inclusion. For every element a of a lattice L, the set $\uparrow a$ of all $x \geqslant a$ in L is a filter in L, called the *principal filter* generated by a, and sometimes denoted by $[a]$. We note that we regard the set L as a filter in L. This filter, the principal filter $\uparrow \bot$ for the least element $\bot$ of L, is sometimes called the *null filter* in L. A filter in the dual lattice L^{op} of a lattice L is called a *dual filter* or an *ideal* in L.

We order filters in a lattice L dually to inclusion, *i.e.* we put $\varphi \leqslant \psi$ for filters φ and ψ in L if $\psi \subset \varphi$. This order has many technical advantages over the order by inclusion used by many authors. Filters in L form a complete lattice, with meets $\varphi \cap \psi$ consisting of all meets $a \cap b$ in L with $a \in \varphi$ and $b \in \psi$, and with set intersections as suprema. The mapping $a \mapsto \uparrow a$, from L to the lattice of filters in L, preserves finite meets and all suprema.

An element x of a lattice L is called *prime* in L if the decreasing set $L \setminus \uparrow x$ is a dual filter in L. This is the case iff (i) $x \neq \perp$, and (ii) for a and b in L such that $a \cup b \leqslant x$, always $a \leqslant x$ or $b \leqslant x$. A *dually prime* or *coprime* element of L is a prime element of L^{op}. A *prime filter* in a lattice L is a filter φ in L which is prime in the complete lattice of filters in L. It ie easily seen that this is the case iff

$$a \cup b \in \varphi \qquad \Longleftrightarrow \qquad a \in \varphi \quad \text{or} \quad b \in \varphi,$$

for elements a and b of L.

We say that a filter φ in a complete lattice L is *completely prime* if for a subset S of L, the join $\bigcup S$ is in φ iff $S \cap \varphi \neq \emptyset$. It follows that if $s = \bigcup(L \setminus \varphi)$, then $\varphi = L \setminus \downarrow s$. Thus this element s is coprime. Conversely, if s is a coprime element in a complete lattice and $\varphi = L \setminus \downarrow s$, then it is easily seen that the filter φ in L is completely prime, and $s = \bigcup(L \setminus \varphi) \notin \varphi$. Thus we have a bijection between coprime elements and completely prime filters in a complete lattice L.

It should be noted that some authors use "prime" and "coprime" dually to the sense given above, and they may even consider the greatest element $\top$ of L as "prime".

24.9. Examples. Natural numbers form a complete lattice with the divisibility order, *i.e.* $m \leqslant n$ iff n is a multiple of m. The least element of this lattice is the number 1, and the greatest element is 0. For positive integers, order in the divisibility lattice is compatible with the usual order.

The divisibility lattice $\mathbb{N}$ thus defined is distributive and complete, with $\sup S = 0$ for every infinite subset S of $\mathbb{N}$, but it does not satisfy the infinite distributive law 24.6.(1). Meets and joins in $\mathbb{N}$ are greatest common divisors and least common multiples. It is easily seen that the prime elements of $\mathbb{N}$ are 0 and the powers p^k of prime numbers p, with $k > 0$, and that $\mathbb{N}$ has no coprime elements. We note that the prime numbers are the atoms of $\mathbb{N}$, *i.e.* natural numbers $p \neq 1$ such that $1 \mid x$ and $x \mid p$ only for $x = 1$ and $x = p$. We also note that every element of $\mathbb{N}$ is a supremum of prime elements, *i.e.* powers of prime numbers.

A linearly ordered set with a least element 0 and a greatest element 1 is a Heyting algebra with every element except 1 prime, and every element except 0 coprime.

25. Spectral Theory

25.1. Spectral spaces. For a complete Heyting algebra H, we denote by ΣH the topological space with the completely prime filters in H as points, and with the *hull-kernel topology*, i.e. the coarsest topology for which the sets

$$a^* = \{\varphi \in \Sigma H \mid a \in \varphi\}$$

are open. We note that $\bigcap a_i^* = (\bigcap a_i)^*$ for a finite family of elements of H since the elements of ΣH are filters in H, and that $\bigcup a_i^* = (\bigcup a_i)^*$ for an arbitrary family of elements of H since these filters are completely prime. Thus every open set of ΣH is one of the sets a^*. The space ΣH is called the *spectral space* of H.

Spectral spaces are discussed in some detail in [35], but the notations of [35] are dual to ours, and "prime" (*i.e.* coprime) elements are used instead of completely prime filters.

25.2. We denote by TOP the category of topological spaces and continuous maps, and by TOP_0 the full subcategory of T_0 spaces.

Theorem. *Spectral spaces define a contravariant functor* $\Sigma : \mathrm{CHA}^{\mathrm{op}} \to \mathrm{TOP}$, *adjoint on the right to the functor* $\mathcal{O} : \mathrm{TOP}^{\mathrm{op}} \to \mathrm{CHA}$. *For a topological space X and a complete Heyting algebra H, morphisms $f : X \to \Sigma H$ and $g : H \to \mathcal{O}X$ are adjoint if and only if*

$$(1) \qquad\qquad a \in f(x) \iff x \in g(a),$$

for all $a \in H$ and $x \in X$.

PROOF. If we consider f and g as mappings $X \to \mathrm{P}H$ and $H \to \mathrm{P}X$, then each of the mappings determines the other one by (1). In particular, if $x \in X$, then $1 \in f(x) \iff x \in g(1)$; thus $1 \in f(x)$ for all x iff $g(1) = X$, the greatest element of $\mathcal{O}X$. From this, and from

$$a \cap b \in f(x) \iff x \in g(a \cap b)$$

and $\qquad\qquad a \in f(x) \wedge b \in f(x) \iff x \in g(a) \wedge x \in g(b),$

we see that each $f(x)$ is a filter iff g preserves finite meets. It follows in the same way that each filter $f(x)$ is completely prime iff g preserves all suprema. Finally, $f^\leftarrow(a^*) = g(a)$ by (1); thus f is continuous iff each $g(a)$ is an open set of X. This establishes the desired bijection, which is clearly natural in X and H.

25.3. Units. The contravariant adjunction $\Sigma^{\mathrm{op}} \dashv \mathcal{O}$ has two units which we denote by

$$\varepsilon : \mathrm{Id}_{\mathrm{CHA}} \to \mathcal{O}\Sigma^{\mathrm{op}} \qquad \text{and} \qquad \nu : \mathrm{Id}_{\mathrm{TOP}} \to \Sigma\mathcal{O}^{\mathrm{op}}.$$

For a complete Heyting algebra H and $a \in H$, we have $\varepsilon_H(a) = a^*$; thus ε_H is surjective. If X is a topological space and $x \in X$, then $\nu_X(x)$ is the filter of open neighborhoods of x in X. Thus ν_X is injective if, and only if, X is a T_0 space. The continuous map ν_X is always coarse; we have $\nu_X\,\!\!^{\leftarrow}(U^*) = U$ for an open set U of X.

As for every contravariant adjunction, we always have

$$(1) \qquad \Sigma\nu_X \circ \varepsilon_{\mathcal{O}X} = \mathrm{id}_{\mathcal{O}X}, \qquad \text{and} \qquad \mathcal{O}\varepsilon_H \circ \nu_{\Sigma H} = \mathrm{id}_{\Sigma H}.$$

Since $\varepsilon_{\mathcal{O}X}$ is surjective and $\mathcal{O}\varepsilon_H$ injective, the two morphisms in the first equation are inverse isomorphisms of complete Heyting algebras, and the two maps in the second equation inverse homeomorphisms.

25.4. Duality. We say that a T_0 space X is *sober* if the unit ν_X is a homeomorphism, and that a complete Heyting algebra H is *pointed* if ε_H is an isomorphism.

Theorem. *Every spectral space ΣH is sober, every complete Heyting algebra $\mathcal{O}X$ is pointed, and the adjunction $\Sigma^{\mathrm{op}} \dashv \mathcal{O}$ induces a duality between the categories of pointed complete Heyting algebras and sober topological spaces.*

PROOF. This follows immediately from 25.3 and the definitions.

25.5. Discussion. If φ is a completely prime filter in $\mathcal{O}X$ for a topological space X, then φ consists by 24.8 of all open sets $U \not\subset V$ of X, for some coprime open set V of X. If $F = X \setminus V$, then F is closed, and φ consists of all open sets of X which meet F. The set F is irreducible, *i.e.* if F is the union of two closed sets, then one of these sets is F. Conversely, if F is an irreducible closed set in X, then the open sets of X which meet F form a completely prime filter φ in $\mathcal{O}X$. In this situation, $\varphi = \nu_X(x)$ for a point x of X if and only if F is the closure of $\{x\}$ in X.

We can introduce a pre-order in a topological space X by putting $x \leqslant y$ for x, y in X if $\nu_X(x) \leqslant \nu_X(y)$, *i.e.* if every neighborhood of y is a neighborhood of x, or equivalently if y is in the closure of $\{x\}$. This pre-order is an order if X is a T_0 space, and discrete if X is a T_1 space. Open sets are decreasing and closed sets increasing; the converse need not be true. The closure of $x \in X$ is $\uparrow x$ in this pre-order. Thus X is sober if and only if every irreducible closed set of X is of the form $F = \uparrow x$, for a unique $x \in F$.

25.6. The following result shows that all finite Heyting algebras are isomorphic to algebras $\mathcal{O}X$ for finite T_0 spaces X, with topologies determined by the induced order.

Theorem. *Every finite T_0 space X is sober, and the open sets of X are the decreasing sets for the induced order of X. Every finite distributive lattice is a pointed complete Heyting algebra.*

PROOF. For a point x of X, the points $y \leqslant x$ are the points in every open neighborhood of x, hence in the intersection $\downarrow x$ of all neighborhoods of x. If X is finite, then this intersection is an open neighborhood of x. It follows that every decreasing set is open, as a set union of open sets $\downarrow x$. Thus every increasing set is closed if X is finite. Let now y be a minimal point in a closed set F of X, and form the set union F' of all sets $\uparrow z$ with $z \in F \setminus \uparrow y$. Then F' is closed in the finite space X, and $F = F' \cup \uparrow y$, with $y \notin F'$ since y is minimal in F. Thus $F = \uparrow y$ if F is irreducibe, and X is sober if X is a T_0 space.

A finite distributive lattice H is a complete Heyting algebra by 24.6, and the completely prime filters are the principal filters $[p] = \uparrow p$ in H with p prime in H. For elements a, b of H with $b \not\leqslant a$, let p be a minimal element of H such that $p \leqslant b$ and $p \not\leqslant a$. If $u \cup v \geqslant p$ in H, then $p = u' \cup v'$ for $u' = u \cap p$ and $v' = v \cap p$. It follows that $u' = p$ or $v' = p$, for otherwise $u' \leqslant a$ and $v' \leqslant a$, hence $p \leqslant a$. Thus p is prime, and $\uparrow p$ is in b^*, but not in a^*. This shows that H is pointed.

26. Set-valued Presheaves

26.1. Definition. If $\mathcal{C}$ is a small category, then a contravariant functor $F : \mathcal{C}^{op} \to \mathbf{A}$ is called a *presheaf* on $\mathcal{C}$ with values in $\mathbf{A}$.

In this Section, we consider only set-valued presheaves; these presheaves are the objects of a functor category $[\mathcal{C}^{op}, \mathrm{SET}]$.

26.2. Theorem. *For every small category $\mathcal{C}$, the category $[\mathcal{C}^{op}, \mathrm{SET}]$ of set-valued presheaves on $\mathcal{C}$ is a topos.*

PROOF. This theorem can be viewed as a special case of Theorem 58.2 in Chapter 6, but we prefer to prove it by explicit constructions. We obtain subobjects and finite limits in 26.3, the subobject classifier in 26.4, the cartesian closed structure of $[\mathcal{C}^{op}, \mathrm{SET}]$ in 26.5, and powerset objects in 26.6.

26.3. Subpresheaves. By 7.5, the category $[\mathcal{C}^{\mathrm{op}}, \mathrm{SET}]$ admits all finite limits, indeed all small limits, since the category of sets admits all small limits.

A morphism $\mu : A \to B$ of presheaves clearly is monomorphic if every mapping μ_i, for an object i of $\mathcal{C}$ is injective. Conversely, we form a kernel pair of μ in $[\mathcal{C}^{\mathrm{op}}, \mathrm{SET}]$ by forming a kernel pair of each μ_i in sets. It follows that μ is monomorphic in $[\mathcal{C}^{\mathrm{op}}, \mathrm{SET}]$ only if every μ_i is injective.

We say that a set-valued functor $S : \mathcal{C}^{\mathrm{op}} \to \mathrm{SET}$ is a *subfunctor* of a functor $A : \mathcal{C}^{\mathrm{op}} \to \mathrm{SET}$ if Si is a subset of Ai for every object i of $\mathcal{C}$, and Sf a subset restriction of Af for every morphism f of $\mathcal{C}$. If this is the case, then subset inclusions $Si \to Ai$ define a *subfunctor inclusion* $S \to A$ in $[\mathcal{C}^{\mathrm{op}}, \mathrm{SET}]$. It is easily seen, using the discussion of the preceding paragraph, that every monomorphism in $[\mathcal{C}^{\mathrm{op}}, \mathrm{SET}]$ is equivalent to a unique subfunctor inclusion. Thus subobjects in $[\mathcal{C}^{\mathrm{op}}, \mathrm{SET}]$ can be replaced by subfunctors.

26.4. Subfunctors classified by sieves. For an object a of a category $\mathcal{C}$, a *sieve* on $\mathcal{C}$ at a is defined as a collection S of morphisms of $\mathcal{C}$ with codomain a, such that always $uf \in S$ if $u \in S$ and uf is defined in $\mathcal{C}$.

A sieve S at an object a of $\mathcal{C}$ can be regarded as a subfunctor of the functor Ya, with Si consisting of all morphisms $u : i \to a$ of $\mathcal{C}$ which are in S. We call Ya, regarded as a sieve, the *full sieve* at a, consisting of all morphisms of $\mathcal{C}$ with codomain a. As a sieve, Ya is characterized by the condition $\mathrm{id}_a \in Ya$.

Some authors define sieves at an object a of $\mathcal{C}$ as arbitrary sets of morphisms of $\mathcal{C}$ with codomain a, and they call sieves as defined above *saturated sieves*. If S is a set of morphisms of $\mathcal{C}$, with common codomain a, then the *saturation* of S can be defined as the sieve, in our sense, of all possible compositions uf in $\mathcal{C}$ with $u \in S$. We say that this sieve is *generated* by S.

For a small category $\mathcal{C}$, we denote by Ωa the set of all sieves on $\mathcal{C}$ at an object a of $\mathcal{C}$. For $g : a \to b$ in $\mathcal{C}$, we define a mapping $\Omega g : \Omega b \to \Omega a$ by putting

$$u \in (\Omega g)(T) \quad \Longleftrightarrow \quad gu \in T,$$

for a sieve T at b and a morphism $u : i \to a$ of $\mathcal{C}$. This is clearly functorial, and $(\Omega g)(Yb) = Ya$.

A terminal object 1 of $[\mathcal{C}^{\mathrm{op}}, \mathrm{SET}]$ assigns to every object a of $\mathcal{C}$ a singleton. Thus a morphism $\rho : 1 \to A$ in $[\mathcal{C}^{\mathrm{op}}, \mathrm{SET}]$ is obtained by choosing a member ρ_a of Aa for every object a of $\mathcal{C}$, so that $(Ag)(\rho_b) = \rho_a$ if $g : a \to b$ in $\mathcal{C}$, and morphisms $\rho : 1 \to A$ correspond bijectively to subfunctors R of A with each set Ra a singleton. In particular, we define $\top$ as the subfunctor of Ω with $\top_a = \{Ya\}$ for every object a of $\mathcal{C}$. We claim that this defines a subobject classifier for $[\mathcal{C}^{\mathrm{op}}, \mathrm{SET}]$.

For a presheaf A, we need a bijection between subfunctors S of A and morphisms $\sigma : A \to \Omega$ of presheaves, so that a square

$$
\begin{array}{ccc}
S & \longrightarrow & \top \\
\downarrow & & \downarrow \\
A & \xrightarrow{\ \sigma\ } & \Omega
\end{array}
$$

with subfunctor inclusions as vertical arrows, is a pullback square in $[\mathcal{C}^{\mathrm{op}}, \mathrm{SET}]$ iff σ corresponds to S. This is achieved by putting

(1) $$u \in \sigma_a(x) \quad \Longleftrightarrow \quad (Au)(x) \in Si,$$

for $u : i \to a$ in $\mathcal{C}$ and $x \in Aa$. We have in particular

(2) $$x \in Sa \quad \Longleftrightarrow \quad \sigma_a(x) = Ya$$

for $x \in Aa$. We omit the easy verifications.

26.5. Objects $[A, B]$. We recall from Section 17 that for every object a of $\mathcal{C}$, there is a slice category $\mathcal{C}/a$, and a domain functor $D_a : \mathcal{C}/a \to \mathcal{C}$. For a presheaf A on $\mathcal{C}$ and an object a of $\mathcal{C}$, we denote by $A \downarrow a$ the functor $AD_a{}^{\mathrm{op}} : (\mathcal{C}/a)^{\mathrm{op}} \to \mathrm{ENS}$.

For presheaves A and B on $\mathcal{C}$, and for an object a of $\mathcal{C}$, we denote by $[A, B]a$ the set of all natural transformations $\sigma : A \downarrow a \to B \downarrow a$. Thus a member σ of $[A, B]a$ assigns to every arrow $u : i \to a$ of $\mathcal{C}$ a mapping $\sigma_u : Ai \to Bi$ such that always

$$Bf \circ \sigma_u = \sigma_{uf} \circ Af$$

for $f : i \to j$ in $\mathcal{C}$, or $f : uf \to u$ in $\mathcal{C}/a$.

For $g : a \to b$ in $\mathcal{C}$, we define a mapping

$$[A, B]g : [A, B]b \to [A, B]a : \tau \mapsto \tau g$$

by putting $(\tau g)_u = \tau_{gu}$ for $u : i \to a$ in $\mathcal{C}$. It is easily seen that this works, and that the sets $[A, B]a$ and the mappings $[A, B]g$ define a presheaf $[A, B]$ on $\mathcal{C}$. It remains to prove that presheaves $[A, B]$ provide a cartesian closed structure for $[\mathcal{C}^{\mathrm{op}}, \mathrm{SET}]$.

For a presheaf F on $\mathcal{C}$, a morphism $\varphi : F \times A \to B$ of presheaves assigns to every object i of $\mathcal{C}$ a mapping $\varphi_i : Fi \times Ai \to Bi$ such that always

(1) $$\varphi_j \circ (Ff \times Af) = Bf \circ \varphi_i$$

for $f : j \to i$ in $\mathcal{C}$. A morphism $\hat{\varphi} : F \to [A, B]$ assigns to $x \in Fa$ a member $\hat{\varphi}(x)$ of $[A, B]a$, and hence to $u : i \to a$ in $\mathcal{C}$ and $x \in Fa$ a mapping $\hat{\varphi}(x)_u : Ai \to Bi$. Thus $\hat{\varphi}$ assigns to $u : i \to a$ a mapping $\hat{\varphi}(-)_u : Fa \to (Bi)^{Ai}$. Using exponential adjunction for sets, we can say that $\hat{\varphi}$ is determined by mappings $\tilde{\varphi}_u : Fa \times Ai \to Bi$, one for every morphism $u : i \to a$ of $\mathcal{C}$, so that $\hat{\varphi}(x)_u = \tilde{\varphi}_u(x, -)$ for every $x \in Fa$. It is easily seen that these mappings $\tilde{\varphi}$ must satisfy the naturality requirements

$$(2) \qquad \tilde{\varphi}_{gu} = \tilde{\varphi}_u \circ (Fg \times \mathrm{id}_{Ai}),$$

for $g : a \to b$ in $\mathcal{C}$, and

$$(3) \qquad \tilde{\varphi}_{uf} \circ (\mathrm{id}_{Fa} \times Af) = Bf \circ \tilde{\varphi}_u,$$

for $f : j \to i$ in $\mathcal{C}$.

For the desired bijection between morphisms $\varphi : F \times A \to B$ and morphisms $\hat{\varphi} : F \to [A, B]$ in $[\mathcal{C}^{\mathrm{op}}, \mathrm{SET}]$, we let $\hat{\varphi}$ correspond to φ if always

$$\varphi_i = \tilde{\varphi}_{\mathrm{id}_i} \qquad \text{and} \qquad \tilde{\varphi}_u = \varphi_i \circ (Fu \times \mathrm{id}_{Ai}),$$

for an object i and an arrow $u : i \to a$ of $\mathcal{C}$. We omit the easy verification that this provides the desired bijection between collections of mappings φ_i which satisfy (1), and collections of mappings $\tilde{\varphi}_u$ satisfying (2) and (3), and that this bijection is natural in F.

26.6. Powerset objects. By the general theory, objects $[A, \Omega]$ are powerset objects PA in $[\mathcal{C}^{\mathrm{op}}, \mathrm{SET}]$, with natural transformations $A \downarrow a \to \Omega \downarrow a$ as elements of a set $(PA)a$.

Proposition. *For an object A of $[\mathcal{C}^{\mathrm{op}}, \mathrm{SET}]$, there is a natural bijection between natural transformations $A \downarrow a \to \Omega \downarrow a$ and subfunctors of the functor $A \downarrow a$.*

PROOF. As in 26.5, a natural transformation $\sigma : A \downarrow a \to \Omega \downarrow a$ assigns to every arrow $u : i \to a$ of $\mathcal{C}$ a mapping $\sigma_u : Ai \to \Omega i$, so that always

$$(1) \qquad \sigma_{uf} \circ Af = \Omega f \circ \sigma_u$$

for $f : j \to i$ in $\mathcal{C}$. A subfunctor S of $A \downarrow a$ is obtained by assigning to every arrow $u : i \to a$ of $\mathcal{C}$ a subset Su of the set Ai, so that Af always maps Su into $S(uf)$ for $f : j \to i$ in $\mathcal{C}$. We obtain σ corresponding to S by putting

$$(2) \qquad f \in \sigma_u(x) \quad \Longleftrightarrow \quad (Af)(x) \in S(uf),$$

for $j \xrightarrow{f} i \xrightarrow{u} a$ in $\mathcal{C}$ and $x \in Ai$. It is easily verified that each $\sigma_u(x)$ is a sieve if S is a subfunctor of $A \downarrow a$, and that σ satisfies (1). Putting $f = \mathrm{id}_i$ in (2), we get S back from σ by

$$(3) \qquad x \in Su \quad \Longleftrightarrow \quad \mathrm{id}_i \in \sigma_u(x).$$

Then (2) follows from (1), and it is easily verified that (3) defines a subfunctor S of $A \downarrow a$, if $\sigma : A \downarrow a \to \Omega \downarrow a$ in $[\mathcal{C}^{\mathrm{op}}, \mathrm{SET}]$.

27. Examples and Complements

27.1. Examples. We note four examples of topoi $[\mathcal{C}^{\mathrm{op}}, \mathrm{SET}]$.

(1) If $\mathcal{C}$ is discrete, with a set I of objects, then $[\mathcal{C}^{\mathrm{op}}, \mathrm{SET}]$ is the product category SET^I, with families $(A_i)_{i \in I}$ of sets indexed by I as its objects.

(2) If $\mathcal{C}$ has one object, then the morphisms of $\mathcal{C}$ form a monoid M, and $[\mathcal{C}^{\mathrm{op}}, \mathrm{SET}]$ is the category of *right M-sets*.

(3) If $\mathcal{C} = 0 \overset{\alpha}{\longrightarrow} 1$, with two objects and one arrow, then the objects of $[\mathcal{C}^{\mathrm{op}}, \mathrm{SET}]$ are mappings $\alpha_A : A_1 \to A_0$, and morphisms $(f_1, f_0) : \alpha_A \to \alpha_B$ are commutative squares in SET, pairs of mappings such that $f_0 \cdot \alpha_A = \alpha_B \cdot f_1$.

(4) If $\mathcal{C} = 0 \overset{s}{\underset{t}{\rightrightarrows}} 1$, with two objects and two parallel arrows, then objects of $[\mathcal{C}^{\mathrm{op}}, \mathrm{SET}]$ are directed graphs. A *directed graph* A consists of a set A_0 of *vertices*, a set A_1 of *arrows*, and mappings $s_A : A_1 \to A_0$ and $t_A : A_1 \to A_0$ which assign to every arrow α its *source* $s_A \alpha$ and its *target* $t_A \alpha$. In this example, a morphism $f : A \to B$ of $[\mathcal{C}^{\mathrm{op}}, \mathrm{SET}]$ is a homomorphism of directed graphs, consisting of mappings $f_0 : A_0 \to B_0$ and $f_1 : A_1 \to B_1$ such that $f_0 s_A = s_B f_1$ for sources, and $f_0 t_A = t_B f_1$ for targets. This means that graph homomorphisms must map arrows to arrows, vertices to vertices, and preserve sources and targets.

We shall describe the constructions of Section 26 for these examples, omitting the easy proofs.

27.2. Powers of SET. If $\mathcal{C}$ is discrete, with a set I of objects, then $[\mathcal{C}^{\mathrm{op}}, \mathrm{SET}] = \mathrm{SET}^I$, with families $A = (A_i)_{i \in I}$ of sets as objects, and families of mappings $f_i : A_i \to B_i$ as morphisms $f : A \to B$. In this situation, $\Omega = 2^I$ for the set $2 = \{0, 1\}$. A subfamily of a family $(A_i)_{i \in I}$ assigns to every $i \in I$ a subset S_i of A_i, and the characteristic morphism of the subfamily inclusion $S \to A$ is the family of characteristic functions $A_i \to 2$, in the usual sense, of the components S_i of S. If A and B are families of sets, indexed by I, then $[A, B]$ consists of all I-indexed families of mappings $f_i : A_i \to B_i$.

27.3. M-sets. If a monoid M is viewed as a category with one object, then the objects of $[M^{\mathrm{op}}, \mathrm{SET}]$ are right M-sets, *i.e.* sets A with an action of M from the right, with the formal laws

$$x \cdot e = x, \qquad x \cdot (ab) = (x \cdot a) \cdot b,$$

for $x \in A$, the neutral element e of M, and a, b in M. A morphism $f : A \to B$ of right M-sets is a mapping $f : A \to B$ of the underlying sets such that always

$$f(x \cdot a) = f(x) \cdot a,$$

for $x \in A$ and $a \in M$.

For this category, Ω is the set of right ideals of M, *i.e.* of subsets S of M with $ab \in S$ whenever $a \in S$ and $b \in M$, with M acting on sieves by

$$Sg = \{x \in M \mid gx \in S\}.$$

For M-sets A and B, an element of $[A, B]$ is a family of mappings $\sigma_a : A \to B$, one for each $a \in M$, so that always

$$\sigma_a(x) \cdot b = \sigma_{ab}(x \cdot b),$$

and M acts on these families by $(\sigma \cdot a)_b = \sigma_{ab}$.

In the particular case that M is a group G, then $\Omega = \{\emptyset, G\}$ with the trivial action of G, and an element σ of $[A, B]$ is given by a single mapping $\sigma_e : A \to B$, for the neutral element e of G, with

$$(\sigma_e \cdot g)(x) = \sigma_g(x) = \sigma_e(x \cdot g^{-1}) \cdot g$$

for $x \in A$ and $g \in G$.

27.4. Mappings and commutative squares. If $\mathcal{C}$ is the category $0 \overset{\alpha}{\longrightarrow} 1$, then $\Omega_0 = \{\emptyset, \{0\}\}$, and $\Omega_1 = \{\emptyset, \{\alpha\}, \{\alpha, 1\}\}$, with $\alpha_\Omega(\emptyset) = \emptyset$, and $\alpha_\Omega(S) = \{0\}$ otherwise. For an object $\alpha_A : A_1 \to A_0$, a subfunctor is a restriction $\alpha_S : S_1 \to S_0$ of α_A to sets $S_1 \subset A_1$ and $S_0 \subset A_0$, with $\alpha_A^{\to}(S_1) \subset S_0$.

For objects A and B of $[\mathcal{C}^{op}, \mathrm{SET}]$, the set $[A, B]_0$ is the set of all mappings $\sigma_0 : A_0 \to B_0$, and $[A, B]_1$ the set of all commutative squares $(\sigma_1, \sigma_0) : A \to B$, with $\alpha_{[A,B]}(\sigma_1, \sigma_0) = \sigma_0$.

27.5. Directed graphs. In this example, the graph Ω of sieves has two vertices $0 = \emptyset$ and $1 = \{0\}$, and five arrows

$$\emptyset : 0 \to 0, \qquad \{s\} : 1 \to 0, \qquad \{t\} : 0 \to 1,$$
$$\{s, t\} : 1 \to 1, \qquad \{s, t, 1\} : 1 \to 1.$$

For a directed graph A, the graph PA has as vertices all subsets of A_0, and as arrows $S : S_s \to S_t$, for sets S_s and S_t of vertices of A, all sets S of arrows of A such that

$$s_A^{\to}(S) \subset S_s \qquad \text{and} \qquad t_A^{\to}(S) \subset S_t.$$

For directed graphs A and B, the graph $[A, B]$ has as vertices all mappings $A_0 \to B_0$, from the set A_0 of vertices of A to the set B_0 of vertices of B. For vertices σ_s and σ_t of $[A, B]$, an arrow $\sigma : \sigma_s \to \sigma_t$ is a mapping $\sigma : A_1 \to B_1$, from the arrows of A to the arrows of B, which preserves sources and targets, i.e. σ satisfies the conditions

$$s_B \circ \sigma = \sigma_s \circ s_A \qquad \text{and} \qquad t_B \circ \sigma = \sigma_t \circ t_A.$$

27.6. Functors $[\mu, B]$ and $[A, \mu]$. We return now to the general theory of presheaf categories $[\mathcal{C}^{\mathrm{op}}, \mathrm{SET}]$.

Natural transformations $\mu : C \to A$ and $\nu : B \to D$ in $[\mathcal{C}^{\mathrm{op}}, \mathrm{SET}]$ define natural transformations

$$[\mu, B] : [A, B] \to [C, B] \qquad \text{and} \qquad [A, \nu] : [A, B] \to [A, D],$$

with $[\mu, B] \circ \hat{\varphi}$ adjoint to $\varphi \circ (\mathrm{id}_F \times \mu)$, and $[A, \nu] \circ \hat{\varphi}$ adjoint to $\nu \circ \varphi$, if $\hat{\varphi} : F \to [A, B]$ is adjoint to $\varphi : F \times A \to B$. We show that $[\mu, B]$ and $[A, \nu]$ are defined by compositions.

Proposition. *For natural transformations* $\mu : C \to A$ *and* $\nu : B \to D$ *in* $[\mathcal{C}^{\mathrm{op}}, \mathrm{SET}]$, *and for* $\sigma : A \downarrow a \to B \downarrow a$ *in* $[A, B]a$ *with* a *an object of* $\mathcal{C}$, *we have*

$$[\mu, B]_a(\sigma) = \sigma \cdot \mu \qquad \text{and} \qquad [A, \nu]_a(\sigma) = \nu \cdot \sigma,$$

with $\qquad (\sigma \cdot \mu)_u = \sigma_u \circ \mu_i \qquad \text{and} \qquad (\nu \cdot \sigma)_u = \nu_i \circ \sigma_u,$

for $u : i \to a$ *in* $\mathcal{C}$.

PROOF. If we replace $\hat{\varphi}$ by $[\mu, B] \circ \hat{\varphi}$, then we replace the corresponding $\tilde{\varphi}_u$ in 26.5 by $\tilde{\varphi}_u \circ (\mathrm{id}_{Fa} \times \mu_i)$, for $u : i \to a$ in $\mathcal{C}$. Thus if $\hat{\varphi}_a(x) = \sigma$, for some $x \in FA$, then $[\mu, B]_a(\sigma) = \tau$, with

$$\tau_u = \tilde{\varphi}_u(x, -) \circ \mu_i = \sigma_u \circ \mu_i.$$

Used for $\hat{\varphi} = \mathrm{id}_{[A, B]}$, this proves the lefthand part of the Proposition. The righthand part is proved in the same way; we omit the details.

27.7. Inverse images in $[\mathcal{C}^{\mathrm{op}}, \mathrm{SET}]$. For $\mu : A \to B$ in $[\mathcal{C}^{\mathrm{op}}, \mathrm{SET}]$, we obtain the inverse image morphism $\mathsf{P}\mu : \mathsf{P}B \to \mathsf{P}A$ from the morphism $[\mu, \Omega]$ by putting

$$(\mathsf{P}\mu)_a(T) = S \qquad \Longleftrightarrow \qquad [\mu, \Omega]_a(\tau) = \sigma,$$

if S and T are the subfunctors of $A \downarrow a$ and $B \downarrow a$ corresponding to $\sigma \in [A, \Omega]_a$ and $\tau \in [A, \Omega]_a$ by 26.6.

Proposition. *If* $\mu : A \to B$ *in* $[\mathcal{C}^{\mathrm{op}}, \mathrm{SET}]$, *and if* $S = (\mathrm{P}\mu)_a(T)$, *for an object* a *of* $\mathcal{C}$ *and subfunctors* S *of* $A \downarrow a$ *and* T *of* $B \downarrow a$, *then*

$$x \in Su \quad \Longleftrightarrow \quad \mu_i(x) \in Tu,$$

for $u : i \to a$ *in* $\mathcal{C}$ *and* $x \in Ai$.

PROOF. By 27.6, we have $\sigma = \tau \cdot \mu$. Thus

$$x \in Su \quad \Longleftrightarrow \quad \mathrm{id}_i \in (\tau_u \circ \mu_i)(x) \quad \Longleftrightarrow \quad \mu_i(x) \in Tu,$$

with 26.6.(3) used twice.

28. Sheaves for a Complete Heyting Algebra

28.1. Notations. If H is an ordered set, regarded as a category, then a set-valued presheaf A on H assigns to every element i of H a set Ai, and to every pair of elements i, j of H with $i \leqslant j$ a mapping $\alpha_{i,j} : Aj \to Ai$. These mappings must satisfy the coherence conditions

$$\alpha_{i,i} = \mathrm{id}_{Ai}, \qquad \text{and} \qquad \alpha_{i,j} \circ \alpha_{j,k} = \alpha_{i,k},$$

if $i \in H$ and $i \leqslant j \leqslant k$ in H.

The mappings $\alpha_{i,j}$ are usually called the *restriction mappings* of the presheaf A, and the notation

$$\alpha_{i,j}(x) = x|i$$

is used for $i \leqslant j$ in H and $x \in Aj$.

We note that for an ordered set H and $a \in H$, the slice category H/a is simply the set $\downarrow a$ of all $i \leqslant a$ in H. Thus if A is a presheaf on H and $a \in H$, then $A \downarrow a$ is the restriction of A to the set $\downarrow a$.

28.2. Sheaves. From now on, H shall be a complete Heyting algebra. Then a presheaf A on H is called a *sheaf* on H if for a subset I of H with $\sup I = s$ in H, and for elements x_i of A_i, one for each $i \in I$, such that

(1) $$x_i|(i \cap j) = x_j|(i \cap j),$$

for every pair i, j in H, there is always exactly one element x of Hs such that $x_i = x|i$ for every $i \in I$. We denote by $\mathrm{Sh}\,H$ the full subcategory of $[H^{\mathrm{op}}, \mathrm{SET}]$ with sheaves as its objects.

28.3. Example and discussion. The sets $Ya = \downarrow a$ for $a \in H$, with restrictions $x|b = x \cap b$ if $x \leqslant a$ and $b \leqslant a$, define a sheaf Y on H.

To see this, let $I \subset H$, and let $x_i \leqslant i$ for $i \in I$. If $x_i = s \cap i$ for $i \in I$, with $s \leqslant \sup i$, then

$$s = s \cap (\sup i) = \sup(s \cap i) = \sup x_i \,.$$

If also $x_i \cap j = x_j \cap i$ for i, j in I, then

$$j \cap (\sup x_i) = \sup(j \cap x_i) = \sup(i \cap x_j) = x_j$$

for $j \in I$. Thus we have indeed a sheaf.

If A is a presheaf on H, and $I \subset H$, then consider the diagram D_I of sets and mappings, with vertices Ai for $i \in I$, and with restriction mappings

$$Ai \quad \longrightarrow \quad A(i \cap j) \quad \longleftarrow \quad Aj$$

as its arrows. The set of all families $x = (x_i)_{i \in I}$ with $x_i \in Ai$ for each $i \in I$, and with 28.2.(1) satisfied, is a limit of this diagram, with a limit cone consisting of the projections $x \mapsto x_i$. If A is a sheaf, then we can replace this limit by $A(\sup I)$, with restriction mappings for A as the projections of a limit cone.

If we replace I by the decreasing subset $J = {\downarrow} I$ of H, consisting of all $j \in H$ with $j \leqslant i$ for some $i \in I$, then the diagram D_I is replaced by a diagram D_J with the sets Aj, for $j \in J$ as vertices, and with all restriction mappings $Aj \to Aj'$, for $j' \leqslant j$ in J, as arrows. However, the limit of the diagram does not change. If $x \in \lim D_J$, with projections $x_j \in Aj$, then the projections x_i of x with $i \in I$ clearly satisfy 28.2.(1). Conversely, if $x \in \lim D_I$ with projections x_i, then put $x_j = x_i|j$ for $j \leqslant i$ in J, with $i \in I$. If x satisfies 28.2.(1), this does not depend on a particular choice of i. If $j' \leqslant j$, then clearly $x_{j'} = x_j|j'$; thus x with projections x_j is in $\lim D_J$. This proves the following result.

Proposition. *A presheaf A on H is a sheaf if and only if for every decreasing subset J of H, the set $A(\sup J)$ is a limit of the diagram D_J constructed above, with restriction mappings as projections.*

28.4. Theorem. *For a complete Heyting algebra H, the category $\mathrm{Sh}\, H$ of set-valued sheaves on H is a topos.*

PROOF. Since sheaves satisfy categorical limit conditions, every limit of sheaves in $[H^{\mathrm{op}}, \mathrm{SET}]$ is a sheaf, and hence a limit in $\mathrm{Sh}\, H$. Thus $\mathrm{Sh}\, H$ has small limits, not only finite limits.

We complete the proof by describing subsheaves and their characteristic morphisms in 28.5, the cartesian closed structure of $\mathrm{Sh}\, H$ in 28.6, and powerset objects in $\mathrm{Sh}\, H$ in 28.7, with easily supplied details omitted.

28.5. Subsheaf classifier. We note first that we can identify sieves on an element a of H with decreasing subsets of $\downarrow a$ in H. If we do this, then 26.4.(1), for the characteristic morphism $\sigma : A \to \Omega$ of a subfunctor S of a presheaf A, becomes

$$i \in \sigma_a(x) \quad \Longleftrightarrow \quad x \mid i \in Si,$$

for $i \leqslant a$ in H and $x \in Aa$.

For $x \in Aa$ and a subset I of $\downarrow a$ in H with $x \mid i \in Si$ for every $i \in I$, we must have $x \mid s \in Ss$ for $s = \sup I$ if S is a sheaf. If A is a sheaf, then it is easily seen that this necessary condition for S to be a sheaf is also sufficient. For the characteristic morphism σ of S, this means that if $I \subset \sigma_a(x)$, then $\sup I \in \sigma_a(x)$. It follows that $\sigma_a(x) = \downarrow s$ for some $s \leqslant a$.

Thus we obtain a subobject classifier $\top$ for sheaves by putting $\Omega_{\text{sh}} a = \downarrow a$, and $\top_a = \{a\}$, for $a \in H$, with $\Omega_{\text{sh}} = \mathsf{Y}$ a sheaf by 28.3. The characteristic morphism $s : A \to \Omega_{\text{sh}}$ of a subsheaf S of a sheaf A is given by

$$x \mid i \in Si \quad \Longleftrightarrow \quad i \leqslant s_a(x), \quad \text{and} \quad s_a(x) = \sup\{i \leqslant a \mid x \mid i \in Si\},$$

for $i \leqslant a$ in H and $x \in Ai$.

28.6. The cartesian closed structure of $\text{Sh}\, H$ is inherited from the cartesian closed structure for presheaves, by the following result.

Proposition. *If B is a sheaf on H, then every presheaf $[A, B]$, for a presheaf A on H, is a sheaf.*

PROOF. For a presheaf A on H and $a \in H$, the functor $A \downarrow a$ is simply the restriction of A to $\downarrow a$. If $b \leqslant a$ in H and $\sigma : A \downarrow a \to B \downarrow a$ in $[A, B]a$, then $\sigma \mid b : A \downarrow b \to B \downarrow b$ is the restriction of σ to $\downarrow b$. Now let $I \subset H$ be decreasing, and assume that for each $i \in I$, a natural transformation $\sigma(i) : A \downarrow i \to B \downarrow i$ is given, with $\sigma(i) = \sigma(j) \mid i$ for $i \leqslant j$. This means that mappings $\sigma_i : Ai \to Bi$ are given, one for each $i \in I$, with $s_i(x \mid i) = s_j(x) \mid i$ if $i \leqslant j$ and $x \in Aj$. We must show that these mappings form the restriction to I of a natural transformation $\sigma : A \downarrow a \to B \downarrow a$, for $a = \sup I$. If $b \leqslant a$ and $x \in Ab$, then we must have $\sigma_i(x \mid i) = \sigma_b(x) \mid i$ for each $i \leqslant b$ in I. Since H is a complete Heyting algebra, b is the supremum of these i. Since B is a sheaf, the equations determine $\sigma_b(x)$ in Bb uniquely. It is easily verified that the mappings σ_b thus obtained define the desired natural transformation σ.

28.7. Sheaves $\mathsf{P}A$. We define a *subsheaf* of a sheaf A as a subfunctor of A which is a sheaf. For $a \in H$, we denote by $(\mathsf{P}A)a$ the set of all subsheaves of the sheaf $A \downarrow a$ on $\downarrow a$. If $b \leqslant a$ and $S \in \mathsf{P}A$, then $S \mid b$ is the restriction of S to $\downarrow b$. It is easily seen that PA then is a sheaf, isomorphic to the sheaf $[A, \Omega_{sh}]$. A subsheaf S of $A \downarrow a$ corresponds to $\sigma : A \downarrow a \to \Omega_{sh} \downarrow a$ by the isomorphism if, equivalently,

$$x \in Si \quad \Longleftrightarrow \quad \sigma_i(x) = i, \qquad \text{or} \qquad j \leqslant \sigma_i(x) \quad \Longleftrightarrow \quad x \,|\, j \in Sj \,,$$

for $j \leqslant i \leqslant a$ in H and $x \in Ai$. We omit details, noting only that a subfunctor S of $A \downarrow a$ is a subsheaf iff for a subset I of $\downarrow a$ and an element $x \in A(\sup I)$ such that $x \,|\, i \in Si$ for all $i \in I$, we always have $x \in S(\sup I)$.

28.8. Change of base. A morphism $\mu : H \to K$ of complete Heyting algebras (see 24.6) induces a functor $[\mu^{op}, \text{SET}]$ from presheaves on K to presheaves on H, with $A \mapsto A\mu$ given by $(A\mu)a = A(\mu a)$ for $a \in H$ and a presheaf A on K. This functor is called a *change of base*. We note the following result.

Theorem. *A change of base functor*

$$[\mu^{op}, \text{SET}] : [K^{op}, \text{SET}] \to [H^{op}, \text{SET}] \,,$$

for a morphism $\mu : H \to K$ of complete Heyting algebras, preserves sheaves.

PROOF. For a decreasing subset I of K, assume that $x_i \in (A\mu)i$ is given for each $i \in I$, with $x_i = x_j \,|\, i$ if $j \in I$ and $i \leqslant j$. If A is a sheaf, there is a unique x in $A(\sup(\mu i)) = (A\mu)(\sup i)$ with $x_i = x \,|\, i$ in $A\mu$, i.e. $x_i = x \,|\, \mu(i)$ in A, for all $i \in I$. Thus $A\mu$ is a sheaf.

29. Separated Presheaves

29.1. Definitions. We say that a presheaf A on a complete Heyting algebra H is *separated* if for a subset I of H and elements $x_i \in Ai$, one for each $i \in I$, there is at most one x in $A(\sup I)$ such that $x_i = x \,|\, i$ for each $i \in I$. It is clear that such an element exists only if the x_i satisfy 28.2.(1); it follows that sheaves are separated. We denote by $\text{Sep}\,H$ the category of separated presheaves on H and their morphisms.

As in 28.3, we can restrict our attention to decreasing subsets of H, i.e. to sieves, and to families $(x_i)_{i \in I}$ such that always $x_i = x_j \,|\, i$ if $j \in I$ and $i \leqslant j$. It will be convenient to say that I is a *dense sieve* at a in H if I is a decreasing subset of H, and $\sup I = a$.

29.2. Theorem. *Separated presheaves on a Heyting algebra H are the objects of a reflective full subcategory $\text{Sep}\,H$ of the category $[H^{op}, \text{SET}]$ of presheaves on H, with epimorphic reflections.*

PROOF. For a presheaf A on A and $a \in H$, we put $x \equiv y$ if there is a dense sieve I at a such that $x \mid i = y \mid i$ for all $i \in I$. Since the intersection of two dense sieves at a is dense at a, in the complete Heyting algebra H, this defines an equivalence relation in Aa. Let $(QA)a$ be the set of equivalence classes in Aa, and $q_a : Aa \to (QA)a$ the quotient mapping. If $x \equiv y$ in Aa, with $x \mid i = y \mid i$ for $b \in I$, a dense sieve at a, then $x \mid b \equiv y \mid b$ in Ab since $I \cap \downarrow b$ is a dense sieve at b. Thus we can define restrictions $(q_a(x)) \mid b = q_b(x \mid b)$ of equivalence classes, so that QA is a presheaf and $q : A \to QA$ a morphism of presheaves. We claim that this is a reflector for separated presheaves.

If $(q_a(x)) \mid i = (q_a(y)) \mid i$, for all i in a dense sieve I at $a \in H$, then for each $i \in I$ there is a dense sieve J_i at i with $x \mid j = y \mid j$ for all $j \in J_i$. The set union $J = \bigcup J_i$ is clearly a dense sieve at a, with $x \mid j = y \mid j$ for each $j \in J$. But then $q_a(x) = q_a(y)$, so that QA is separated. If $\mu : A \to B$ in $[\mathcal{C}^{\mathrm{op}}, \mathrm{SET}]$ and $x \equiv y$ in Aa, then clearly $\mu_a(x) \equiv \mu_a(y)$ in Ba. If B is separated, then $\mu_a(x) = \mu_a(y)$ follows. Thus $\mu_a = \nu_a q_a$ for a unique $\nu_a : (QA)a \to Ba$. It is easily seen that the mappings ν_a define a morphism $\nu : QA \to B$ of presheaves. This proves our claim.

Since every presheaf admits a reflection for separated presheaves, these reflections define a functor, a reflector for separated presheaves.

29.3. Proposition. *The reflector for separated presheaves preserves fibred products $A \times_C B$ with C separated.*

PROOF. We consider pullbacks

$$
\begin{array}{ccc}
S & \longrightarrow & B \\
\downarrow & & \downarrow \beta \\
A & \xrightarrow{\alpha} & C
\end{array}
\quad \text{and} \quad
\begin{array}{ccc}
T & \longrightarrow & QB \\
\downarrow & & \downarrow \bar{\beta} \\
QA & \xrightarrow{\bar{\alpha}} & C
\end{array}
$$

of presheaves. Up to isomorphisms, Sa consists of pairs (x, y) in $Aa \times Ba$ with $\alpha(x) = \beta(y)$, and Ta consists of pairs $(\bar{x}, \bar{y})$ of equivalence classes of pairs (x, y) in $A \times B$, with $\bar{\alpha}(\bar{x}) = \bar{\beta}(\bar{y})$, hence $\alpha(x) \equiv \beta(y)$. Since C is separated, this means that $\alpha(x) = \beta(y)$, so that $(\bar{x}, \bar{y})$ is the equivalence class of a pair (x, y) in S. But then $T = QS$ up to isomorphism.

29.4. Closed and dense subpresheaves. We say that a subfunctor S of a presheaf A on H is *dense* if for $a \in H$ and $x \in Aa$, there is always a dense sieve I at a such that $x \mid i \in Si$ for all $i \in I$, and we say that S is a *closed* subfunctor of A if for $x \in Aa$ and a dense sieve I at a such that $x \mid i \in Si$ for all $i \in I$, we always have $x \in Sa$.

Proposition. *Subfunctors of presheaves on* H *have the following properties.*

(i) Every subfunctor inclusion $S \to A$ *factors* $S \to T \to A$ *with* S *dense in* T *and* T *closed in* A.

(ii) If A *is separated, then every subfunctor of* A *is separated.*

(iii) If A *is separated and* S *dense in* A, *then the inclusion* $S \to A$ *is epimorphic in* Sep H.

(iv) A subfunctor S *of a sheaf* A *is a sheaf if and only if* S *is closed in* A.

PROOF. We sketch the proof; details are easily filled in.

For (i), and for $a \in H$, let Ta consist of all $x \in Aa$ such that every $x \,|\, i$ with $i \in I$ is in Si, for some dense sieve I at a. Then T clearly is a closed subfunctor of A, and S a dense subfunctor of T.

(ii) follows immediately from the definitions, and (iv) has already been proved in 28.5.

For (iii), let $\mu : A \to B$ and $\nu : A \to B$, with B separated. If $x \in Aa$, and $\mu_i(x \,|\, i) = \nu_i(x \,|\, i)$ for all i in a dense sieve I at a, then $\mu_a(x) = \nu_a(x)$ follows. Thus $\mu = \nu$ if μ and ν have the same restriction to a dense subfunctor of A.

29.5. Theorem. *The category of separated presheaves on a complete Heyting algebra* H *is a quasitopos.*

PROOF. The reflector for separated presheaves clearly induces a reflector for Sep H/C in $[\mathcal{C}^{\mathrm{op}}, \mathrm{SET}]/C$, for every separated presheaf C. By 29.3, the reflector for Sep H/C preserves finite products. Thus Sep H is locally cartesian closed, by 9.4.

By 29.4.(i), a monomorphism μ in Sep H factors $\mu = \nu\delta$ with δ equivalent to a dense subfunctor inclusion, hence epimorphic in Sep H by 29.4.(iii), and ν a closed subfunctor inclusion. Thus strong monomorphisms in Sep H are equivalent to closed subfunctor inclusions. Conversely, it is easily seen that closed subfunctor inclusions in Sep H are strong monomorphisms.

As already noted in 28.5, a subfunctor S of a presheaf A, with characteristic morphism $\sigma : A \to \Omega$, is a closed subfunctor of A if and only if every sieve $\sigma_a(x)$ is of the form $\downarrow s$. It follows that the subobject classifier $\top$ for sheaves also classifies closed subfunctors in $[H^{\mathrm{op}}, \mathrm{SET}]$, and hence strong monomorphisms in Sep H.

29.6. Proposition. *The coarse objects of the quasitopos* Sep H *are the sheaves over* H.

PROOF. For a presheaf A, putting $(x, y) \in Da \iff x = y$, for $a \in H$ and x, y in Aa, defines a diagonal subfunctor D of $A \times A$. It is easily seen that this subfunctor is closed if, and only if, A is separated. Thus if A is separated, there is in $\operatorname{Sep} H$ a monomorphism $s_A : A \to \mathsf{P}A$, by 14.4 and the proof of 29.5. Since $\operatorname{Sep} H$ and $\operatorname{Sh} H$ have the same subobject classifier, they also have the same powerset objects. Now A is a sheaf iff s_A is closed, by 29.4.(iv), and coarse iff s_A is a strong monomorphism, by 21.6. Since strong and closed monomorphisms are the same in $\operatorname{Sep} H$, this proves the result.

30. Sheaves on Topological Spaces

30.1. Definition and examples. If $H = \mathcal{O}X$, the complete Heyting algebra of open sets of X for a topological space X, then a sheaf on H is also called a *sheaf on the space* X. We give two examples.

(1) For a topological space X, let AU be the set of all continuous functions $u : U \to X$ for an open set U of X, with $u|V$ the usual restriction if $V \subset U$ and $u \in AU$.

(2) For a continuous map $p : Y \to X$, we define the *sheaf of sections* Γp on X as follows. Elements of $\Gamma(p, U)$, for U open in X, are all continuous functions $u : U \to Y$ such that $p \circ u$ is the subspace inclusion. These functions are called *sections* of p. We shall see in 30.5 that every sheaf on a topological space is isomorphic to a sheaf of sections, with p a local homeomorphisms.

For open sets U_i and functions $s_i : U_i \to Y$, with $ps_i : U_i \to X$ the inclusion map for each U_i, let $U = \bigcup U_i$. Then $s_i = s|U_i$, for all i, for at most one $s : U \to X$, with $u(x) = u_i(x)$ for $x \in U_i$, and with $ps : U \to X$ the inclusion map if s exists. If x is also in U_j, we have $s_i(x) = s_j(x)$ if s_i and s_j have the same restriction to $U_i \cap U_j$. Thus s is well defined, and continuous on U as each $s|U_i$ is continuous, if the maps s_i satisfy 28.2.(1).

The proof that Example (1) defines a sheaf is analogous; we omit it.

30.2. Total spaces. The *total space* LA of a presheaf A on X is obtained as the colimit of a diagram D of topological spaces and continuous maps, constructed as follows. For every pair (U, s) with U open in X and $s \in AU$, the diagram D has a vertex $D_s = U$. For every restriction $t = s \mid V$ in A, there is an arrow $D_t \to D_s$ in D, the inclusion map $V \to U$. Inclusion maps $i_s : D_s \to X$ then define a cone $i : D \to X$, and D with this cone becomes a diagram in TOP/X.

The colimit cone $D \to LA$ in TOP consists of mappings $\hat{s} : D_s \to LA$, with a mapping $\ell A : LA \to X$ such that $\ell A \circ \hat{s}$ is the inclusion map $U \to X$ for every pair (U, s) with $s \in U$. The mappings $\hat{s}$ are collectively surjective, and we have $\hat{s}(x) = \hat{t}(x)$, for $s \in AU$, $t \in AV$, and $x \in U \cap V$, iff $s|W = t|W$ for some open neighborhood W of x contained in $U \cap V$. The topology of LA

is the final topology for the maps $\hat{s}$, with $Z \subset LA$ opem iff the inverse image $\hat{s}^{\leftarrow}(Z)$ is open in U, for every pair (U, s). It follows that $\ell A : LA \to X$ is continuous, and the colimit of D in TOP/X.

For a pair (U, s) with U open in X and $s \in AU$, we claim that the set $\hat{s}^{\rightarrow}(U)$ is open in LA. If $\hat{t}(x) = \hat{s}(x)$, with $t \in AV$ and $x \in V$, then $s \,|\, W = t \,|\, W$ for an open neighborhood W of x, contained in $U \cap V$, and then $W \subset \hat{t}^{\leftarrow}(\hat{s}^{\rightarrow}(U))$, so that $\hat{t}^{\leftarrow}(\hat{s}^{\rightarrow}(U))$ is open in U. Conversely, if $\hat{s}(x) \in Z$, with $s \in AU$, $x \in U$, and Z open in LA, then $\hat{s}^{\leftarrow}(Z)$ is a neighborhood V of x, and $\hat{t}^{\rightarrow}(V) \subset Z$, with $\hat{s}(x) = \hat{t}(x)$, for $t = s \,|\, V$. Thus sets $\hat{s}^{\rightarrow}(U)$ form a basis of open sets in LA.

30.3. Functors Γ and ℓ. In 30.1.(2), we have defined for every topological space Y over a space X, i.e. for every object $p : Y \to X$ of TOP/X, a sheaf of sections $\Gamma(p, -)$ on the space X. If $f : p \to q$ in TOP/X, then mappings

$$\Gamma(f, U) : \Gamma(p, U) \to \Gamma(q.U) : s \mapsto fs$$

cleary define a morphism $\Gamma(f, -) : \Gamma(p.-) \to \Gamma(q, -)$ of sheaves on X. Thus we have a sheaf of sections functor $\Gamma : \mathrm{TOP}/X \to \mathrm{Sh}\,X$. We may also view Γ as a functor from TOP/X to presheaves over X.

Total spaces also define a functor, from presheaves over X to TOP/X. If $\mu : A \to B$ is a morphism of presheaves and $t = \mu_U(s)$, with $s \in AU$, and if $V \subset U$ in $\mathcal{O}_X$, then $\hat{t}\,|\,V = \hat{r}$ for $r = \mu_V(s \,|\, V)$. Thus maps $\hat{t}$ form a cone for the diagram D used to construct LA, and there is a unique map $\ell\mu : \ell A \to \ell B$ in TOP/X, given by $\ell\mu \circ \hat{s} = \hat{t}$, for $s \in AU$ and $t = \mu_U(s)$.

30.4. Theorem. *The total space functor ℓ is left adjoint to the sheaves of sections functor Γ.*

PROOF. For a presheaf A and an object $p : Y \to X$ of TOP/X, we let $\varphi : A \to \Gamma p$ correspond to $f : LA \to Y$, with $p \circ f = \ell A$, if

(1) $f \circ \hat{s} = \varphi_U(s)$,

for every pair (U, s) with U open in X and $s \in AU$. If f is given, then $p \circ f \circ \hat{s} = \ell A \circ \hat{s}$ is the inclusion $s : U \to X$; thus $\varphi_U(s)$ is a section of p. Naturality of φ is easily verified. If φ is given, then the sections $\varphi_U(s)$ form a cone for the diagram D of 30.2 in TOP. Thus (1) is satisfied for a unique continuous map $f : LA \to Y$, with $pf = \ell A$ since each $\varphi_U(s)$ is a section for p.

30.5. Proposition. *For a presheaf A over X, the unit $\eta_A : A \to \Gamma\ell A$ is an isomorphism if and only if A is a sheaf.*

PROOF. If η_A is an isomorphism, then A is a sheaf since $\Gamma L A$ is a sheaf by 30.2. In general, the unit η_A corresponds to $f = \mathrm{id}_{LA}$ by 30.4.(1); thus $\eta_{A,U}(s) = \hat{s}$ for $s \in AU$. If $\hat{s} = \hat{t}$ for s, t in AU, then U has a cover of open sets W with $s \,|\, W = t \,|\, W$. But then $s = t$ if A is a sheaf.

If $u : U \to LA$ is a section and $x \in U$, let $\hat{i}^{\to}(V)$ be a basic open set of LA containing $u(x)$, hence with $x \in V$. If $u^{\leftarrow}(\hat{i}^{\to}(V)) = W$, then W is open in U, hence in X, and $u \,|\, W = \hat{i} \,|\, W$ since u and $\hat{i}$ are sections. Thus U has a covering by open subsets U_i with $u \,|\, U_i = \hat{s}_i$ for elements $s_i \in AU_i$. The restrictions $\hat{s}_i$ of u satisfy 28.2.(1). Since the mappings $\eta_{A,U}$ are injective and preserve restrictions, the elements $s_i \in AU_i$ also satisfy 28.2.(1). As A is a sheaf, it follows that $s_i = s \,|\, U_i$ for a (unique) $s \in AU$. But then $\hat{s}_i = \hat{s} \,|\, U_i$ for each U_i, and $u = \hat{s}$ follows. Thus each $\eta_{A,U}$ is bijective, and η_A is an isomorphism of sheaves.

30.6. Local homeomorphisms. We say that a map $p : Y \to X$ of topological spaces is a *local homeomorphism* if Y has a cover by open sets $s^{\to}(U)$ with U open in X and $s : U \to Y$ a section for p. The restrictions of s and p to U and $s^{\to}(U)$ then become inverse homeomorphisms. The maps $\ell A : LA \to X$ are local homeomorphisms; we show now that every local homeomorphism is isomorphic to a map ℓA.

Proposition. *For an object* $p : Y \to X$ *of* TOP$/X$, *the counit* $\varepsilon_p :$ $\ell\Gamma p \to p$ *of* $\ell \dashv \Gamma$ *is an isomorphism of* TOP$/X$ *if and only if* p *is a local homeomorphism.*

PROOF. Since $\varepsilon_p : \ell\Gamma p \to p$ corresponds by 30.4 to $\mathrm{id}_{\Gamma p}$, it is defined by compositions

$$\varepsilon_p \circ \hat{s} = s,$$

one for each section $s : U \to Y$ of p. If ε_p is an isomorphism of TOP$/X$, then p is a local homeomorphism since $\ell\Gamma p$ is one. Conversely, if p is a local homeomorphism, then we show that the sections $s : D_s \to Y$ form a colimit cone for the diagram D of 30.2, as do the sections $\hat{s}$, and it follows that ε_p is an isomorphism.

Since p is a local homeomorphism, the sections s are collectively surjective. If $s(x) = t(x')$, for sections s and t on open sets U and V of X, then $x = x'$. If $s(x) \in u^{\to}(W)$, for a basic open set of Y, then $s^{\leftarrow}(u^{\to}(W))$ and $t^{\leftarrow}(u^{\to}(W))$ are open neighborhoods of x in U and V. If W' is the intersection of these sets and of W, then W' is open with $x \in W'$, and $s \,|\, W' = u \,|\, W' = t \,|\, W'$ since s, t, u are sections. Thus the maps $s : D_s \to Y$ form a colimit cone at the set level. It is easily seen that Y has the final topology for these sections, with Z open in Y iff $s^{\leftarrow}(Z)$ is open in X for every section s. Thus the maps s form a colimit cone in TOP, as claimed.

30.7. Discussion. By the results of 30.4–30.6, sheaves over a topological space X define a reflective full subcategory of the category of presheaves over X, with morphisms $\eta_A : A \to \Gamma \ell A$ as reflections, and local homeomorphisms define a coreflective full subcategory of TOP/X, with maps $\varepsilon_p : \ell \Gamma p \to p$ as coreflections. The functors Γ and ℓ induce inverse equivalences between the categories of sheaves and of local homomorphisms.

Every continuous map $f : X \to Y$ induces a morphism $\mathcal{O}f : \mathcal{O}Y \to \mathcal{O}X$ of complete Heyting algebras, and hence by 28.8 a change of base functor $\mathrm{Sh}\,X \to \mathrm{Sh}\,Y$ for sheaves. If Y is sober (see 25.4 and 25.5), then it can be shown that these are the only change of base functors $\mathrm{Sh}\,X \to \mathrm{Sh}\,Y$.

31. Examples of Topological Quasitopoi

31.1. Generalities. We consider in this Section a topological category $(\mathbf{T}, p)$ over sets. Objects and morphisms of $\mathbf{T}$ will be called *spaces* and *continuous maps*. By 11.7, we may assume that limits and colimits in $\mathbf{T}$ are obtained by lifting limits and colimits in SET.

We note first that trivial objects (11.5) of $\mathbf{T}$ are coarse objects in the sense of Section 21. Thus they define a full subcategory $\mathbf{T}^{\mathrm{gr}}$ of $\mathbf{T}$ which is a topos isomorphic to sets, and has the properties stated in Section 21. It follows that relations in $\mathbf{T}$ are represented; a powerset object $\mathrm{P}X$ in $\mathbf{T}$ is obtained by providing the powerset $\mathrm{P}pX$ with the trivial structure.

Subobjects in $\mathbf{T}$ correspond bijectively to subspaces, where Y is called a *subspace* of X if pY is a subset of pX, and Y is the initial lift for X and the subset inclusion $pX \to pY$. Subspaces are always represented, by the set $\{0, 1\}$ with the trivial structure.

31.2. Function spaces. Let 1 be a terminal object of $\mathbf{T}$, with $p1 = \{*\}$. If $\mathbf{T}$ is cartesian closed, with exponential objects $[Y, Z]$, then we have natural bijections between morphisms

$$f : Y \to Z, \qquad f_1 : 1 \times Y \to \ , \qquad f_2 : 1 \to [Y, Z]$$

of $\mathbf{T}$, with $f_1 = \mathrm{ev} \cdot (f_2 \times \mathrm{id}_Y)$. If $\mathbf{T}$ has constant maps, then morphisms f_2 correspond bijectively to elements of $p[Y, Z]$, and $f \mapsto f_2(*)$ defines a bijection from $\mathbf{T}(Y, Z)$ to $p[Y, Z]$. We use this to replace $p[Y, Z]$ by $\mathbf{T}(Y, Z)$, and then

$$\mathrm{ev}(f, y) = \mathrm{ev}(f_2(*), y) = f_1(*, y) = f(y)$$

for $y \in pY$. It follows that $\hat{\varphi}(x) = \varphi(x, -)$ for $x \in pX$ if $\hat{\varphi} : X \to [Y, Z]$ is exponentially adjoint to $\varphi : X \times Y \to Z$.

More generally, if $\mathbf{T}$ has constant maps and $\varphi : X \times Y \to Z$ in $\mathbf{T}$, then $\varphi(x, -) : Y \to Z$ is a map of $\mathbf{T}$ for every $x \in pX$. It follows that exponential adjunction at the set level assigns to every $\varphi : X \times Y \to Z$ in $\mathbf{T}$ a mapping $\hat{\varphi} : pX \to \mathbf{T}(Y, Z)$. A structure of $\mathbf{T}(X, Y)$ is then called *proper* if $\hat{\varphi}$ is continuous for every φ. We denote by $[Y, Z]_c$ the final lift of $\mathbf{T}(Y, Z)$ for all mappings $\hat{\varphi}$ for maps $\varphi : X \times Y \to Z$ in $\mathbf{T}$; this defines the finest proper structure of $\mathbf{T}(Y, Z)$ in $\mathbf{T}$.

Proposition. *The category* $\mathbf{T}$ *is cartesian closed if and only if evaluation* $\mathrm{ev} : [Y, Z]_c \times Y \to Z$ *is continuous in* $\mathbf{T}$.

PROOF. The category $\mathbf{T}$ is cartesian closed with function spaces $[Y, Z]$ iff the spaces $[Y, Z]$ are proper, with $\mathrm{ev} : [Y, Z] \times Y \to Z$ continuous, since $\varphi = \mathrm{ev} \cdot (\hat{\varphi} \times \mathrm{id}_Y)$. In particular, $\mathrm{id} : [Y, Z] \to [Y, Z]_c$ is then continuous since $\hat{\varphi} = \mathrm{id}$ for $\varphi = \mathrm{ev}$. Thus the function spaces must be the spaces $[Y, Z]_c$, and the Proposition follows.

31.3. Representing partial morphisms. If partial morphisms in a topological category $\mathbf{T}$ over sets are represented, then they are typically represented by one-point extensions $\vartheta_Y : Y \to \tilde{Y}$, with

$$p\tilde{Y} = \{\emptyset\} \cup \{\{x\} \mid x \in pY\},$$

and $\vartheta_Y(x) = \{x\}$ for $x \in pY$. Putting $pX = E$, we have for every partial morphism $X \xleftarrow{m} \cdot \xrightarrow{f} Y$ a unique pullback square

$$
\begin{array}{ccc}
\cdot & \xrightarrow{f} & E \\
{\scriptstyle m}\downarrow & & \downarrow{\scriptstyle \vartheta_E} \\
pX & \xrightarrow{\bar{f}} & \tilde{E}
\end{array}
$$

in SET. If $\tilde{Y}$ is the final lift of $\tilde{E}$ for the mappings $\bar{f}$ thus obtained, and the resulting map $\vartheta_E : Y \to \tilde{Y}$ is an embedding in $\mathbf{T}$, then this embedding clearly represents partial morphisms in $\mathbf{T}$ with codomain Y. If every empty space in $\mathbf{T}$ is a subspace of a non-empty space, and in particular if $\mathbf{T}$ has constant maps, then it follows from the results of Section 65 that all representations of partial morphisms in $\mathbf{T}$ are obtained in this way.

31.4. Convergence spaces. We define a *convergence space* as a pair (S, q) consisting of a set S and a *convergence relation* $q : \mathsf{F}S \to S$, from the set $\mathsf{F}S$ of filters on S to S, subject to two conditions.

31.4.1. The point filter $[\{x\}]$ converges to x for every $x \in S$.

31.4.2. If $\varphi q x$ and ψ is finer than φ, then $\psi q x$.

We include the null filter $[\emptyset] = \mathsf{P}S$ on S in $\mathsf{F}S$; this filter converges to every point of S.

We may denote a convergence space X by $(|X|, q_X)$. A *continuous map* $f : X \to Y$ of convergence spaces is a mapping $f : |X| \to |Y|$ which preserves convergence, i.e. we always have $(\mathsf{F}f)(\varphi) \, q_Y \, f(x)$ if $\varphi q_X x$. It is easily verified that convergence spaces and their continuous maps form a topological category over sets, with constant maps. We note that $\mathsf{F}f$ is defined by putting

$$T \in (\mathsf{F}f)(\varphi) \quad \Longleftrightarrow \quad f^{\leftarrow}(T) \in \varphi,$$

for $T \subset Y$ and a filter φ on X. It follows that $(\mathsf{F}f)\varphi$ is generated by the sets $f^{\to}(S)$ with $S \in \varphi$.

The category of convergence spaces is cartesian closed. For convergence spaces X and Y, the function space $[X, Y]$ consists of all continuous maps $f : X \to Y$ with *continuous convergence*. This means that a filter Φ on the function space converges to a map f if and only if always

$$\varphi \, q_X \, x \quad \Longrightarrow \quad (\mathsf{F}\mathrm{ev}_{X,Y})(\Phi \times \varphi) \, q_Y \, f(x).$$

Here $\mathrm{ev}_{X,Y} : (f, x) \mapsto f(x)$ is evaluation, and $\Phi \times \varphi$ is generated by the sets $F \times A$ with $F \in \Phi$ and $A \in \varphi$.

Partial morphisms are represented; the construction of 31.3 provides the representations. Every filter on $\tilde{X}$ converges to the added point of $\tilde{X}$, and a filter φ on $\tilde{X}$ converges to a point $\{x\}$ with $x \in X$ iff the filter of sets $\vartheta_X^{\leftarrow}(S)$ with $S \in \varphi$ converges to x in X.

Thus convergence spaces and their continuous maps form a quasitopos.

31.5. Remarks.

The category of convergence spaces defined in 31.4 is too large for many purposes. We obtain better behaved full subcategories by imposing one of the following axioms.

31.5.1. If $\varphi q_X x$, then $(\varphi \cup [\{x\}]) q_X x$.

31.5.2. If $\varphi q_X x$ and $\psi q_X x$, then $(\varphi \cup \psi) q_X x$.

31.5.3. If $\psi q_X x$ for every ultrafilter ψ finer than φ, then $\varphi q_X x$.

We note that 31.5.3 $\Longrightarrow$ 31.5.2 $\Longrightarrow$ 31.5.1, and that $\varphi \cup \psi$ in 31.5.1 and 31.5.2 consists of all sets $S \cup T$ with $S \in \varphi$ and $T \in \psi$. Spaces which satisfy 31.5.2 are called *limit spaces* [66, 29], and spaces which satisfy 31.5.3 are called *pseudotopological spaces* [17]. Axiom 31.5.1 was introduced by D.C. KENT [62]. We note that pseudotopological spaces can be characterized by a convergence relation from ultrafilters to points which must satisfy 31.4.1.

It is easily seen that the axioms listed above carry over from a space X to all function spaces $[Z, X]$, and to the space $\tilde{X}$. Thus the categories of Kent convergence spaces, of limit spaces and of pseudotopological spaces also are quasitopoi.

31.6. Bounded set structures. We define a *B-set* as a pair (S, S) consisting of a set S and a set S of non-empty subsets of S, subject to the following conditions.

31.6.1. Every set $\{x\}$ with $x \in S$ is bounded.

31.6.2. If $A \in S$, then every non-empty subset of A is in S.

A *bounded map* $f : (S, S) \to (T, \mathcal{T})$ of B-sets is a mapping $f : S \to B$ of the underlying sets such that $f^{\to}(A) \in \mathcal{T}$ for every set $A \in S$. By the usual *abus de langage*, we shall use the same symbol for a B-set and its underlying set.

For B-sets S and T the function B-set $[S, T]$ has as elements all bounded maps $f : S \to T$, and as bounded sets all sets $F \subset [S, T]$ such that the set $\mathrm{ev}_{S,T}^{\to}(F \times A)$ is bounded in T for every bounded set A of S, with evaluation $\mathrm{ev}_{S,T}$ defined as for sets.

Partial morphisms with codomain a B-set S are represented by a one-point extension $\vartheta_S : S \to \tilde{S}$, with $A \subset \tilde{S}$ bounded in $\tilde{S}$ iff $\vartheta_S^{\leftarrow}(A)$ is bounded in S or empty.

A B-set S is called a *bornological set* [53] if it satisfies the condition

31.6.3. If A and B are bounded in S, then $A \cup B$ is bounded in S.

Bornological sets also form a quasitopos; function sets $[T, S]$ and the one-point extension $\tilde{S}$ are bornological sets if S is a bornological set.

31.7. Pairs of sets. This example does not have constant maps. Objects of **T** are now pairs (X, A) with $A \subset X$, and with underlying set $p(X, A) = X$. Maps $f : (X, A) \to (Y, B)$ are mappings $f : X \to Y$ such that $f^{\to}(A) \subset B$. Initial objects for sources and final objects for sinks are easily constructed. If $f : X \to Y$ is a mapping, then

$$f_*(X, A) = (Y, f^{\to}(A)) \quad \text{and} \quad f^*(Y, B) = (X, f^{\leftarrow}(B)),$$

for $A \subset X$ and $B \subset Y$.

Function spaces $[(X, A), (Y, B)]$ are pairs (Y^X, F), with F the set of all $f : (X, A) \to (Y, B)$ in **T**, and partial morphisms with codomain (Y, B) are represented by the one-point extension $(\tilde{Y}, \tilde{B})$.

31.8. Subsequential spaces. A *subsequential space* X [59] consists of a set $|X|$ with a convergence relation for sequences on $|X|$, satisfying the following three axioms.

(i) For every $x \in |X|$, the constant sequence at x converges to x.

(ii) If a sequence (x_n) in $|X|$ converges to $x \in |X|$, then every subsequence of (x_n) converges to x.

(iii) If every subsequence of a sequence (x_n) in $|X|$ has a subsequence which converges to $x \in |X|$, then (x_n) converges to x.

A continuous map of subsequential space is a mapping of the underlying sets which preserves convergence of sequences.

It is easily verified that subsequential spaces form a topological category over sets, with constant maps. For subsequential spaces Y and Z, we let $[Y, Z]$ be the set of continuous functions $f : Y \to Z$, with continuous convergence, *i.e.* (f_n) converging to f if and only if the sequence of values $f_n(y_n)$ converges to $f(y)$ in Z whenever (y_n) converges to y in Y. The one-point extension $\tilde{Y}$ of a subsequential space has every sequence converging to the added point $\emptyset$, and a sequence (z_n) converging to a point $\{y\}$ if and only if it has a subsequence of points $z_{n_i} = \{y_i\}$ with (y_i) converging to y in Y.

31.9. Partial examples. The category of *uniform limit spaces* is topological over sets, with constant maps. It is cartesian closed, but does not have representation of partial maps. We refer to [68] or [106] for a description of this category and its cartesian closed structure. The category of *merotopic spaces* of M. KATĚTOV [61] is another cartesian closed category which contains uniform spaces as a full and dense subcategory.

The category of *neighborhood spaces* has partial maps represented, but it is not cartesian closed. A neighborhood space (E, ν) consists of a set E and a mapping ν which assigns to every $x \in E$ a filter ν_x of neighborhoods of x, with x in every neighborhood of x. A continuous map $f : X \to Y$ of neighborhood spaces is a mapping $f : pX \to pY$ of the underlying sets, with the property that for $x \in pX$ and a neighborhood V of $f(x)$ in Y, the set $f^{\leftarrow}(V)$ always is a neighborhood of x in X. Partial morphisms with codomain Y are represented by a one-point extension $\tilde{Y}$, with $\tilde{Y}$ the only neighborhood of the added point.

Neighborhood spaces have also been called *pretopological spaces* or *closure spaces*; they were called *gestufte Räume* by F. Hausdorff [42]. A neighborhood space can be characterized by its closure operation on subsets, which preserves finite set unions but need not be idempotent.

Chapter 3

LOGIC IN A QUASITOPOS

Every quasitopos $\mathbf{E}$, and in particular every topos, has an internal language and an internal logic, which can be used to obtain properties of $\mathbf{E}$ and to carry out constructions in $\mathbf{E}$. This logic is the topic of the present chapter.

We first introduce propositional connectives and generalized quantifiers, for monomorphisms and for strong monomorphisms. Connectives and quantifiers for strong monomorphisms can be internalized as morphisms of the quasitopos. We do this in a two-step process which always works, and we construct the internal versions of the most important propositional connectives.

We associate with every quasitopos a formal language. Terms and statements in this language can be interpreted in the quasitopos; thus the quasitopos is a model of its language. This correspondence allows us to describe constructions in the quasitopos in the language of the quasitopos; we give some examples. The language of a quasitopos is typed; types are associated with the objects of the quasitopos. The language includes the usual propositional connectives, universal and existential quantifiers, and set and function formation, and it assigns unary operators to the morphisms of the quasitopos.

A quasitopos and its language have an internal logic, with interpretations of statements as internal truth-value tables. This logic is intuitionistic, with some restrictions. These restrictions are due to the presence of types A such that the statement $(\exists x)(x = x)$ is not internally true for variables x of type A. Laws for the internal logic will also be discussed.

The internal language of a quasitopos $\mathbf{E}$ is an extension of the internal language of the topos $\mathbf{E}^{\mathrm{gr}}$ of coarse objects of $\mathbf{E}$, and the two languages have the same internal propositional connectives and quantifiers. Thus the internal logic of $\mathbf{E}$ is essentially equivalent to the internal logic of $\mathbf{E}^{\mathrm{gr}}$.

For topoi, the internal language and its logic were introduced by J. BÉNA-BOU, W. MITCHELL [74], M. FOURMAN [30], G. OSIUS, and others. Our presentation is based on the work of G. OSIUS [80].

In the last two sections of this chapter, we use the internal language and logic of a quasitopos $\mathbf{E}$ to construct internal union and intersection operators for subobjects of an object of $\mathbf{E}$, and to construct a category of internal relations in $\mathbf{E}$.

32. Propositional Connectives

32.1. We recall from 20.1 that for an object A of a quasitopos $\mathbf{E}$, we denote by $\operatorname{Mon} A$ and $\mathcal{M} A$ the full subcategories of $\mathbf{E}/A$ with monomorphisms and with strong monomorphisms of $\mathbf{E}$, with codomain A, as their objects. Both subcategories are preordered classes, and morphisms of $\operatorname{Mon} A$ and of $\mathcal{M} A$ are also monomorphic or strongly monomorphic in $\mathbf{E}/A$. For $f : A \to B$ in $\mathbf{E}$, we denote by $f^{\leftarrow} : \mathcal{M} B \to \mathcal{M} A$ and by $f^{\leftarrow} : \operatorname{Mon} B \to \operatorname{Mon} A$ the restrictions of the pullback functor $f^* : \mathbf{E}/B \to \mathbf{E}/A$.

Proposition. *$\operatorname{Mon} A$ is a full reflective subcategory of $\mathbf{E}/A$, with strong epimorphisms as reflections, and $\mathcal{M} A$ is a full reflective subcategory of $\mathbf{E}/A$ and of $\operatorname{Mon} A$, with epimorphic reflections.*

PROOF. If we factor f with codomain A as $f = m_1 e_1$ by 23.3, with e_1 epimorphic and m_1 strongly monomorphic, then $e_1 : f \to m_1$ in $\mathbf{E}/A$. If $u : f \to m$ in $\mathbf{E}/A$, with m strongly monomorphic, then $mu = m_1 e_1$ in $\mathbf{E}$, and it follows that $u = t e_1$ for a unique morphism $t : m_1 \to m$ of $\mathcal{M} A$. Thus $e_1 : f \to m_1$ is a reflection for $\mathcal{M} A$ in $\mathbf{E}/A$.

The proof for $\operatorname{Mon} A$ is analogous, using the factorization $f = m_2 e_2$ in 23.3.

32.2. Propositional connectives. We define a *binary propositional connective* in $\mathbf{E}$ as a binary operation $\odot$, defined on $\operatorname{Mon} A$ or on $\mathcal{M} A$, for every object A of $\mathbf{E}$, such that always

$$(1) \qquad m_1' \simeq m_1 \text{ and } m_2' \simeq m_2 \quad \Longrightarrow \quad m_1' \odot m_2' \simeq m_1 \odot m_2 ,$$

for morphisms in $\operatorname{Mon} A$ or in $\mathcal{M} A$, and

$$(2) \qquad (f^{\leftarrow} m_1) \odot (f^{\leftarrow} m_2) \simeq f^{\leftarrow}(m_1 \odot m_2) ,$$

for $f : A \to B$ in $\mathbf{E}$, and for m_1 and m_2 in $\operatorname{Mon} B$ or $\mathcal{M} B$. Nullary, unary, ternary, ... propositional connectives are defined similarly, and our results apply also, *mutatis mutandis*, to these.

Propositional connectives can also be defined for relations. We observe, however, that we get the same connectives. The preordered class $\operatorname{Rel}(X, A)$ of relations $(u, v) : X \to A$ is isomorphic to $\mathcal{M}(X \times A)$ for objects X and A of $\mathbf{E}$, and $\mathcal{M} A$ is isomorphic to $\operatorname{Rel}(A, 1)$.

32.3. Theorem. *Meets, joins and implications*

$$m_1 \cap m_2 , \qquad m_1 \cup m_2 , \qquad m_1 \to m_2 ,$$

define binary propositional connectives, for monomorphisms and for strong monomorphisms of a quasitopos **E***, and greatest and least objects*

$$\text{id}_A \quad \text{and} \quad o_A : 0 \to A$$

define nullary propositional connectives "true" and "false".

We note that monomorphism o_A are in general not strong; they have to be replaced by their reflections in $\mathcal{M}A$, in order to obtain "false" for strong monomorphisms.

PROOF. Fibred products $a \times_A b$, and the terminal object id_A of **E**$/A$, provide finite meets in Mon A and $\mathcal{M}A$. The unique morphism $o_A : 0 \to A$ is a least element of Mon A by the Corollary of 23.5, and the monomorphism $m : X \cup Y \to A$ of 23.8 provides a join $a \cup b$ in Mon A. Finite joins in $\mathcal{M}A$ are obtained as reflections of finite joins in Mon A.

For objects a and m of Mon A, we have $m \cap a \simeq a_> a^* m$. As the functors $a_>$ and a^* have right adjoints a^* and a_*, their composition $- \cap a$ also has a right adjoint $a \to -$. This functor on **E**$/A$ maps Mon A into itself; thus Mon A is cartesian closed. The same argument shows that $\mathcal{M}A$ is cartesian closed.

Pullbacks by $f : A \to B$ preserve meets $m_1 \cap m_2$, which are pullbacks, and $f^\leftarrow \text{id}_B \simeq \text{id}_A$. Finite joins in Mon B are colimits, in **E**$/B$ and in **E**, and thus preserved by $f^\leftarrow$, by 23.4. For finite joins in $\mathcal{M}B$, we note that the square

$$
\begin{array}{ccc}
\text{Mon} B & \xrightarrow{f^\leftarrow} & \text{Mon} A \\
\downarrow & & \downarrow \\
\mathcal{M}B & \xrightarrow{f^\leftarrow} & \mathcal{M}A
\end{array}
,
$$

with reflectors as vertical arrows, commutes up to equivalence, by 32.1 and 23.1. It follows that $f^\leftarrow : \mathcal{M}B \to \mathcal{M}A$ also preserves finite joins.

We have $a \simeq f^\leftarrow m_1$ if there is a pullback square

$$
\begin{array}{ccc}
\cdot & \xrightarrow{r} & \cdot \\
a \downarrow & & \downarrow m_1 \\
A & \xrightarrow{f} & B
\end{array}
$$

in **E**. In this situation, we get

$$f^\leftarrow m_1 \to f^\leftarrow m_2 \simeq a_* a^* f^* m_2 \simeq a_* r^* m_1^* m_2$$

$$\simeq f^* m_{1*} m_1^* m_2 \simeq f^\leftarrow (m_1 \to m_2),$$

using 18.3.

32.4. Internal propositional connectives. We obtain internal propositional connectives for a quasitopos $\mathbf{E}$ in two steps.

The first step is to replace morphisms m in $\mathcal{M}A$, for an object A of $\mathbf{E}$, by morphisms $u : A \to \Omega$, by the correspondance

$$(1) \qquad\qquad u = \operatorname{ch} m \quad \Longleftrightarrow \quad m \simeq u^{\leftarrow}\top,$$

which defines a bijection between subobjects of A and morphisms in $\mathbf{E}(A, \Omega)$. We use this to replace a propositional connective $\odot$ defined on $\mathcal{M}$ by a connective, also denoted $\odot$ for convenience, on morphisms of $\mathbf{E}$ with codomain Ω, by putting

$$(2) \qquad\qquad u \odot v = \operatorname{ch}(u^{\leftarrow}\top \odot v^{\leftarrow}\top),$$

for u and v in a set $\mathbf{E}(A, \Omega)$. This is always possible iff 32.2.(1) is valid for $\odot$. The correspondence (1) replaces a pullback $f^{\leftarrow}m$ by $\operatorname{ch} m \circ f$; thus 32.2.(2) becomes

$$(3) \qquad\qquad (u \odot v) \circ f = (u \circ f) \odot (v \circ f)$$

in the new setting.

In the second step, we define the *internal propositional connective* $\otimes :$ $\Omega \times \Omega \to \Omega$ corresponding to a binary propositional connective $\odot$ by putting

$$(4) \qquad\qquad \otimes = p \odot q = \operatorname{ch}(p^{\leftarrow}\top \odot q^{\leftarrow}\top) : \Omega \times \Omega \to \Omega,$$

for the projections $\Omega \xleftarrow{p} \Omega \times \Omega \xrightarrow{q} \Omega$ of the product $\Omega \times \Omega$.

This definition and the following result can obviously be generalized to connectives of any arity, *i.e.* with any number of arguments, with $\otimes = \operatorname{ch}(\odot\top)$ for a unary connective, and $\otimes = \operatorname{ch}\odot_1$ for a nullary connective.

Proposition. *If $\odot$ is a binary propositional connective for strong monomorphisms in a quasitopos $\mathbf{E}$, and $\otimes$ the corresponding internal propositional connective, then*

$$(5) \qquad\qquad u \odot v = \otimes \circ \langle u, v \rangle,$$

for characteristic morphisms of strong monomorphisms with the same codomain.

PROOF. This follows immediately from (3) and (4).

32.5. Internal true and false. A nullary propositional connective assigns to every object A of $\mathbf{E}$ a monomorphism $\odot_A$, with $\odot_A \simeq \tau_A^{\leftarrow}\odot_1$ for the morphism $\tau_A : A \to 1$. If $\odot_1$ is strong, then $\otimes = \operatorname{ch}\odot_1 : 1 \to \Omega$ defines an internal connective, with $\text{ch} \odot_A = \otimes \circ \tau_A$ for every object A. We have in particular the following result.

Proposition. *If* $o_A : 0 \to A$ *factors* $o_A = \bar{o}_A \, u$, *with* u *monomorphic and epimorphic, and with* $\bar{o}_A$ *strongly monomorphic, then*

$$\top = \text{ch id}_1 \quad \text{and} \quad \bot = \text{ch} \, \bar{o}_1$$

define internal "true" and internal "false".

PROOF. This follows from the preceding discussion, since "true" and "false" correspond to the greatest element id_A and the least element $\bar{o}_A$ of $\mathcal{M}A$, for every object A of $\mathbf{E}$.

32.6. Internal conjunction. Internal "and" is easily obtained:

Proposition. $\wedge = \text{ch} \langle \top, \top \rangle$ *defines internal conjunction in* $\mathbf{E}$.

PROOF. It is easily verified that we have pullbacks

$$
\begin{array}{ccc}
\Omega & \xrightarrow{\tau\Omega} & 1 \\
{\scriptstyle \langle \top \tau\Omega, \, \text{id}_\Omega \rangle} \downarrow & & \downarrow {\scriptstyle \top} \\
\Omega \times \Omega & \xrightarrow{p} & \Omega
\end{array}
\qquad
\begin{array}{ccc}
\Omega & \xrightarrow{\tau\Omega} & 1 \\
{\scriptstyle \langle \text{id}_\Omega, \, \top \tau\Omega \rangle} \downarrow & & \downarrow {\scriptstyle \top} \\
\Omega \times \Omega & \xrightarrow{q} & \Omega
\end{array}
\;,
$$

and

$$
\begin{array}{ccc}
1 & \xrightarrow{\quad \top \quad} & \Omega \\
{\scriptstyle \top} \downarrow & & \downarrow {\scriptstyle \langle \top \tau\Omega, \, \text{id}_\Omega \rangle} \\
\Omega & \xrightarrow{\langle \text{id}_\Omega, \, \top \tau\Omega \rangle} & \Omega \times \Omega
\end{array}
$$

in $\mathbf{E}$. Thus $p^{\leftarrow}\top \cap q^{\leftarrow}\top \simeq \langle \top, \top \rangle$, and the Proposition follows from 32.4.

32.7. Lemma. *For strong monomorphisms* a *and* b *with the same codomain,* $a \to b$ *is an equalizer of*

$$\text{ch} \, a = p \circ \langle \text{ch} \, a, \text{ch} \, b \rangle \quad \text{and} \quad \text{ch}(a \cap b) = \wedge \circ \langle \text{ch} \, a, \text{ch} \, b \rangle .$$

PROOF. For a strong monomorphism m with the same codomain, we have

$$m \leqslant a \to b \iff m \cap a \leqslant b \iff m \cap a \cap b \simeq m \cap a .$$

Since $m \cap a \simeq m \circ m^* a$ and $\text{ch} \, m^* a = \text{ch} \, a \circ m$, and similarly for $a \cap b$, we have

$$m \leqslant a \to b \iff (\text{ch} \, a) \circ m = \text{ch}(a \cap b) \circ m ,$$

and the Lemma follows.

32.8. Internal order and implication. For strong monomorphisms m_1 and m_2 of $\mathbf{E}$ with the same codomain, we have $m_1 \leqslant m_2$ iff $m_1 \cap m_2 \simeq m_1$. This is the case iff

$$\wedge \circ \langle \operatorname{ch} m_1, \operatorname{ch} m_2 \rangle = \operatorname{ch} m_1 \,,$$

and hence iff $\langle \operatorname{ch} m_1, \operatorname{ch} m_2 \rangle$ factors through an equalizer of $\wedge$ and the first projection p of $\Omega \times \Omega$. Because of this, we denote this equalizer by $\leq$ and call it the *internal order* of Ω.

Proposition. $\rightarrow\ = \operatorname{ch} \leq$ *defines internal implication.*

PROOF. We use 32.4.(4) for $p^\leftarrow\top \rightarrow q^\leftarrow\top$. By 32.7, this is internal order, an equalizer of p and of $\operatorname{ch}(p^\leftarrow\top \cap q^\leftarrow\top) = \wedge$.

32.9. Internal negation. We define *negation* in $\operatorname{Mon} A$ and in $\mathcal{M} A$, for an object A of $\mathbf{E}$, by putting

$$\neg m \simeq m \rightarrow o_A \,,$$

with o_A the least element of $\operatorname{Mon} A$ or of $\mathcal{M} A$. If m is strong, then

$$\operatorname{ch}(\neg m) = \neg \circ \operatorname{ch} m \,,$$

with internal negation $\neg$ on the righthand side, by 32.4 for unary connectives.

Proposition. $\neg = \operatorname{ch}(\neg\top) = \operatorname{ch} \bot$ *defines internal negation.*

PROOF. Using 32.4 for the unary operation $\neg$, we have

$$\neg = \operatorname{ch}(\neg \top) = \operatorname{ch}(\top \rightarrow o_\Omega) \,,$$

with $o_\Omega \simeq \tau_\Omega{}^\leftarrow o_1$. By 32.7 and 32.5, $\top \rightarrow o_\Omega$ is an equalizer of $\operatorname{ch} \top = \operatorname{id}_\Omega$ and of

$$\operatorname{ch}(\top \cap o_\Omega) = \operatorname{ch} o_\Omega = \bot \circ \tau_\Omega \,.$$

This equalizer is clearly $\bot$, up to equivalence, and the result follows.

32.10. Internal disjunction. We construct in $\mathbf{E}$ commutative squares

$$
\begin{array}{ccc}
1 & \xrightarrow{\ \top\ } & \Omega \\
{\scriptstyle \top}\downarrow & & \downarrow{\scriptstyle q^\leftarrow\top} \\
\Omega & \xrightarrow{\ p^\leftarrow\top\ } & \Omega \times \Omega
\end{array}
\qquad \text{and} \qquad
\begin{array}{ccc}
1 & \xrightarrow{\ \top\ } & \Omega \\
{\scriptstyle \top}\downarrow & & \downarrow{\scriptstyle v} \\
\Omega & \xrightarrow{\ u\ } & \Omega \vee \Omega
\end{array} \,,
$$

for the projections p and q of $\Omega \times \Omega$, with a pullback square at left by 32.6, and a pushout square at right. By 23.8, we have

$$p^\leftarrow\top = d \circ u, \qquad q^\leftarrow\top = d \circ v,$$

in $\mathbf{E}$ for a monomorphism

$$d = p^{\leftarrow}\top \vee q^{\leftarrow}\top : \Omega \vee \Omega \to \Omega \times \Omega$$

of $\mathbf{E}$. If $\mathbf{E}$ is a quasitopos, then d is in general not a strong monomorphism. Thus we must put

$$d = m\,e, \qquad \vee = \operatorname{ch} m,$$

with e monomorphic and epimorphic, and m a strong monomorphism, to obtain internal disjunction.

33. Quantifiers

33.1. Notations. For $f : A \to B$ in $\mathbf{E}$, we saw in Section 20 that the pullback functor $f^{\leftarrow} : \mathcal{M}A \to \mathcal{M}B$ has a right adjoint $\forall_f$, for which the two squares in the diagram

(1)

$$
\begin{array}{ccc}
\mathcal{M}A & \longrightarrow & \mathbf{E}/A \\
f^{\leftarrow} \big\uparrow \big\downarrow \forall_f & & f^* \big\uparrow \big\downarrow f_* \\
\mathcal{M}B & \longrightarrow & \mathbf{E}/A
\end{array}
$$

with inclusion functors as horizontal arrows, are commutative up to equivalence. We now construct a left adjoint of the functor $f^{\leftarrow}$. We discuss only classes $\mathcal{M}A$ in this Section; statements and results which do not involve relations or characteristic morphisms remain valid for classes $\operatorname{Mon} A$.

33.2. Theorem. *For $f : A \to B$ in a quasitopos $\mathbf{E}$, the inverse image functor $f^{\leftarrow} : \mathcal{M}B \to \mathcal{M}A$ has a left adjoint functor $\exists_f : \mathcal{M}A \to \mathcal{M}B$ for which the two squares in the diagram*

(1)

$$
\begin{array}{ccc}
\mathbf{E}/A & \xrightarrow{R_A} & \mathcal{M}A \\
f_> \big\downarrow \big\uparrow f^* & & \exists_f \big\downarrow \big\uparrow f^{\leftarrow} \\
\mathbf{E}/B & \xrightarrow{R_B} & \mathcal{M}B
\end{array}
$$

commute up to equivalence, with reflectors as horizontal arrows.

The functor $\exists_f$ thus defined is called an *external existential quantifier* in $\mathbf{E}$.

PROOF. By 32.1, R_B is given by reflections $e : v \to m$ with e epimorphic and m strongly monomorphic in **E**. Pullbacks preserve this situation; thus $f^{\leftarrow} R_B \simeq R_A f^*$.

For an object m of $\mathcal{M} A$, we put $\exists_f m \simeq m_1$ if $fm = m_1 e_1$ with e_1 epimorphic and m_1 strongly monomorphic. If u is an object of $\mathbf{E}/A$, with reflection $e : u \to m$ for $\mathcal{M} A$ and $fm = m_1 e_1$ as above, then $f_> u = m_1 e_1 e$, and

$$R_B f_> u \simeq m_1 \simeq \exists_f R_A u$$

by the definitions. Thus both squares in (1) commute up to equivalence.

For objects m of $\mathcal{M} A$ and m' of $\mathcal{M} B$, we have $f^{\leftarrow} m' \simeq f^* m'$, and we note that $\exists_f m \simeq R_B f_> m$ for the reflector $R_B : \mathbf{E}/B \to \mathcal{M} B$. Thus there is $r : \exists_f m \to m'$ iff there is $s : f_> m \to m'$ iff there is $t : m \to f^* m' \simeq f^{\leftarrow} m'$. As these morphisms are unique if they exist, and clearly natural in m', this establishes the desired adjunction $\exists_f \dashv f^{\leftarrow}$.

33.3. Example. For sets, we replace subobjects by subsets, so that $f^{\leftarrow} : \mathcal{M} B \to \mathcal{M} A$, for a mapping $f : A \to B$, becomes the inverse image mapping $f^{\leftarrow} : PB \to PA$. The left adjoint $\exists_f$ then becomes a mapping $\exists_f : PA \to PB$, with

$$S \subset f^{\leftarrow}(T) \iff \exists_f(S) \subset T,$$

for $S \subset A$ and $T \subset B$. It follows that $\exists_f$ is the usual image mapping,

$$y \in \exists_f(S) \iff (\exists x \in A)(x \in S \text{ and } y = f(x))$$

for $S \subset A$ and $y \in B$. In particular, if f is a projection $A \times B \to B$ of a product, then

$$y \in \exists_f(S) \iff (\exists x \in A)((x, y) \in S)$$

for $S \subset A \times B$ and $y \in B$. Thus operators $\exists_f$ generalize existential quantifiers.

33.4. Proposition. *For a pullback square*

$$
\begin{array}{ccc}
\cdot & \xrightarrow{\ t\ } & C \\
\downarrow{\scriptstyle s} & & \downarrow{\scriptstyle g} \\
A & \xrightarrow{\ f\ } & B
\end{array}
$$

in **E**, *the functors* $\exists_t s^{\leftarrow}$ *and* $g^{\leftarrow} \exists_f$ *are equivalent.*

PROOF. For an object m of $\mathcal{M}A$, we can form pullback squares

$$
\begin{array}{ccccc}
\cdot & \xrightarrow{m'} & \cdot & \xrightarrow{t} & \cdot \\
\downarrow & & \downarrow{\scriptstyle s} & & \downarrow{\scriptstyle g} \\
\cdot & \xrightarrow{m} & A & \xrightarrow{f} & B
\end{array}
\qquad \text{and} \qquad
\begin{array}{ccccc}
\cdot & \xrightarrow{e_2} & \cdot & \xrightarrow{m_2} & C \\
\downarrow & & \downarrow & & \downarrow{\scriptstyle g} \\
\cdot & \xrightarrow{e_1} & \cdot & \xrightarrow{m_1} & B
\end{array} ,
$$

with $m_1 e_1 = fm$ and $m_2 e_2 = tm'$, and with e_1 is epimorphic and m_1 strongly monomorphic. Then e_2 is epimorphic and m_2 strongly monomorphic, and thus

$$
g^{\leftarrow}\exists_f\, m \simeq g^{\leftarrow} m_1 \simeq m_2 \simeq \exists_t\, m' \simeq \exists_t\, s^{\leftarrow} m
$$

as claimed.

33.5. Definition. For $g : B \to C$ in $\mathbf{E}$ and an object A of $\mathbf{E}$, we use the isomorphisms of 20.1 to define a composition $g \circ (u, v)$ in $\mathrm{Rel}\,(A, C)$, for a relation $(u, v) : A \to B$, by putting

$$
(r, s) \simeq g \circ (u, v) \quad \Longleftrightarrow \quad \langle r, s \rangle \simeq \exists_{\mathrm{id} \times g}\langle u, v \rangle .
$$

We then proceed exactly as in Section 20, replacing $\forall_{\mathrm{id} \times g}(u, v)$ by $g \circ (u, v)$, and using 33.4 instead of 18.3. We state definitions and results, omitting the proofs.

Proposition. *If $f : X \to A$ and $g : B \to C$ in $\mathbf{E}$, then*

$$
g \circ ((u, v) \circ f) \simeq (g \circ (u, v)) \circ f
$$

for a relation $(u, v) : A \to B$ in $\mathbf{E}$.

33.6. Internal existential quantifiers. For $g : B \to C$ in $\mathbf{E}$, we put

$$
\exists g = \chi(g \circ \exists_B) : \mathsf{P}B \to \mathsf{P}C .
$$

This defines an *internal existential quantifier* $\exists g$ in $\mathbf{E}$.

Proposition. *If $(u, v) : A \to B$ is a relation in $\mathbf{E}$, then*

$$
\chi(g \circ (u, v)) = \exists g \circ \chi(u, v)
$$

for a morphism $g : B \to C$ of $\mathbf{E}$.

33.7. Proposition. *Internal existential quantifiers in $\mathbf{E}$ define a functor $\exists : \mathbf{E} \to \mathbf{E}$.*

33.8. Theorem. *If*

$$\begin{array}{ccc} \cdot & \xrightarrow{\ t\ } & C \\ {\scriptstyle s}\downarrow & & \downarrow{\scriptstyle g} \\ A & \xrightarrow{\ f\ } & B \end{array}$$

is a pullback square in a quasitopos $\mathbf{E}$, *then* $\exists s \circ \mathsf{P}i = \mathsf{P}f \circ \exists g$ *in* $\mathbf{E}$.

Corollary. *If* $f : A \to B$ *is monomorphic in* $\mathbf{E}$, *then* $\mathsf{P}f \circ \exists f = \mathrm{id}_{\mathsf{P}A}$ *in* $\mathbf{E}$.

33.9. We note an additional result for external and internal quantifiers. The first part of this result is valid for $\mathrm{Mon}\,A$ and $\mathrm{Mon}\,B$ if f is a strong epimorphism.

Proposition. *If* $f : A \to B$ *is epimorphic in* $\mathbf{E}$, *then the functors* $\exists_f\, f^\leftarrow$ *and* $\forall_f\, f^\leftarrow$ *for* $\mathcal{M}A$ *and* $\mathcal{M}B$ *are equivalences, and*

$$\exists f \circ \mathsf{P}f = \mathrm{id}_{\mathsf{P}A} = \forall f \circ \mathsf{P}f.$$

PROOF. For a pullback square

$$\begin{array}{ccc} \cdot & \xrightarrow{\ m'\ } & A \\ {\scriptstyle f'}\downarrow & & \downarrow{\scriptstyle f} \\ \cdot & \xrightarrow{\ m\ } & B \end{array}$$

with m strongly monomorphic, we have $m' \simeq f^\leftarrow m$, and f' is epimorphic. But then $\exists_f\, m' \simeq m$, and $\exists_f\, f^\leftarrow$ is an equivalence. It follows that the right adjoint $\forall_f\, f^\leftarrow$ is also an equivalence. Since a morphism $\mathrm{id}_X \times f$ is epimorphic, we also have

$$f \circ \left(f^{\mathrm{op}} \circ (u, v) \right) \simeq (u, v) \simeq \forall_{\mathrm{id} \times f}\left(f^{\mathrm{op}} \circ (u, v) \right)$$

for a relation (u, v). Now the displayed internal version of the Proposition follows immediately from 33.6, 13.8 and 20.5.

33.10. Remark. For $f : A \to B$ in $\mathbf{E}$, the functors $\exists_f$ and $\forall_f$ for strong monomorphisms can be replaced by operators

$$\exists_f : \mathbf{E}(A, \Omega) \to \mathbf{E}(B, \Omega) \qquad \text{and} \qquad \forall_f : \mathbf{E}(A, \Omega) \to \mathbf{E}(B, \Omega),$$

using the technique of 32.4. As already pointed out in 32.4, this also replaces the functor $f^\leftarrow$ by a composition $- \circ f$. If we combine this with exponential adjunction, then we have the following result.

Proposition. *If $\hat{\varphi} : X \to \mathsf{P}A$ is exponentially adjoint to $\varphi : X \times A \to \Omega$, then the following are exponentially adjoint:*

$$
\begin{array}{ccc}
\exists f \circ \hat{\varphi} & to & \exists_{\mathrm{id} \times f} \, \varphi \,, \\
\mathsf{P}g \circ \hat{\varphi} & to & \varphi \circ (\mathrm{id}_X \times g) \,, \\
\forall f \circ \hat{\varphi} & to & \forall_{\mathrm{id} \times f} \, \varphi \,,
\end{array}
$$

for $f : A \to B$ and $g : C \to A$ in $\mathbf{E}$.

PROOF. Since $\hat{\varphi} = \chi(u, v)$ for $\varphi = \mathrm{ch}\langle u, v \rangle$, this follows immediately from 13.8, 20.5, and 33.6.

34. The Language of a Quasitopos

34.1. Ingredients. The language of a quasitopos $\mathbf{E}$ is essentially a typed first-order language, but it allows set and function formation and thus must be regarded as a higher-order language. As usual, the language has *terms* and *statements*. We shall use the word *formula* to include terms and statements. Every formula has a *type*. Types are objects of $\mathbf{E}$, with Ω as the type of statements. Symbols of the language are as follows.

(1) For every type A, there is an adequate supply of *variables* of type A.

(2) All morphisms of $\mathbf{E}$ are *unary operators* in the language.

(3) For every pair of types A, B, there is a binary *pair builder* $\langle -, - \rangle_{A,B}$ or $\diamondsuit_{A,B}$.

(4) For every type A, there are *binary predicates* $=_A$ (equality) and $\in_A$ (membership).

(5) The language has statements $\top$ (true) and $\bot$ (false), and *propositional connectives* $\wedge$ (and), $\vee$ (or), and $\to$ (implies). Other connectives could be added to this list, but the usual ones can be constructed from the ones listed. We note only that $\Phi \to \bot$ replaces negation $\neg \Phi$.

(6) For every variable x, the language has a *universal quantifier* $(\forall x)$ or $\forall_x$, and an *existential quantifier* $(\exists x)$ or $\exists_x$.

(7) For every variable x, there is a *set builder* $\{x \mid -\}$ or S_x. For every variable x and every type B, there is a *function builder* $\Lambda_{x,B}$. We shall use the generic term *quantifier* for these operators as well as for universal and existential quantifiers.

If two versions of a symbol are given, the first one is for traditional notation, the second one for Polish notation.

34.2. Terms and statements. Terms and statements of the language of $\mathbf{E}$ are constructed recursively by the following rules. We denote by $\tau(F)$

the type of a formula F, with $\tau(F) = \Omega$ if F is a statement. Note that there can be terms of type Ω; we do not make a sharp distinction between terms of type Ω and statements.

(1) Every variable is a term; the type of this term is the type of the variable.

(2) If T is a term of type A and $f : A \to B$ in $\mathbf{E}$, then fT is a term of type B.

(3) If S is a term of type A and T a term of type B, then $\langle S, T \rangle_{A,B}$ is a term of type $A \times B$.

(4) If S and T are terms of the same type A, then $S =_A T$ is a statement. If t is a term of type A and T a term of type $\mathsf{P}A$, then $t \in_A T$ is a statement.

(5) $\top$ and $\perp$ are statements. If Φ and Ψ are statements, then $\Phi \wedge \Psi$, $\Phi \vee \Psi$, and $\Phi \to \Psi$ are statements.

(6) If Φ is a statement and x a variable, then $(\forall x)\Phi$ and $(\exists x)\Phi$ are statements. In these statements, every occurrence of x in Φ is replaced by a link to the quantifier $(\forall x)$ or $(\exists x)$, so that x does not occur in the quantified statements.

(7) If Φ is a statement, and x a variable of type A, then $\{x \mid \Phi\}$ is a term of type $\mathsf{P}A$. If x is a variable of type A and T a term of type B, then $\Lambda_{x,B} T$ is a term of type B^A. In these terms, every occurrence of x in Φ or in T is replaced by a link to the set builder $\{x \mid -\}$ or to the function builder $\Lambda_{x,B}$, so that x does not occur in the resulting term.

Formulas are non-empty, and every formula is obtained by Rules (1)–(7). Type subscripts of pair builders, predicates and function builders are usually omitted; variable subscripts of quantifiers cannot be omitted unless explicit links are used.

34.3. Subformulas and links.

Every symbol of the language represents a nullary, unary or binary operation, with operands and results of specified types. Variables, and the statements $\top$ and $\perp$, are nullary symbols. In the recursive process of building a formula F, every unary or binary symbol of F generates a *subformula* of F from smaller, previously constructed, subformulas, with nullary symbols as atomic one-symbol subformulas. We call a symbol in F the *leading symbol* of the subformula which it generates in the construction of F. Subformulas are nested; if s is a symbol in a subformula G of F, then the subformula of F with leading symbol s is a subformula of G.

In our formulation of formula building, a variable x in a formula F is replaced by a link if x is inside the subformula generated by a quantifier with x as subscript, i.e. by one of the symbols $(\forall x)$, $(\exists x)$, $\{x \mid -\}$, or $\Lambda_{x,B}$. Thus x does not occur in a subformula with one of these leading symbols.

In practice, links are not shown. If a variable x occurs inside a subformula generated by a quantifier $(\forall x), (\exists x), \{x \mid -\}$, or $\Lambda_{x,B}$, then it is linked to one of these quantifiers. A particular occurrence of x may be inside several subformulas of this type. If this is the case, then these subformulas and their generators are nested, and the symbol is linked to the innermost quantifier. The rules in 34.2 allow nested quantifiers using the same variable; they also allow linked and unlinked occurrences of the same variable in a formula. However, these situations can and should be avoided.

34.4. Polish notation. In traditional notation, parentheses and other delimiters for subformulas must often be used to make the meaning of formulas clear and unambiguous. In fact, the pair builder and set builder symbols have built-in delimiters in traditional notation. There are rules and conventions for reducing the number of delimiter pairs needed for formulas, but we need not discuss these rules.

Polish notation needs no delimiters. Every operation symbol immediately precedes its operands, and every symbol in a formula is the first symbol of the subformula which it generates. Thus a formula becomes a non-empty string $s_1, s_2, \ldots, s_k$ of symbols, and subformulas become substrings. A string F of symbols is a formula iff F is *well-formed* and *well-typed*.

There is a simple algorithm which determines whether a string of symbols is well-formed. This algorithm also determines the subformula generated by each symbol, and hence the operand or operands of each operation symbol. The algorithm assigns to the string $s_1, s_2, \ldots, s_k$ a string $n_0, n_1, n_2, \ldots, n_k$ of integers as follows. (i) $n_0 = 1$. (ii) If n_i is defined and $i < k$, then $n_{i+1} = n_i + r - 1$, where $r = 0$ if s_{i+1} is nullary, $r = 1$ if s_{i+1} is unary, and $r = 2$ if s_{i+1} is binary. The string of symbols is *well-formed* if $n_i > 0$ for $i < k$, and $n_k = 0$.

In a well-formed formula F, the subformula generated by a symbol s_i in F is the substring of F beginning with s_i and ending with the first symbol s_j, with $j \geq i$, such that $n_j = n_{i-1} - 1$. A subformula of a well-formed formula is again well-formed. If s_i is nullary, a variable or one of the statements $\top$ and $\bot$, then the subformula generated by s_i has just one symbol s_i. If s_i is unary, then s_i generates a subformula $s_i G$, with G the subformula generated by s_{i+1}. If s_i is binary, then the generated subformula is $s_i G H$ for the two operands of s_i, with G the subformula generated by s_{i+1}, and H generated by the symbol immediately following G in F. In this way, a well-formed formula can be parsed completely and unambiguously.

A formula in the typed language must also be *well-typed*. Every symbol in a formula determines the type of the subformula which it generates, and the types of its operand or operands. Operands are obtained by the rules of the preceding paragraph, and their types must satisfy the conditions of 34.2. Type subscripts can be omitted in writing down a formula; they can always be reconstructed from the parsing of the formula. Variable subscripts could also be omitted, but then variables linked to a quantifier or a set builder or a function builder must be replaced by explicit links to the operator.

EXAMPLE. The formula $s_A x =_{PA} \{y \mid y =_A x\}$, with variables x, y of type A becomes $=_{PA} s_A x S_y =_A x y$ in Polish notation. This is parsed as

$$
\begin{array}{ccccccc}
=_{PA} & s_A & x & S_y & =_A & x & y \\
1 & 2 & 2 & 1 & 1 & 2 & 1 & 0
\end{array}
$$

with the integers n_i shown below the symbols s_i. The non-trivial subformulas are

$$ s_A x, \quad S_y =_A xy, \quad =_A xy, $$

and the whole formula. The first two displayed subformulas are the operands of $=_{PA}$, and y is linked to S_y.

34.5. Substitution. We denote by $F[\Sigma/y]$ (read: "F with Σ for y") the result of substituting a formula Σ of type $\tau(y)$ for a variable y, with Σ replacing y for every unlinked occurrence of y in F. We note that substitution of Σ for y is only allowed if Σ has the same type as y.

It is clear from this definition that substitution has the following properties.

(i) If y does not occur in F, then substituting Σ for y does not change F.

(ii) If v is a variable, of the same type as y and not occurring in F, then $F[\Sigma/y]$ is the same formula as $(F[v/y])[\Sigma/v]$.

34.6. Theorem. *If F is a formula, y a variable, and Σ a formula of type $\tau(y)$, then $F[\Sigma/y]$ is a formula, of the same type as F.*

PROOF. If F is y, then $F[\Sigma/y]$ is Σ. Other variables, and the statements $\top$ and $\bot$, are not changed by substitutions for y.

If F is fT, then $F[\Sigma/y]$ is $f(T[\Sigma/y])$.

If F is $\langle S, T \rangle_{A,B}$, then $F[\Sigma/y]$ is $\langle S[\Sigma/y], T[\Sigma/y] \rangle_{A,B}$.

If F is $\Phi \wedge \Psi$, then $F[\Sigma/y]$ is $(\Phi[\Sigma/y]) \wedge (\Psi[\Sigma/y])$. Other logical connectives, and predicates $=_A$ and $\in_A$, are treated in the same way.

If F is $(\forall x)\Phi$, then a substitution for x does not change F. If y is distinct from x and x does not occur in Σ, then $F[\Sigma/y]$ is $(\forall x)(\Phi[\Sigma/y])$. If x occurs in Σ, then $F[\Sigma/y]$ is $(\forall v)(\Psi[\Sigma/y])$, where Ψ is $\Phi[v/x]$, for a variable v which does not occur in Φ or Σ. Existential quantifiers $(\exists x)$, set builders $\{x \mid -\}$, and function builders $\Lambda_{x,B}$ are treated in the same way as universal quantifiers.

35. Interpretations of Formulas

35.1. Products of types. For a list $L = (x_1, x_2, \ldots, x_n)$ of distinct variables of the language of a quasitopos $\mathbf{E}$, we form in $\mathbf{E}$ a finite product $P_L = \prod \tau(x_i)$, with projections $\pi_i^L : P_L \to \tau(x_i)$. If L is empty, then we put $P_L = 1$, a terminal object of $\mathbf{E}$.

If $M = (y_1, y_2, \ldots, y_m)$ is a list of distinct variables which contains the list L, i.e. each x_i is y_{μ_i} for some subscript μ_i, then putting $\pi_i^L \circ \pi_L^M = \pi_{\mu_i}^M$, for $1 \leqslant i \leqslant n$, defines a projection $\pi_L^M : P_M \to P_L$. We note that π_L^L and π_M^M are inverse isomorphisms if L and M contain the same variables, i.e. if the list M is a permutation of the list L. More generally, $\pi_L^L = \mathrm{id}_{P_L}$, and $\pi_K^M = \pi_K^L \circ \pi_L^M$ if M contains L and L contains K. If L is empty, then $\pi_L^M = \tau_{P_M} : P_M \to 1$.

In particular, if L is a list of distinct variables and x a variable which does not occur in L, then we have a list Lx, obtained by adding x to the list L, for which $P_{Lx} = P_L \times \tau(x)$. This list has projections $\pi_i^{Lx} = \pi_i^L \circ \pi_L^{Lx}$, and an added projection π_x^{Lx}, with π_L^{Lx} and π_x^{Lx} the projections of $P_L \times \tau(x)$. This construction can be used for a recursive definition of products P_L of types and their projections π_i^L; we omit the details.

We allow substitutions t for x_i in a list L of variables if t has the same type as x_i and does not occur in L. Substitutions of this type do not change P_L and its projections π_i^L.

35.2. Interpretations of Formulas. For a formula F of the language of $\mathbf{E}$, and for a list L of variables which contains all variables which occur in F, we define an *interpretation* $|F|_L : P_L \to \tau(F)$ of F in $\mathbf{E}$ recursively as follows.

(1) $|x_i|_L = \pi_i^L$.

(2) $|fT|_L = f \circ |T|_L$.

(3) $|\langle S, T \rangle|_L = \langle |S|_L, |T|_L \rangle$.

(4) $|S =_A T|_L = \delta_A \circ \langle |S|_L, |T|_L \rangle$ and $|t \in T|_L = \varepsilon_A \circ \langle |T|_L, |t|_L \rangle$, with $\delta_A = \mathrm{ch}\langle \mathrm{id}_A, \mathrm{id}_A \rangle : A \times A \to \Omega$, and for $\varepsilon_A : PA \times A \to \Omega$ exponentially adjoint to id_{PA}.

(5) $|\top|_L = \top \circ \tau_{P_L}$, $|\bot|_L = \bot \circ \tau_{P_L}$, and $|\Phi \wedge \Psi|_L = \wedge \circ \langle |\Phi|_L, |\Psi|_L \rangle$. The connectives $\vee$ and $\to$ are treated in the same way as $\wedge$.

(6) If L does not contain x, then

$$|(\forall x)\Phi|_L = \forall_p |\Phi|_{Lx} \qquad \text{and} \qquad |(\exists x)\Phi|_L = \exists_p |\Phi|_{Lx},$$

for $p = \pi_L^{Lx}$.

(7) If L does not contain x, then $|\{x \mid \Phi\}|_L : P_L \to PA$ is exponentially adjoint to $|\Phi|_{Lx} : P_L \times A \to \Omega$, and $|\Lambda_{x,B} T|_L : P_L \to B^A$ exponentially adjoint to $|T|_{Lx} : P_L \times A \to B$.

We put $|(\forall x)\Phi|_L = |(\forall x)\Phi|_{L'} \circ \pi_{L'}^L$, if L contains x, with L' obtained by omitting x from L. Interpretations $|(\exists x)\Phi|_L$, $|\{x \mid \Phi\}|_L$, and $|\Lambda_x T|_L$ are defined in the same way if L contains x.

35.3. We need some rules which facilitate computations of interpretations. We begin with some useful reductions.

Lemma. (i) *If* $|S|_L$ *and* $|T|_L$ *are defined, then* $|S =_A T|_L = \operatorname{ch} m$ *for an equalizer* m *of* $|S|_L$ *and* $|T|_L$.

(ii) *If* $|\Phi|_L$ *and* $|\Psi|_L$ *are defined, then* $|\Phi \to \Psi|_L = \operatorname{ch} m$ *for an equalizer* m *of* $|\Phi|_L$ *and* $|\Phi \wedge \Psi|_L$.

(iii) *If* Q_x *is one of the operators* $(\forall x)$, $(\exists x)$, $\{x \mid -\}$, $\Lambda_{x,B}$, *and if*

$$|G|_{Mx} = |F|_{Lx} \circ (f \times \operatorname{id}_A)$$

for formulas F *and* G, *with* $f : P_M \to P_L$ *in* $\mathbf{E}$ *and* $A = \tau(x)$, *then*

$$|Q_x G|_M = |Q_x F|_L \circ f.$$

PROOF. (i) follows immediately from the fact that m is an equalizer of r and s in a diagram

$$
\begin{array}{ccccc}
\cdot & \longrightarrow & A & \longrightarrow & 1 \\
\downarrow{\scriptstyle m} & & \downarrow{\scriptstyle \langle \operatorname{id}_A, \operatorname{id}_A \rangle} & & \downarrow{\scriptstyle \top} \\
\cdot & \xrightarrow{\langle r, s \rangle} & A & \xrightarrow{\delta_A} & \Omega
\end{array}
$$

of pullback squares, and (ii) follows from 32.7.

(iii) is valid for $\{x \mid -\}$ and Λ_x, since this replaces $|F|_{Lx}$ by the exponentially adjoint morphism, and $\hat{\varphi} \circ f$ is exponentially adjoint to $\varphi \circ (f \times \operatorname{id}_A)$ if $\hat{\varphi} : P_L \to B^A$ is exponentially adjoint to $\varphi : P_L \times A \to B$.

For $(\forall x)$, we note that

(1)
$$
\begin{array}{ccc}
P_M \times A & \xrightarrow{q} & P_M \\
\downarrow{\scriptstyle f \times \operatorname{id}_A} & & \downarrow{\scriptstyle f} \\
P_L \times A & \xrightarrow{p} & P_L
\end{array}
$$

with $p = \pi_L^{Lx}$ and $q = \pi_M^{Mx}$, is a pullback square. Since functors $f^{\leftarrow}$ become compositions $- \circ f$ if we replace classes $\mathcal{M}-$ by sets $\mathbf{E}(-, \Omega)$, we have

$$|(\forall x) G|_M = \forall_q |G|_{Mx} = \forall_q (|F|_{Lx}(f \times \operatorname{id}_A))$$

$$= (\forall_p |F|_{Lx}) \circ f = |(\forall x) F|_L \circ f,$$

by 20.7. This proves (iii) for $(\forall x)$. The same proof works for $(\exists x)$, using 33.4.

35.4. Proposition. *The interpretations defined above satisfy the following rules.*

(i) $|F|_L$ *is not changed by a substitution* u *for* x, *in* F *and in* L, *for a variable* u *of the same type as* x *which does not occur in* L.

(ii) $|F|_M = |F|_L \circ \pi_L^M$ *if* M *contains all variables in* L.

(iii) $|F[\Sigma/y]|_L = |F|_{Ly} \circ \langle \mathrm{id}_{P_L}, |\Sigma|_L \rangle$ *if* y *does not occur in* L, *and* $|F|_{Ly}$ *and* $|\Sigma|_L$ *are defined.*

PROOF. We prove this by recursion over the number of symbols in F.

We note first that Rule (i) follows from Rules (ii) and (iii). Denote by M the list obtained by changing x to u, with $P_M = P_L$. From (ii) and (iii), we get

$$|F[u/x]|_M = |F|_L \circ \pi_L^{Mx} \circ \langle \mathrm{id}_{P_M}, \pi_u^M \rangle .$$

It is easily seen that

$$\pi_i^L \circ \pi_L^{Mx} \circ \langle \mathrm{id}_{P_M}, \pi_u^M \rangle = \pi_i^M = \pi_i^L$$

for a variable x_i distinct from x in L, and

$$\pi_x^L \circ \pi_L^{Mx} \circ \langle \mathrm{id}_{P_M}, \pi_u^M \rangle = \pi_u^M = \pi_x^L$$

for the remaining variable x. Thus we get $|F[u/x]|_M = |F|_L \circ \mathrm{id}_{P_L}$.

Interpretations $|x_i|_L = \pi_i^L$, and $|\top|_L = \top \circ \tau_{P_L}$ and $|\bot|_L$ clearly satisfy Rule (ii). Rule (iii) is valid for y since $y[\Sigma/y]$ is Σ, and $|y|_{Ly} = \pi_y^{Ly}$. If u is a variable distinct from y, or one of the symbols $\top$ or $\bot$, then $u[\Sigma/y]$ is u, and $|u|_{Ly} = |u|_L \circ \pi_L^{Ly}$; thus (iii) is again valid.

It is clear that the rules remain valid if we build formulas by (2)–(5) in 34.2, from operands which satisfy the rules. Only cases (6) and (7), for $G = Q_x F$ in the notation of 35.3.(iii), with the rules valid for F, need discussion.

For Rule (ii), we consider first the case that x does not occur in L. If x does not occur in M, then we use 35.3.(iii) with $G = F$ and $f = \pi_L^M$, with $f \times \mathrm{id}_A = \pi_{Lx}^{Mx}$, to get Rule (ii) for F. If x occurs in M, then

$$|F|_M = |F|_L \circ \pi_L^{M'} \circ \pi_{M'}^M = |F|_L \circ \pi_L^M ,$$

with M' obtained by omitting x from M. If x occurs in L, and if L' is obtained by omitting x from L, then again

$$|F|_M = |F|_{L'} \circ \pi_{L'}^M = |F|_{L'} \circ \pi_{L'}^L \circ \pi_L^M = |F|_L \circ \pi_L^M .$$

If y is x for Rule (iii), then $G[\Sigma/y]$ is G, and

$$|G|_{Ly} \circ \langle \mathrm{id}_{P_L}, |\Sigma|_L \rangle = |G|_L \circ \pi_L^{Ly} \circ \langle \mathrm{id}_{P_L}, |\Sigma|_L \rangle = |G|_L ;$$

thus Rule (iii) is valid. If y is distinct from x and x occurs in L, then we can replace $G = Q_x F$ by $G = Q_u F'$, for $F' = F[u/x]$, by the proof of 34.6, with u distinct from y and not occurring in L, and with Rule (iii) valid for F' which has the same length as F. Thus we may assume that x does not occur in L and is distinct from y. We then use 35.3.(iii) with L and M replaced by Ly and L, and with

$$f = \langle \mathrm{id}_{P_L}, |\Sigma|_L \rangle \qquad \text{and} \qquad f \times \mathrm{id}_A = \pi^{Lxy}_{Lyx} \circ \langle \mathrm{id}_{P_{Lx}}, |\Sigma|_{Lx} \rangle,$$

to get Rule (iii) for G.

35.5. Supports and internal interpretations. Every formula F of the language of $\mathbf{E}$ has infinitely many interpretations, one for every list L of variables which contains all variables occurring in F. If the variables listed in L are exactly the variables which occur in F, then we say that L is a *support* of F, and the interpretation $|F|_L$ is called an *internal interpretation* of F. We denote by $\sigma(F)$ a support of F, and an internal interpretation is denoted by $\|F\|_L$, or just by $\|F\|$ if L is given by the context.

Supports are determined up to a permutation of variables, and if L and M are supports of a formula F, then $\|F\|_M = \|F\|_L \circ \pi^M_L$ for an isomorphism π^M_L, by Rule (ii) of 35.4. Every interpretation of a formula F can be obtained from an internal interpretation of F by Rule 35.4.(ii). Supports and internal interpretations can be made definite by listing variables in a specified order, e.g. by their first appearance in F, in Polish notation.

The constructions of interpretations $|F|_L$ listed in 35.2 could easily have been restricted to internal interpretations. However, if F results from a binary operation, as in cases (3), (4) and (5), then the construction of $\|F\|$ may well require non-internal interpretations of the constituents of F.

36. Internal Validity

36.1. Definition. A statement Φ of the language of a quasitopos $\mathbf{E}$ is called *internally valid* if an internal interpretation $\|\Phi\| = |\Phi|_L$ of Φ factors $\|\Phi\| = \top \circ \tau_{P_L}$. It is clear from 35.5 that this is the case if and only if $|\Phi|_L = \top \circ \tau_{P_L}$ for every list L of variables such that $|\Phi|_L$ is defined. It is not enough to require this for one interpretation $|\Phi|_M$ unless this interpretation is internal, since a projection π^M_L need not be epimorphic in $\mathbf{E}$.

We write $\models \Phi$ to indicate that Φ is internally valid, and we obtain some properties of internal validity.

36.2. Proposition. (i) $\models \Phi \to \Psi$ *for statements* Φ *and* Ψ *if and only if always* $|\Phi|_L \le |\Psi|_L$, *whenever both interpretations are defined.*

(ii) $\models S = T$ *for terms* S *and* T *if and only if always* $|S|_L = |T|_L$, *whenever both interpretations are defined.*

PROOF. This follows immediately from 35.3.(i) and 35.3.(ii).

36.3. Logical and internal equivalence. We define *logical equivalence* by

$$a \leftrightarrow b \simeq (a \to b) \wedge (b \wedge a),$$

for strong monomorphisms a, b with the same codomain. It follows from 32.7 that this is an equalizer of $\mathrm{ch}\, a, \mathrm{ch}\, b$, and $\mathrm{ch}(a \cap b) = \mathrm{ch}\, a \cap \mathrm{ch}\, b$, and hence an equalizer of $\mathrm{ch}\, a$ and $\mathrm{ch}\, b$. Logical equivalence is a propositional connective. By 32.4.(4), the internal version of this connective is

$$\leftrightarrow \; = \; \delta_\Omega \; = \; \mathrm{ch}\langle \mathrm{id}_\Omega, \mathrm{id}_\Omega \rangle,$$

for the equalizer $\langle \mathrm{id}_\Omega, \mathrm{id}_\Omega \rangle$ of the projections of $\Omega \times \Omega$.

We define logical equivalence for statements Φ and Ψ by

$$\Phi \leftrightarrow \Psi \quad \Longleftrightarrow \quad (\Phi \to \Psi) \wedge (\Psi \to \Phi),$$

and we say that Φ and Ψ are *internally equivalent* if $\models \Phi \leftrightarrow \Psi$. By 36.2, this is the case iff $|\Phi|_L = |\Psi|_L$, for every list L of variables which includes all variables occurring in Φ or in Ψ. It follows that

$$\models \Phi \leftrightarrow \Psi \quad \Longleftrightarrow \quad \models \Phi \to \Psi \quad \text{and} \quad \models \Psi \to \Phi.$$

We shall obtain some basic internal equivalences in the next Section.

36.4. Internal equality. We say that terms S and T of the same type are *internally equal* if $\models S = T$. This means by 36.2 that $|S|_L = |T|_L$ whenever both interpretations are defined.

For statements Φ and Ψ, regarded as formulas of type Ω, we have

$$\models (\Phi \leftrightarrow \Psi) \leftrightarrow (\Phi =_\Omega \Psi);$$

thus internal equivalence can be considered as a special case of internal equality.

Internal equality and internal equivalence have all the properties one expects; we list the most important ones without proof.

Proposition. *The following statements are internally valid, for variables and terms of the same type.*

$\models x = x$.

$\models x = y \rightarrow y = x$.

$\models (x = y \wedge y = z) \rightarrow x = z$.

$\models S = T \rightarrow (\Phi[S/x] \leftrightarrow \Phi[T/x])$.

$\models S = T \rightarrow (R[S/x] = R[T/x])$.

These rules are also valid, mutatis mutandis, for internal equivalence.

36.5. Deduction rules. A *deduction rule* is a statement of the form

$$(\models \Phi_1 \text{ and } \ldots \text{ and } \models \Phi_k) \implies \models \Psi.$$

We write this usually in the concise form

$$(1) \qquad \frac{\Phi_1, \ldots, \Phi_k}{\Psi}.$$

A deduction rule of this form is related to the assertion

$$(2) \qquad \models (\Phi_1 \wedge \ldots \wedge \Phi_k) \rightarrow \Psi,$$

but (1) does not always follow from (2). It follows from (2) and the hypothesis of (1) that $|\Psi|_L = \top \circ \tau_{P_L}$, for a list L of variables which includes all variables of $\Phi_1, \ldots, \Phi_k$ and of Ψ. If N is a support of Ψ, then $|\Psi|_L = |\Psi|_N \circ \pi_N^L$. We can conclude from this that $|\Psi|_N = \top \circ \tau_{P_N}$ only if π_N^L is epimorphic.

This state of affairs leads to the following discussion.

36.6. Inhabited and witnessed types. We say that a type A is *inhabited* if

$$\models (\exists x)(x = x)$$

for a variable x of type A.

We say more generally that a type A is *witnessed* by a list L of distinct variables if for a variable x of type A, but not in L, we have

$$\models (\exists x) \, \Phi$$

for some statement Φ with $|\Phi|_{Lx}$ defined.

We note that an inhabited type is witnessed by an empty list L.

36.7. Proposition. *The following statements are valid for a list L of variables.*

(i) *Every inhabited type is witnessed by L.*

(ii) *If a type A is witnessed by L, then A is witnessed by every list of variables which contains L.*

(iii) *A type A is witnessed by L if and only if the projection π_L^{Lx} is epimorphic in $\mathbf{E}$, for a variable x of type A and not in L. In particular, a type A is inhabited if and only if $\tau_A : A \rightarrow 1$ is epimorphic in $\mathbf{E}$.*

(iv) *If there is a term R of type A with all variables of R in L, then A is witnessed by L. In particular, the type of every variable in L is witnessed by L.*

(v) *A projection π_L^M is epimorphic if and only if the type of every variable in M is witnessed by L.*

PROOF. (i) and (ii) are obvious from the definitions.

For (iii), we note that $|(\exists x)\Phi|_L = \mathrm{ch}(\exists_p m)$, for $p = \pi_L^{Lx}$ and $|\Phi|_L = \mathrm{ch}\, m$. This factors through $\top$ iff $\exists_p m$ is isomorphic, *i.e.* pm epimorphic. But then $p = \pi_L^{Lx}$ is epimorphic. Conversely, if $R = \pi_x^{Lx}\langle x_1, \ldots x_k, x\rangle$ for the variables $x_1, \cdots, x_k$ of L, with the $(k+1)$-tuple obtained by repeated pair-building, then $\|R\| = \pi_x^{Lx} = |x|_{Lx}$, and $m \simeq \mathrm{id}_{P_{Lx}}$ if $\|R = x\| = \mathrm{ch}\, m$. Thus $\exists_p m$ is isomorphic, and $\models (\exists x)(R = x)$ for this R, if $p = \pi_L^{Lx}$ is epimorphic.

For (iv), we have $m = \langle \mathrm{id}_{P_L}, |R|_L\rangle$ in 35.3.(i) for $S = R$ and $T = x$, with $\pi_L^{Lx} m = \mathrm{id}_{P_L}$, if x is a variable of type A which does not occur in L. Thus A is witnessed by L, by (iii).

For (v), we note that π_L^M is, up to isomorphism, a composite of morphisms π_N^{Nx}, with L contained in N and x not in N. If all types of variables in M are witnessed by L, and hence by N containing L, then each π_N^{Nx} is epimorphic, and thus π_L^M is epimorphic. Conversely, $\pi_L^M = \pi_L^{Lx} \circ \pi_{Lx}^M$ for x in M, but not in L, and π_L^{Lx} is epimorphic if π_L^M is epimorphic.

37. Rules of Internal Logic

37.1. Tautologies. An *intuitionistic tautology* can be defined as a formal law for Heyting algebras (see Section 24), *i.e.* an equation $\varphi = \psi$ or an inequality $\varphi \leqslant \psi$ which is valid in every Heyting algebra. Since the operations of Heyting algebras are logical connectives, we can translate an intuitionistic tautology into a statement $\varphi \leftrightarrow \psi$ or $\varphi \rightarrow \psi$, either by considering the variables of φ and ψ as statements, or by substituting statements for these variables. We then have the following result.

Proposition. *Every intuitionistic tautology is internally valid.*

PROOF. This follows immediately from the definitions, using 32.8 for inequalities, and the fact that every set $\mathbf{E}(P_L, \Omega)$ is a Heyting algebra.

37.2. Proposition. *If x occurs in Φ, then the statements*

$$(\forall x)\Phi \rightarrow \Phi \qquad \text{and} \qquad \Phi \rightarrow (\exists x)\Phi$$

are internally valid, and $\models \Phi$ if and only if $\models (\forall x)\Phi$.

PROOF. By assumption, Φ has a support Lx with x not listed in L. It follows from the definitions that

$$|(\forall x)\Phi \to \Phi|_{Lx} = (p^{\leftarrow}\forall_p|\Phi|_{Lx}) \to |\Phi|_{Lx},$$

for the projection $p = \pi_L^{Lx}$. Since $p^{\leftarrow} \dashv \forall_p$, we have $p^{\leftarrow}\forall_p|\Phi|_{Lx} \leq |\Phi|_{Lx}$, and the first claim follows with 36.2. The second claim is verified similarly, using the adjunction $\exists_p \dashv p^{\leftarrow}$. If $\models \Phi$, then $\models (\forall x)\Phi$ since the operator $\forall_p$ preserves greatest elements of lattices $\mathcal{M}A$ or $\mathbf{E}(A,\Omega)$. The converse rule follows from the validity of $(\forall x)\Phi \to \Phi$ by 37.4; it can also be verified directly.

37.3. Proposition. *If Σ is a term of type A, for a variable x of type A which occurs in Φ, but not in Σ, then*

$$\frac{\Phi}{\Phi[\Sigma/x]}, \quad \text{and} \quad \models \Phi[\Sigma/x] \to (\exists x)\Phi.$$

PROOF. If L is a list which does not contain x, but contains all other variables of Φ and all variables of Σ, construct a pullback square

$$
\begin{array}{ccc}
\cdot & \xrightarrow{m_1} & \cdot \\
\downarrow{m'} & & \downarrow{m} \\
P_L & \xrightarrow{\langle \mathrm{id}, s\rangle} & P_L \times A
\end{array},
$$

with $|\Phi|_{Lx} = \mathrm{ch}\, m$ and $s = |\Sigma|_L$. Then $|\Phi[\Sigma/x]|_L = \mathrm{ch}\, m'$, and it follows that $\mathrm{ch}\, m'$ factors through $\top$ if $\mathrm{ch}\, m$ does. This proves the deduction rule. For $p = \pi_L^{Lx}$, we have

$$m' = p\langle \mathrm{id}, s\rangle m' = p\, m\, m_1 \leqslant \exists_p\, m;$$

this shows that $\models \Phi[\Sigma/x] \to (\exists x)\Phi$.

37.4. Restricted modus ponens. The most important deduction rule,

(MP) $$\frac{\Phi, \quad \Phi \to \Psi}{\Psi},$$

is called *modus ponens*. Unfortunately, this rule is not always internally valid. For example, if Φ is $x = x$ and Ψ is $(\exists x)(x = x)$, for a variable x of a type A with $\tau_A : A \to 1$ not epimorphic, then $\models \Phi$, and $\Phi \to \Psi$ is internally valid by 37.3, but Ψ is not internally valid.

Because (MP) is not internally valid in general, internal equality and internal equivalence are not fully transitive. They are internally transitive by 36.4, but external transitivity of these relations is subject to the same restrictions as modus ponens.

The following seems to be a "best possible" internally valid form of modus ponens; we call it *restricted modus ponens*.

Theorem. *Modus ponens* (MP) *is internally valid if every type of a variable occurring in* Φ *is witnessed by a list of variables which occur in* Ψ, *and in particular if every variable of* Φ *occurs in* Ψ.

PROOF. Let L be a list of all variables which occur in Φ or Ψ. If the premises of (MP) are valid, then $|\Phi|_L \le |\Psi|_L$, and $|\Phi|_L = \top \circ \tau_{P_L}$. It follows that also $|\Psi|_L = \top \circ \tau_{P_L}$. If N is a support of Ψ, then

$$|\Psi|_N \circ \pi_N^L = |\Psi|_L = \top \circ \tau_{P_L} = \top \circ \tau_{P_N} \circ \pi_N^L.$$

If every variable in L is witnessed by a list of variables in N, then π_N^L is epimorphic by 36.7, and $\models \Psi$ follows.

37.5. Transitivity of inference. Since $\Phi \leftrightarrow (\top \to \Phi)$, modus ponens can be viewed as a special case of the syllogistic deduction rule

(SL) $$\frac{\Sigma \to \Phi, \quad \Phi \to \Psi}{\Sigma \to \Psi}.$$

This rule is not universally valid, but we have the following result.

Theorem. *The deduction rule* (SL) *is internally valid if every type of a variable occurring in* Φ *is witnessed by a list of variables which occur in* Σ *or* Ψ, *and in particular if every variable of* Φ *occurs in* Σ *or* Ψ.

The proof of this result is similar to that of 37.4; we omit it.

37.6. Proposition. *The following quantification rules are internally valid.*

$$(\forall x)(\forall y)\Phi \ \leftrightarrow \ (\forall y)(\forall x)\Phi,$$

$$(\exists x)(\exists y)\Phi \ \leftrightarrow \ (\exists y)(\exists x)\Phi,$$

$$(\forall x)(\Phi \wedge \Psi) \ \leftrightarrow \ (\forall x)\Phi \wedge (\forall x)\Psi,$$

$$(\exists x)(\Phi \vee \Psi) \ \leftrightarrow \ (\exists x)\Phi \vee (\exists x)\Psi.$$

We omit the easy proofs of these rules.

37.7. Proposition. *If the variable* x *does not occur in the term* T, *then a substitution statement* $\Phi[T/x]$ *is internally equivalent to the two statements*

$$(\exists x)((T = x) \wedge \Phi) \quad \text{and} \quad (\forall x)((T = x) \to \Phi),$$

and to $T \in \{x \mid \Phi\}$.

PROOF. For a list L of variables which does not contain x, but all other variables in Φ and in T, let $t = |T|_L$. Then $\langle \mathrm{id}_{P_L}, t \rangle$ is an equalizer of $|T|_{Lx} = t\pi_L^{Lx}$ and $|x|_{Lx} = \pi_x^{Lx}$, and

$$|T = x|_{Lx} = \mathrm{ch}\langle \mathrm{id}_{P_L}, t \rangle \qquad \text{and} \qquad |\Phi[T/x]|_L = |\Phi|_{Lx} \circ \langle \mathrm{id}_{P_L}, t \rangle,$$

follow by 35.3 and 35.4. If $p = \pi_L^{Lx}$, then $p \circ \langle \mathrm{id}_{P_L}, t \rangle = \mathrm{id}_{P_L}$, and

$$|(\exists x)(T = x \wedge \Phi)|_L = \exists_p \exists_{\langle \mathrm{id}, t \rangle} \langle \mathrm{id}, t \rangle^{\leftarrow} |\Phi|_{Lx}$$

$$= \langle \mathrm{id}, t \rangle^{\leftarrow} |\Phi|_{Lx} = |\Phi|_{Lx} \circ \langle \mathrm{id}, t \rangle = |\Phi[T/x]|_L,$$

so that $\Phi[T/x]$ is equivalent to the first displayed statement. Equivalence to the second statement is obtained in the same way, using universal quantifiers. Finally,

$$|T \in \{x \mid \Phi\}|_L = \varepsilon_A \circ (|\{x \mid \Phi\}|_L \times \mathrm{id}_A) \circ \langle \mathrm{id}_{P_L}, t \rangle$$

$$= |\Phi|_{Lx} \circ \langle \mathrm{id}_{P_L}, t \rangle = |\Phi[T/x]|_L,$$

by the definitions and rules, and we are done.

37.8. Proposition. *If the variable x does not occur in Φ, then the following equivalences are internally valid.*

$$(\exists x)(\Phi \wedge \Psi) \;\leftrightarrow\; \Phi \wedge (\exists x)\Psi,$$

$$(\forall x)(\Phi \rightarrow \Psi) \;\leftrightarrow\; \Phi \rightarrow (\forall x)\Psi,$$

$$(\forall x)(\Psi \rightarrow \Phi) \;\leftrightarrow\; (\exists x)\Psi \rightarrow \Phi.$$

PROOF. For a list L which contains all variables occurring in the three statements, but not x, let $|\Phi|_L = \mathrm{ch}\,\varphi$ and $|\Psi|_{Lx} = \mathrm{ch}\,\psi$, and let $p = \pi_L^{Lx} : P_L \times A \rightarrow P_L$, with A the type of x. Then

$$|(\exists x)(\Phi \wedge \Psi)|_L = \mathrm{ch}(\exists_p \exists_{\varphi \times \mathrm{id}} (\varphi \times \mathrm{id})^{\leftarrow} \psi)$$

$$= \mathrm{ch}(\exists_\varphi \exists_{p'}(\varphi \times \mathrm{id})^{\leftarrow} \psi) = \mathrm{ch}(\exists_\varphi \varphi^{\leftarrow} \exists_p \psi) = |\Phi \wedge (\exists x)\Psi|_L,$$

using 33.4 for a pullback square $p\,(\varphi \times \mathrm{id}_A) = \varphi\,p'$.

The second equivalence is proved in the same way, using universal quantifiers.

For the third equivalence, we note that the two sides have interpretations

$$\mathrm{ch}(\forall_p(\psi \rightarrow p^{\leftarrow} \varphi)) \qquad \text{and} \qquad \mathrm{ch}((\exists_p \psi) \rightarrow \varphi),$$

and that

$$m \leqslant \forall_p(\psi \rightarrow p^{\leftarrow} \varphi) \quad \Longleftrightarrow \quad p^{\leftarrow} m \wedge \psi \leqslant p^{\leftarrow} \varphi)$$

$$\Longleftrightarrow \quad \psi \leqslant p^{\leftarrow}(m \rightarrow \varphi) \quad \Longleftrightarrow \quad \exists_p \psi \leqslant m \rightarrow \varphi$$

$$\Longleftrightarrow \quad m \wedge \exists_p \psi \leqslant \varphi \quad \Longleftrightarrow \quad m \leqslant \exists_p \psi \rightarrow \varphi,$$

for a strong monomorphism m with codomain P_L.

37.9. Proposition. *If the variable x occurs in Ψ, but not in Φ, then the deduction rules*

$$\frac{\Psi \to \Phi}{(\exists x)\Psi \to \Phi} \quad \text{and} \quad \frac{\Phi \to \Psi}{\Phi \to (\forall x)\Psi} \ ,$$

and their converses, are internally valid.

PROOF. We use the notations of the preceding proof, with L minimal. Then

$$\models \Psi \to \Phi \iff |\Psi|_{Lx} \leqslant p^{\frown}|\Phi|_L \,,$$

and

$$\models (\exists x)\Psi \to \Phi \iff \exists_p|\Psi|_{Lx} \leq |\Phi|_L \,,$$

in the first deduction rule. The two inequalities are equivalent since $\exists_p \dashv p^{\frown}$; thus the two internal validity assertions are equivalent. The second rule is obtained in the same way.

38. Some Constructions in a Quasitopos

38.1. We assume again that $\mathbf{E}$ is a quasitopos, provided with its internal language.

Proposition. *There is a constant term $\star$ of type 1, with $\|\star\| = \mathrm{id}_1$, and any two terms of type 1 are internally equal.*

PROOF. The term $\star$ can be defined as $\tau_{P0}\{x \mid x = x\}$, for a variable x of type 0 and the isomorphism $\tau_{P0} : P0 \to 1$. If S and T are terms of type 1, then $|S|_L = \tau_{P_L} = |T|_L$ whenever the two interpretations are defined.

38.2. Proposition. *If T is a term of type A, then*

$$\models s_A T = \{x \mid T = x\} \,,$$

for a variable x of type A which does not occur in T.

This justifies the notation $\{T\}$ for $s_A T$ which is often used.

PROOF. As s_A is exponentially adjoint to δ_A, an interpretation $|s_A T|_L :$ $P_L \to PA$ is exponentially adjoint to

$$\delta_A(|T|_L \times \mathrm{id}_A) = |T = x|_{Lx} \,,$$

if x does not occur in the list L. The result now follows immediately from the definitions.

38.3. Proposition. *If* $f : A \to B$ *and* $g : B \to C$ *in* **E**, *then the equations*

$$\mathrm{id}_A\, T = T \qquad \text{and} \qquad g(f\, T) = (g \circ f)\, T$$

are internally valid for a term T *of type* A.

If $f : A \to A'$ *and* $g : B \to B'$ *in* **E**, *then the equation*

$$(f \times g)\langle S, T \rangle = \langle fS, gT \rangle$$

is internally valid for terms S *of type* A *and* T *of type* B.

PROOF. Both sides of each equation have the same interpretations.

38.4. Proposition. *If* $A \xleftarrow{p} A \times B \xrightarrow{q} B$ *are the projections of a product in* **E**, *then*

$$\models\; U = \langle S, T \rangle \;\leftrightarrow\; S = pU \wedge T = qU\,,$$

and

$$\models\; S = p\langle S, T \rangle\,, \qquad \models\; T = q\langle S, T \rangle\,,$$

for terms S *of type* A, T *of type* B, *and* U *of type* $A \times B$.

PROOF. Put $|S|_L = s, |T|_L = t, |U|_L = u$, for a suitable list L of variables. Then

$$\mathrm{ch}\, m \leqslant |U = \langle S, T \rangle|_L \quad \Longleftrightarrow \quad um = \langle s, t \rangle\, m\,,$$

$$\mathrm{ch}\, m \leqslant |pU = S|_L \quad \Longleftrightarrow \quad pum = sm\,,$$

$$\mathrm{ch}\, m \leqslant |qU = T|_L \quad \Longleftrightarrow \quad qum = tm\,,$$

for a strong monomorphism m with codomain P_L, by 35.3.(i) and the definitions, and

$$|U = \langle S, T \rangle|_L \;=\; |pU = S|_L \wedge |qU = T|_L$$

follows. The other validity assertions follow immediately from the definitions.

38.5. Proposition. *Let* $f : A \to B$ *in* **E**. *For statements* Φ *and* Ψ, *a variable* x *of type* A *which does not occur in* Ψ, *and a variable* y *of type* B *which does not occur in* Φ, *we have the following.*

$$\models\; \exists f\, \{x \mid \Phi\} = \{y \mid (\exists x)(fx = y \wedge \Phi)\}\,.$$

$$\models\; Pf\, \{y \mid \Psi\} = \{x \mid \Psi[fx/y]\}\,.$$

$$\models\; \forall f\, \{x \mid \Phi\} = \{y \mid (\forall x)(fx = y \to \Phi)\}\,.$$

PROOF. Let L be a list containing all variables in Φ except x, or all variables in Ψ except y, let $A \xleftarrow{p} A \times B \xrightarrow{q} B$ be the projections of the product, and put

$$m = \mathrm{id}_{P_L} \times \langle \mathrm{id}_A, f \rangle\,, \qquad \text{with} \qquad |fx = y|_{Lxy} = \mathrm{ch}\, m$$

by 35.3.(i). Then

$$|(\exists x)(fx = y \wedge \Phi)|_{Ly} = \exists_{\mathrm{id} \times q} \exists_m m^{\leftarrow} |\Phi|_{Lxy}$$

$$= \exists_{\mathrm{id} \times q} \exists_m m^{\leftarrow} (\mathrm{id} \times p)^{\leftarrow} |\Phi|_{Lx} = \exists_{\mathrm{id} \times f} |\Phi|_{Lx},$$

using $m \wedge m_1 = \exists_m m^{\leftarrow} m_1$ and the definitions. Taking exponential adjoints, and using 33.10 and 36.2.(ii), we get the first part of the result.

For the second part, we note that

$$|\Psi[fx/y]|_{Lx} = |\Psi|_{Lxy} \circ \langle \mathrm{id}_{P_{Lx}}, f \pi_x^{Lx} \rangle$$

$$= |\Psi|_{Ly} \circ \pi_{Ly}^{Lxy} \circ \langle \mathrm{id}_{P_{Lx}}, f \pi_x^{Lx} \rangle = |\Psi|_{Ly} (\mathrm{id}_{P_L} \times f).$$

Now take exponential adjoints on both sides and use again 33.10.

The third part is obtained in the same way as the first part, with universal instead of existential quantifiers. We omit the details.

38.6. Predicates. We may regard a morphism $\rho : A \to \Omega$ of **E** as a *predicate* of type A, with a statement ρT defined, by 34.2.(2), for every term T of type A.

Proposition. *If $\rho = \mathrm{ch}\, m$ with $m : X \to A$ in **E**, then*

$$\rho T \leftrightarrow (\exists x)(T = mx)$$

is internally valid, for a variable x of type X which does not occur in T.

PROOF. If L is a list of variables which contains all variables in T, then

$$|(\exists x)(T = mx)|_L = \mathrm{ch}(\exists_p \langle a, b \rangle),$$

for an equalizer $\langle a, b \rangle$ of $|T|_{Lx}$ and $|mx|_{Lx} = m \circ \pi_x^{Lx}$, and for the projection $p : P_L \times A \to P_L$. The equalizer is obtained from a diagram

$$
\begin{array}{ccccc}
\cdot & \xrightarrow{\ b\ } & X & \longrightarrow & 1 \\
{\scriptstyle a}\downarrow & & {\scriptstyle m}\downarrow & & \downarrow{\scriptstyle T} \\
P_L & \xrightarrow{\ |T|_L\ } & A & \xrightarrow{\ \rho\ } & \Omega
\end{array}
$$

(1)

of pullback squares in **E**, and

$$\mathrm{ch}(\exists_p \langle a, b \rangle) = \mathrm{ch}\, a = \rho \circ |T|_L = |\rho T|_L$$

follows since a in (1) is a strong monomorphism.

38.7. Proposition. *If x does not occur in S or in T, for a variable x
of type A and terms S and T of type PA, then the rule*

$$\frac{x \in S \;\leftrightarrow\; x \in T}{S = T}$$

and its converse are internally valid.

PROOF. If a list L contains all variables of S, but not x, then

$$|x \in S|_{Lx} = \varepsilon_A \circ (|S|_L \times \mathrm{id}_A)$$

and $|S|_L$ are exponentially adjoint; this is also true for T. Thus

$$|x \in S|_{Lx} = |x \in T|_{Lx} \quad \Longleftrightarrow \quad |S|_L = |T|_L$$

if the interpretations are defined, and the result follows.

39. Internal Unions and Intersections

39.1. Definitions. For a type A, and variables x, X, Z of types A, PA,
and PPA, the internal interpretations

$$\bigcup\nolimits_A = \|\{x \mid (\exists X)(X \in Z \wedge x \in X)\}\| : PPA \to PA,$$

and

$$\bigcap\nolimits_A = \|\{x \mid (\forall X)(X \in Z \to x \in X)\}\| : PPA \to PA,$$

are called an *internal union* and an *internal intersection* in **E**.

39.2. Lemma. *For a term T of type PPA, the equations*

$$\bigcup\nolimits_A T = \{x \mid (\exists X)(X \in T \wedge x \in X)\},$$

and

$$\bigcap\nolimits_A T = \{x \mid (\forall X)(X \in T \to x \in X)\},$$

are internally valid if x does not occur in T.

PROOF. Let L be a list of all variables of T; we can assume that Z does
not occur in L. Then

$$|\{x \mid (\exists X)(X \in T \wedge x \in X)\}|_L = |\{x \mid (\exists X)(X \in Z \wedge x \in X)\}|_{LZ} \circ \langle \mathrm{id}_{P_L}, |T|_L \rangle$$

$$= \bigcup\nolimits_A \circ \, \pi_Z^{LZ} \circ \langle \mathrm{id}_{P_L}, |T|_L \rangle = \bigcup\nolimits_A \circ \, |T|_L ,$$

using 35.4.(iii). The same proof works for intersections.

39.3. Theorem. *If* $f : A \to B$ *in* **E**, *then*

$$\exists f \circ \bigcup_A = \bigcup_B \circ \exists \exists f, \qquad \mathsf{P}f \circ \bigcup_B = \bigcup_A \circ \exists \mathsf{P}f,$$

$$\forall f \circ \bigcap_A = \bigcap_B \circ \exists \forall f, \qquad \mathsf{P}f \circ \bigcap_B = \bigcap_A \circ \exists \mathsf{P}f$$

in a quasitopos **E**.

PROOF. All four equations are obtained by similar techniques; we show the proofs of the second and the third equation.

For variables x of type A and $\mathcal{T}$ of type $\mathsf{PP}B$, we have

$$x \in (\mathsf{P}f \circ \bigcup_B)\mathcal{T} \;\leftrightarrow\; fx \in \bigcup_B \mathcal{T}$$

$$\leftrightarrow\; (\exists Y)(fx \in Y \wedge Y \in \mathcal{T}) \;\leftrightarrow\; (\exists Y)(x \in (\mathsf{P}f)Y \wedge Y \in \mathcal{T})$$

$$\leftrightarrow\; (\exists Y)((\exists X)(x \in X \wedge X = (\mathsf{P}f)Y) \wedge Y \in \mathcal{T})$$

$$\leftrightarrow\; (\exists X)(x \in X \wedge (\exists Y)(X = (\mathsf{P}f)Y \wedge Y \in \mathcal{T}))$$

$$\leftrightarrow\; (\exists X)(x \in X \wedge X \in (\exists \mathsf{P}f)\mathcal{T}) \;\leftrightarrow\; x \in (\bigcup_A \circ \exists \mathsf{P}f)\mathcal{T},$$

by 39.2 and the laws of the internal logic. Thus all these internally equivalent expressions have the same internal interpretation, and taking exponential adjoints of this interpretation for the first and the last expression, we get the second equation.

For the third equation, we note that

$$y \in (\forall f \circ \bigcap_A)\mathcal{S} \;\leftrightarrow\; (\forall x)(y = fx \to x \in \bigcap_A \mathcal{S})$$

$$\leftrightarrow\; (\forall x)(y = fx \to (\forall X)(X \in \mathcal{S} \to x \in X))$$

$$\leftrightarrow\; (\forall X)(X \in \mathcal{S} \to (\forall x)(y = fx \to x \in X))$$

$$\leftrightarrow\; (\forall X)(X \in \mathcal{S} \to y \in (\forall f)X)$$

$$\leftrightarrow\; (\forall X)(X \in \mathcal{S} \to (\forall Y)(Y = (\forall f)X \to y \in Y))$$

$$\leftrightarrow\; (\forall Y)((\exists X)(X \in \mathcal{S} \wedge Y = (\forall f)X) \to y \in Y))$$

$$\leftrightarrow\; (\forall Y)(Y \in (\exists \forall f)\mathcal{S} \to y \in Y) \;\leftrightarrow\; y \in (\bigcap_B \circ \exists \forall f)\mathcal{S},$$

and we then proceed as for the second equation.

39.4. Theorem. *For a quasitopos* **E**, *the functor* $\exists : \mathbf{E} \to \mathbf{E}$, *and the morphisms* $s_A : A \to \mathsf{P}A$ *and* $\bigcup_A : \mathsf{PP}A \to \mathsf{P}A$ *of* **E**, *define a monad* $(\exists, s, \bigcup)$ *on* **E**.

PROOF. We have shown in 39.3 that internal unions define a natural transformation $\bigcup : \exists \exists \to \exists$. We show next that s_A is natural in A. For

$f : A \to B$, we have

$$y \in (\exists f)(s_A x) \quad \leftrightarrow \quad (\exists u)(y = fu \wedge u \in s_A x)$$

$$\leftrightarrow \quad (\exists u)(y = fu \wedge u = x) \quad \leftrightarrow \quad y = fx \quad \leftrightarrow \quad y \in s_B fx$$

for variables x, u, y, by 38.5, 37.7 and 38.2, and $\exists f \circ s_A = s_B \circ f$ follows with 38.7.

It remains to show that

$$\bigcup_A \circ s_{PA} = \mathrm{id}_{PA} = \bigcup_A \circ \exists s_A \quad \text{and} \quad \bigcup_A \circ \bigcup_{PA} = \bigcup_A \circ \exists \bigcup_A$$

for every object A of $\mathbf{E}$. We have

$$y \in (\bigcup_A \circ s_{PA})X \quad \leftrightarrow \quad (\exists Y)(Y \in s_A X \wedge y \in Y)$$

$$\leftrightarrow \quad (\exists Y)(Y = X \wedge y \in Y) \quad \leftrightarrow \quad y \in X \quad \leftrightarrow \quad y \in \mathrm{id}_{PA} X \,;$$

the first equation follows by 38.7. The other equations are obtained similarly.

39.5. Internal complete semilattices. For sets, the category of algebras for the monad $(\exists, s, \bigcup)$ has as objects $(A, \sup)$ complete lattices, with $\sup : PA \to A$ defined by suprema, and as morphisms mappings which preserve all suprema. Thus we say that the category of algebras for the monad $(\exists, s, \bigcup)$ is the category of *internal complete semilattices*, not only for sets, but for any quasitopos $\mathbf{E}$. The free algebras are the pairs $(PA, \bigcup_A)$, and all morphisms $\exists A$ and Pf of $\mathbf{E}$ are homomorphisms of these algebras, by the general theory and 39.3. Internal intersections also provide complete semilattice structures.

Proposition. *For every object A of a quasitopos $\mathbf{E}$, the pair $(PA, \bigcap_A)$ is a complete semilattice. For every morphism f of $\mathbf{E}$, the morphisms Pf and $\forall f$ of $\mathbf{E}$ are homomorphisms of these complete semilattices.*

PROOF. We must show for the first part that the equations

$$\bigcap_A \circ s_{PA} = \mathrm{id}_{PA} \quad \text{and} \quad \bigcap_A \circ \bigcup_{PA} = \bigcap_A \circ \exists \bigcap_A$$

are valid; we omit the details. The second part follows immediately from 39.3.

40. Composition of Relations

40.1. Internal relations. In Section 13, we defined a relation $(u, v) : A \to B$ in a quasitopos $\mathbf{E}$ as a span $A \xleftarrow{u} \cdot \xrightarrow{v} B$ in $\mathbf{E}$ for which $\langle u, v \rangle$ is a strong monomorphism in $\mathbf{E}$. A relation $(u, v) : A \to B$ has two characteristic morphisms,

$$\mathrm{ch}\langle u, v \rangle : A \times B \to \Omega \quad \text{and} \quad \chi(u, v) : A \to PB \,,$$

with $\chi(u,v)$ exponentially adjoint to $\mathrm{ch}\langle u,v\rangle$. In this Section, $\hat{\rho} : A \to PB$ shall always denote the exponential adjoint of $\rho : A \times B \to \Omega$. Either ρ or $\hat{\rho}$ may be considered as an *internal relation* in **E**, with *domain* A and *codomain* B, representing (u,v) if $\rho = \mathrm{ch}(u,v)$.

In the following, we shall use morphisms $\rho : A \times B \to \Omega$ as internal relations $\rho : A \to B$ in **E**; the use of morphisms $\hat{\rho} : A \to PB$ will be discussed in 40.5–40.7.

Proposition. *If $\rho = \mathrm{ch}\langle u,v\rangle$ and $\hat{\rho} = \chi(u,v)$, for a strong monomorphism $\langle u,v\rangle : X \to A \times B$, then*

$$S \in \hat{\rho}R \;\leftrightarrow\; \rho\langle R,S\rangle \;\leftrightarrow\; (\exists x)(R = ux \wedge S = vx)$$

is internally valid, for terms R of type A and S of type B, and a variable x of type X which does not occur in R or S.

PROOF. For a list L of all variables of R and S, we have

$$|S \in \hat{\rho}R|_L \;=\; \varepsilon_B \,\langle \hat{\rho}|R|_L, |S|_L\rangle \;=\; \rho\langle |R|_L, |S|_L\rangle \;=\; |\rho\langle R,S\rangle|_L \,,$$

by the definitions. We also have internal equivalences

$$\rho\langle R,S\rangle \;\leftrightarrow\; (\exists x)(\langle R,S\rangle = \langle u,v\rangle x) \;\leftrightarrow\; (\exists x)(R = ux \wedge S = vx) \,,$$

by 38.6 and 38.4.

40.2. Composition. We define the *composition* $\sigma \circ \rho : A \to C$ of relations $\rho : A \to B$ and $\sigma : B \to C$ by putting

$$\sigma \circ \rho \;=\; \|(\exists y)(\rho\langle x,y\rangle \wedge \sigma\langle y,z\rangle)\| \;:\; A \times C \to \Omega$$

in **E**, for variables x,y,z of types A,B,C.

Lemma. *If R and T are terms of types A and C, then*

$$(\sigma \circ \rho)\langle R,T\rangle \;\leftrightarrow\; (\exists y)(\rho\langle R,y\rangle \wedge \sigma\langle y,T\rangle)$$

is internally valid if the variable y of type B does not occur in R or T.

PROOF. The righthand side of this equivalence is

$$((\exists y)(\rho\langle x,y\rangle \wedge \sigma\langle y,z\rangle)[T/z])[R/x] \,,$$

for variables x of type A and z of type C which do not occur in R or T. By 35.4 and the definitions, this formula has interpretations

$$(\sigma \circ \rho) \circ \pi_{xz}^{Lxz} \circ \langle \mathrm{id}_{P_L} \times \mathrm{id}_A, |T|_L \circ \pi_L^{Lx}\rangle \circ \langle \mathrm{id}_{P_L}, |R|_L\rangle \,,$$

and the Lemma follows.

40.3. External composition. If $\rho = \mathrm{ch}\langle u, v\rangle$ and $\sigma = \mathrm{ch}\langle r, s\rangle$ in 40.2, then

$$\sigma \circ \rho = \mathrm{ch}(\exists_q\,((\langle u, v\rangle \times \mathrm{id}_C) \cap (\mathrm{id}_A \times \langle r, s\rangle)))\,,$$

for the projection $q : A \times B \times C \to A \times C$. It is easily seen that

$$((\langle u, v\rangle \times \mathrm{id}_C) \cap (\mathrm{id}_A \times \langle r, s\rangle) = \langle ur', vr', sv'\rangle$$

for a pullback square

$$
\begin{array}{ccc}
\cdot & \xrightarrow{\;v'\;} & \cdot \\
{\scriptstyle r'}\big\downarrow & & \big\downarrow{\scriptstyle r} \\
\cdot & \xrightarrow[\;v\;]{} & B
\end{array}
$$

in **E**. Thus

$$\sigma \circ \rho = \mathrm{ch}\langle a, b\rangle \qquad \text{for} \qquad \langle ur', sv'\rangle = \langle a, b\rangle \circ e\,,$$

with e epimorphic and $\langle a, b\rangle$ strongly monomorphic. We put

$$(a, b) \simeq (r, s) \circ (u, v)\,,$$

and we call (a, b) an *external composition* of the relations (u, v) and (r, s) in **E**.

With this construction, external composition of relation can be defined in any category **E** with finite limits and factorizations $f = me$, with e epimorphic and m strongly monomorphic. A. KLEIN [63] showed that composition of external relations is associative, up to equivalence, iff pullbacks in **E** preserve epimorphisms. For a quasitopos **E**, we can obtain the associativity more easily by using internal logic.

40.4. Theorem. *Internal relations in a quasitopos* **E** *form a category, with the same objects as* **E**, *with identity relations* δ_A *as identity morphisms, and with the composition defined above.*

We denote this category of relations by Rel**E**.

PROOF. For variables x, y, z, u of the appropriate types, we have

$$(\tau \circ (\sigma \circ \rho))\langle x, u\rangle \;\leftrightarrow\; (\exists z)((\sigma \circ \rho)\langle x, z\rangle \wedge \tau\langle z, u\rangle)$$

$$\leftrightarrow\; (\exists z)((\exists y)(\rho\langle x, y\rangle \wedge \sigma\langle y, z\rangle) \wedge \tau\langle z, u\rangle)$$

$$\leftrightarrow\; (\exists y)(\rho\langle x, y\rangle \wedge (\exists z)(\sigma\langle y, z\rangle \wedge \tau\langle z, u\rangle))$$

$$\leftrightarrow\; (\exists y)(\rho\langle x, y\rangle \wedge (\tau \circ \sigma)\langle y, z\rangle) \;\leftrightarrow\; ((\tau \circ \sigma) \circ \rho)\langle x, u\rangle\,,$$

by 40.2 and the laws of internal logic, and $\tau \circ (\sigma \circ \rho) = (\tau \circ \sigma) \circ \rho$ follows.

For $\delta_A = \text{ch}\langle \text{id}_A, \text{id}_A \rangle$, the internal equivalence

$$\delta_A \langle x, y \rangle \ \leftrightarrow \ x = y,$$

for variables x, y of type A is internally valid by 40.1 and 37.7. Thus

$$(\rho \circ \delta_A)\langle x, z \rangle \ \leftrightarrow \ (\exists y)(x = y \wedge \rho\langle y, z \rangle) \ \leftrightarrow \ \rho\langle x, z \rangle,$$

which proves $\rho \circ \delta_A = \rho$. We prove $\delta_B \circ \rho = \rho$ in the same way.

40.5. Proposition. *External composition of relations satisfies*

$$(r, s) \circ (u, v) \ \simeq \ s \circ r^{\text{op}} \circ (u, v)$$

and $\qquad\qquad \chi((r, s) \circ (u, v)) \ = \ \exists s \circ \text{Pr} \circ \chi(u, v),$

for the compositions of 13.2 and 33.5. In particular,

$$\chi(u, v) \ = \ \exists v \circ Pu \circ s_A$$

for a relation $(u, v) : A \to B$ *in* **E**.

PROOF. With the notations of 40.3, we have

$$(ur', v') \ \simeq \ r^{\text{op}} \circ (u, v) \quad \text{and} \quad (a, b) \ \simeq \ s \circ (ur', v')$$

This proves the first formula, and now the second formula follows immediately from 13.8 and 33.6. Applying this to the particular composition

$$(u, v) \ \simeq \ (u, v) \circ (\text{id}_A, \text{id}_A),$$

we obtain the third formula.

40.6. Theorem. *Exponential adjunction defines an isomorphism between the category of internal relations in a quasitopos* **E** *and the Kleisli category of the monad* $(\exists, s, \bigcup)$ *on* **E**.

PROOF. The two categories have the same objects, and the isomorphism is the identity on objects. We recall that morphisms from A to B in the Kleisli category are morphisms $f : A \to PB$ of **E**, with the composition $g * f = \bigcup_C \circ \exists g \circ f$ for $g : B \to PC$ in **E**, and with identity morphisms s_A, for $s_A : A \to PA$ in **E**.

Since s_A is exponentially adjoint to δ_A, the claimed isomorphism preserves identity morphisms. For relations $\rho = \text{ch}\langle u, v \rangle : A \to B$ and $\sigma = \text{ch}\langle r, s \rangle : B \to C$, we have

$$\hat{\sigma} * \hat{\rho} = \bigcup_C \circ \exists \exists s \circ \exists \text{Pr} \circ \exists s_B \circ \hat{\rho}$$

$$= \exists s \circ \text{Pr} \circ \bigcup_B \circ \exists s_B \circ \hat{\rho}$$

$$= \exists s \circ \text{Pr} \circ \hat{\rho} = \chi((r, s) \circ (u, v))$$

in $\mathbf{E}$, by 40.5, 39.3 and 39.4. Thus we have $\hat{\tau} = \hat{\sigma} * \hat{\rho}$ in the Kleisli category if $\tau = \sigma \circ \rho$ in $\mathrm{Rel}\,\mathbf{E}$.

40.7. Discussion. We recall from 39.5 that algebras for the monad $(\exists, s, \bigcup)$ on $\mathbf{E}$ are internal complete semilattices. Since the category $\mathrm{Rel}\,\mathbf{E}$ is, up to isomorphism, the Kleisli category of the monad $(\exists, s, \bigcup)$ on $\mathbf{E}$, there is an adjunction

$$I \dashv E : \mathrm{Rel}\,\mathbf{E} \to \mathbf{E},$$

with $EI = \exists$, and with $\rho : A \to B$ in $\mathrm{Rel}\,\mathbf{E}$ adjoint to $\hat{\rho} : A \to PB$ in $\mathbf{E}$. We describe this adjunction briefly, without proofs.

The two categories have the same objects, and I is the identity on objects. For $f : X \to A$ in $\mathbf{E}$, we have

$$If = \mathrm{ch}\langle \mathrm{id}_X, f \rangle = \delta_A \circ (f \times \mathrm{id}_A) : X \to A$$

in $\mathrm{Rel}\,\mathbf{E}$, with the composition at right performed in $\mathbf{E}$. It follows that a composition $\rho \circ If$ in $\mathrm{Rel}\,\mathbf{E}$ is the internal version of the composition $(u, v) \circ f$ of Section 13. We note that the composition $Ig \circ \rho$ in $\mathrm{Rel}\,\mathbf{E}$, for $g : B \to C$ in $\mathbf{E}$, is the internal version of the composition $g \circ \rho$ of 33.5.

We have $EA = PA$ for an object A, and

$$E\,\mathrm{ch}\langle u, v \rangle = \exists u \circ Pv$$

for a relation.

The unit s of $I \dashv E$ is given by singleton morphisms s_A in $\mathbf{E}$, and the counit is given by membership relations $\varepsilon_A : PA \to A$ in $\mathrm{Rel}\,\mathbf{E}$, with exponential adjoints $\mathrm{id}_{PA} = \chi \ni_A$ in $\mathbf{E}$. It follows that

$$\bigcup A = E\varepsilon_A$$

for an object A of $\mathbf{E}$.

The adjunction $I \dashv E$ induces a comparison functor K, from $\mathrm{Rel}\,\mathbf{E}$ to the category of sup semilattices over $\mathbf{E}$, with $KA = (PA, \bigcup_A)$, the free sup semilattice generated by A, for an object A of $\mathbf{E}$, and $K\rho = E\rho : KA \to KB$ for a relation $\rho : A \to B$. This functor is full and faithful; we have

$$f = E\,\mathrm{ch}\langle u, v \rangle \quad \Longleftrightarrow \quad fs_A = \chi(u, v),$$

for a relation $(u, v) : A \to B$ and a morphism $f : KA \to KB$ of free internal complete semilattices.

Chapter 4

TOPOLOGIES AND SHEAVES

Lawvere-Tierney topologies of a topos **E**, usually just called topologies of **E**, are an important tool for constructing full reflective subcategories of **E** which are again topoi. The theory of topologies was extended to quasitopoi in [104], largely without proofs. J. PENON provided some proofs in [83], and P. JOHNSTONE [59] added the important result that categories of separated objects, for topologies of topoi, are quasitopoi. We give full proofs in this Chapter, and some additional results.

We first define topologies and obtain their basic properties in Section 41. In Sections 42 and 43, we define and study sheaves and separated objects for a topology of a quasitopos **E**. We show that sheaves and separated objects define full reflective subcategories of **E** which are again quasitopoi. If **E** is a topos, then the category of sheaves is also a topos.

We show in Section 44 that the reflector for sheaves preserves finite limits. Conversely, if the reflector for a reflective full subcategory **B** of a quasitopos **E** preserves finite limits, then we shall see in Section 52, in the more general context of geometric morphisms, that **B** is equivalent to the category of sheaves for a topology of **E**. For a topos, the sheaves determine the topology up to equivalence, but we do not know whether this remains true in general for quasitopoi. This will also be discussed in Section 52.

If γ is a topology of a quasitopos **E**, then the restriction of γ to the topos $\mathbf{E}^{\mathrm{gr}}$ of coarse objects of **E** is a topology of $\mathbf{E}^{\mathrm{gr}}$ which we denote by γ^{gr}. The coarse objects of the quasitopoi $\mathrm{Sep}_\gamma\mathbf{E}$ and $\mathrm{Sh}_\gamma\mathbf{E}$ of separated objects and sheaves for γ are the sheaves for γ^{gr}. Coarse objects for quasitopoi of separated objects and of sheaves were studied by G.P. MONRO [75]. We obtain and extend his results in Section 45.

A topology of a topos **E** can be characterized by a single morphism $j : \Omega \to \Omega$ of **E** which satisfies the conditions of 45.5. This is not the case for quasitopoi. The morphism j can be defined and determines the topology for strong monomorphisms, but not for monomorphisms which are not strong.

The last three sections deal with more special topics. Section 46 considers G.P. MONRO's construction [75] of quasitopoi with a coarse initial object; such quasitopoi are called solid. We show that finite sets can be represented in a non-trivial solid quasitopos, by a faithful functor which preserves finite limits. In Sections 47 and 48, we compare topologies for a topos $[\mathcal{C}^{\mathrm{op}}, \mathrm{SET}]$

135

with Grothendieck topologies of the category $\mathcal{C}$, and we obtain the canonical topology of a topos $[\mathcal{C}^{op}, \text{SET}]$.

41. Closed and Dense Monomorphisms

41.1. Topologies of a quasitopos. Without a statement to the contrary, we assume throughout this Chapter that $\mathbf{E}$ is a quasitopos. A *topology* γ of $\mathbf{E}$ assigns to every monomorphism m of $\mathbf{E}$ a monomorphism γm of $\mathbf{E}$, called the *closure* of m for γ, with the same codomain as m and with the following properties

41.1.1. If $m \leqslant m'$, then $\gamma m \leqslant \gamma m'$.

41.1.2. $m \leqslant \gamma m$, and $\gamma\gamma m \simeq \gamma m$, for every monomorphism m of $\mathbf{E}$.

41.1.3. If f is a morphism of $\mathbf{E}$, then $\gamma(f^{\leftarrow} m) \simeq f^{\leftarrow}(\gamma m)$ for every monomorphism m of $\mathbf{E}$ with the same codomain as f.

41.1.4. If m is a strong monomorphism of $\mathbf{E}$, then γm is a strong monomorphism of $\mathbf{E}$.

Thus a topology of a quasitopos $\mathbf{E}$ assigns to every object A of $\mathbf{E}$ an idempotent closure operator for $\text{Mon}\,A$, and these closure operators are natural in A in the sense that they commute with inverse images. Since $\gamma m \simeq \gamma m'$ if $m \simeq m'$, by 41.1.1, we can also consider a topology as a natural closure operator for subobjects.

41.2. Examples. We note the following examples of topologies.

(1) $\gamma m = \text{id}_A$ for every monomorphism m of $\mathbf{E}$ with codomain A. This is the *trivial topology* of $\mathbf{E}$.

(2) $\gamma m = m$ for every monomorphism m of $\mathbf{E}$. This is the *discrete topology* of $\mathbf{E}$.

(3) $\gamma m = m_1$ for a factorization $m = m_1 e$ of m with e monomorphic and epimorphic, and m_1 a strong monomorphism of $\mathbf{E}$.

(4) Factor the monomorphism $o_A : 0 \to A$, for an initial object 0 of $\mathbf{E}$, as $o_A = \bar{o}_A e$ with e monomorphic and epimorphic and $\bar{o}_A$ a strong monomorphism, and put $\neg m = m \to \bar{o}_A$ for a monomorphism m with codomain A. Then $\gamma m = \neg\neg m$ defines the *double negation topology* of $\mathbf{E}$. We note that every monomorphism $\neg m$ is strong, since the functor $m \to -$, right adjoint to $- \cap m$, preserves strong monomorphisms.

(5) If γ_1 and γ_2 are topologies of $\mathbf{E}$, then $\gamma m = \gamma_1 m \cap \gamma_2 m$ defines a topology $\gamma = \gamma_1 \cap \gamma_2$ of $\mathbf{E}$.

(6) If γ is a topology and A an object of $\mathbf{E}$, and if m is a monomorphism of $\mathbf{E}/A$ with codomain u, then we can consider γm, defined in $\mathbf{E}$, as a

morphism of $\mathbf{E}/A$ with codomain u. This defines a topology of $\mathbf{E}/A$ which we denote by γ/A.

41.3. Order and equivalence of topologies. We define a preorder for topologies of $\mathbf{E}$ by putting $\gamma \leqslant \gamma'$, for topologies γ and γ' of $\mathbf{E}$, if $\gamma m \leqslant \gamma' m$ for every monomorphism m of $\mathbf{E}$. We say that γ is a *finer topology* than γ', and γ' a *coarser topology* than γ, if $\gamma \leqslant \gamma'$.

We put $\gamma \simeq \gamma'$, and we say that γ and γ' are *equivalent*, if $\gamma \leqslant \gamma'$ and $\gamma' \leqslant \gamma$. Concepts defined in terms of a topology are usually the same for equivalent topologies; thus equivalent topologies can be regarded as being essentially the same.

In the preorder of topologies of $\mathbf{E}$, the discrete topology of 41.2.(2) is a finest topology. The trivial topology of 41.2.(1) is a coarsest topology of $\mathbf{E}$, and 41.2.(5) defines the intersection $\gamma_1 \cap \gamma_2$ of two topologies. The topology of 41.2.(3) is a finest topology with every monomorphism γm strong.

41.4. Dense and closed monomorphisms. We say that a monomorphism m of $\mathbf{E}$ is *closed* for a topology γ of $\mathbf{E}$ if $\gamma m \simeq m$, and that m is *dense* if $\gamma m \simeq \mathrm{id}_A$, with A the codomain of m.

Equivalent topologies have the same closed monomorphisms and the same dense monomorphisms. More generally, if $\gamma \leqslant \gamma'$, then every closed monomorphism for γ' is closed for γ, and every dense monomorphism for γ is dense for γ'.

For the topology of 41.2.(3), the dense monomorphisms are the monomorphisms which are also epimorphic, and the closed monomorphisms are the strong monomorphisms. For the topology of 41.2.(4) the closed monomorphisms are the negations $\neg m$ of monomorphisms.

41.5. Properties of a topology. We note the following basic properties of a monomorphism m of $\mathbf{E}$, for a topology γ of $\mathbf{E}$.

41.5.1. *If $f^\leftarrow m$ is defined for a morphism f, then $f^\leftarrow m$ is closed if m is closed, and $f^\leftarrow m$ is dense if m is dense.*

PROOF. This follows immediately from 41.1.3 and the definitions.

41.5.2. *If m factors $m = m_1 m'$ for monomorphisms m_1 and m', then $m \leqslant m_1 \cdot \gamma m' \leqslant \gamma m$, and m' is dense if and only if $m_1 \leqslant \gamma m$.*

PROOF. The first inequality follows from $m' \leqslant \gamma m'$. We note that $m' \simeq m_1{}^\leftarrow m$ and hence $\gamma m' \simeq m_1{}^\leftarrow(\gamma m)$. The second inequality follows, and m' is dense iff $m_1{}^\leftarrow(\gamma m)$ is an isomorphism, hence iff $m_1 \leqslant \gamma m$.

41.5.3. *If $m = m_1 m'$ with m' dense, then $\gamma m \simeq \gamma m_1$, and m is dense if and only if m_1 is dense.*

PROOF. By 41.5.2, we have $m \leqslant m_1 \leqslant \gamma m$ in this situation, and thus $\gamma m \simeq \gamma m_1$. The second part follows immediately.

41.5.4. *m factors $m = m_1 d$ with d dense and m_1 closed. This factorization is unique up to isomorphisms, and $m_1 \simeq \gamma m$.*

PROOF. Since $m \leqslant \gamma m$, we can factor $m = \gamma m \cdot d$, with d dense by 41.5.2. If also $m = m_1 m'$ with m' dense and m_1 closed, then $m_1 \simeq \gamma m$ by 41.5.3.

41.5.5. *m is closed if and only if every commutative square $mf = gd$, with d a dense monomorphism, has a (unique) diagonal t such that $f = td$ and $g = mt$.*

PROOF. If $mf = gd$, then $d \leqslant g^{\leftarrow} m$, hence $\gamma d \leqslant g^{\leftarrow} m$ if m is closed, by 41.5.1. If m is closed and d dense, it follows that $g^{\leftarrow} m$ is an isomorphism. Thus $g = mt$ for a morphism t, and $f = td$ follows.

For the converse, factor $m = \gamma m \cdot d$, with d dense by 41.5.1. If the square $m \cdot \mathrm{id} = \gamma m \cdot d$ has a diagonal, then $\gamma m \leqslant m$, and thus m is closed.

41.5.6. *If $m = m_1 m'$ with m_1 a closed monomorphism, then m is closed if and only if m' is closed.*

PROOF. This follows immediately from 41.5.5 and 10.3.(2).

41.6. Proposition. *A commutative square of monomorphisms*

(1)
$$
\begin{array}{ccc}
\cdot & \xrightarrow{\ m\ } & \cdot \\
{\scriptstyle d_1}\downarrow & & \downarrow{\scriptstyle d} \\
\cdot & \xrightarrow[\ m_1\ ]{} & \cdot
\end{array}
,
$$

with d and d_1 dense, and m and m_1 closed, is a pullback square. If m is strong, then m_1 is strong.

PROOF. We have $m_1 \simeq \gamma(dm)$, and hence

$$d^{\leftarrow} m_1 \simeq d^{\leftarrow}(\gamma(dm)) \simeq \gamma(d^{\leftarrow}(dm)) \simeq \gamma m \simeq m.$$

Thus (1) is a pullback square. If we factor $m_1 = m'e'$, with m' strong and e' monomorphic and epimorphic, then we have pullbacks

$$
\begin{array}{ccccc}
\cdot & \xrightarrow{\ e''\ } & \cdot & \xrightarrow{\ m''\ } & \cdot \\
{\scriptstyle d_1}\downarrow & & \downarrow{\scriptstyle d'} & & \downarrow{\scriptstyle d} \\
\cdot & \xrightarrow[\ e'\ ]{} & \cdot & \xrightarrow[\ m'\ ]{} & \cdot
\end{array}
,
$$

with d' dense and $m''e'' = m$, and with e'' monomorphic and epimorphic. If m is a strong monomorphism, then e'' is an isomorphism. But then $d' \leqslant e'$, and e' is dense. It follows that $m' \leqslant \gamma m_1$. Since m_1 is closed and $m_1 \leqslant m'$, we get $m_1 \simeq m'$, and m_1 is strong.

41.7. Closure for monomorphisms in a topos or quasitopos acts in many ways like closure of subsets in a topological space. Here is one way in which the two closures behave quite differently.

Proposition. *If* m_1 *and* m_2 *are monomorphisms of* **E** *with the same codomain, then* $\gamma(m_1 \cap m_2) \simeq \gamma m_1 \cap \gamma m_2$.

PROOF. We factor $m_1 = \gamma m_1 \cdot d_1$ and $m_2 = \gamma m_2 \cdot d_2$, with d_1 and d_2 dense, using 41.5.4, and we construct pullback squares

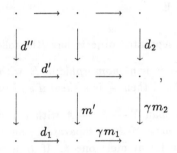

with m' closed, and d' and d'' dense, by 41.5.1. Thus

$$m_1 \cap m_2 \simeq \gamma m_1 \cdot m' \cdot d' \cdot d'',$$

with $\gamma m_1 \cdot m' \simeq \gamma m_1 \cap \gamma m_2$ closed by 41.5.6, and $d' \cdot d''$ dense by 41.5.3. With 41.5.4, the Proposition follows.

42. Separated Objects and Sheaves

42.1. Definitions. We assume in this Section that **E** is a quasitopos provided with a topology γ. We say that an object A of **E** is *separated* for γ if for a span

$$(1) \qquad\qquad Y \xleftarrow{\ d\ } X \xrightarrow{\ f\ } A$$

in **E**, with d a dense monomorphism for γ, there is always at most one morphism $g : Y \to A$ in **E** such that $f = gd$, and we say that A is a *sheaf* for γ if for every span (1) with d a dense monomorphism for γ, there is exactly one morphism $g : Y \to A$ in **E** such that $f = gd$.

We note that every sheaf for γ is a separated object for γ.

We denote by $\mathrm{Sh}_\gamma\,\mathbf{E}$ the full subcategory of $\mathbf{E}$ with the sheaves for γ as objects, and by $\mathrm{Sep}_\gamma\,\mathbf{E}$ the full subcategory with separated objects as its objects. The subscript may be omitted if γ is determined by the context.

42.2. Examples. For the discrete topology of 41.2.(2), every object of $\mathbf{E}$ is a sheaf.

For the trivial topology of 41.2.(1), the monomorphism $0 \to X$, with 0 an initial object of $\mathbf{E}$, is dense for every object X of $\mathbf{E}$. If A is a sheaf, it follows that there is exactly one morphism $X \to A$. Thus only terminal objects are sheaves. The separated objects are the subobjects of a terminal object 1, with the unique morphism $A \to 1$ monomorphic.

For the topology of 41.2.(3), every object of $\mathbf{E}$ is separated, and the sheaves are the coarse objects of $\mathbf{E}$.

42.3. Sheaves and separated objects have the following properties.

42.3.1. *If $m : A \to B$ is monomorphic in $\mathbf{E}$ with B separated, then A is separated. If B is a sheaf, then A is a sheaf if and only if m is closed.*

PROOF. Consider a span $Y \xleftarrow{d} X \xrightarrow{f} A$ with d a dense momomorphism. If $td = f$, then $mtd = mf$. If B is separated, this is possible for at most one morphism mt, hence for at most one t. If B is a sheaf, then $mf = gd$ for exactly one $g : X \to Y$, and then $td = f$ for exactly one t if m is closed, by 41.5.5. Thus A is a sheaf if m is closed. Conversely, if A is a sheaf, then $td = f$, and hence $g = mt$, for exactly one t, and m is closed by 41.5.5.

42.3.2. *If F is separated or a sheaf, then every object $F^A = [A, F]$ is separated or a sheaf.*

PROOF. If $f : X \to F^A$ is given, then we have $f = gd$ for $g : Y \to F^A$ iff $f_1 = g_1 \cdot (d \times \mathrm{id}_A)$ for the exponentially adjoint morphisms f_1 and g_1. If d is dense, then so is the pullback $d \times \mathrm{id}_A$ of d. Thus we get exactly one g_1 and hence exactly one g if F is a sheaf, and at most one g_1, hence at most one g, if F is separated.

42.3.3. *An object A of $\mathbf{E}$ is separated if and only if the morphism $\langle \mathrm{id}_A, \mathrm{id}_A \rangle : A \to A \times A$ is closed.*

PROOF. Consider a span $Y \xleftarrow{d} X \xrightarrow{f} A$ with d a dense monomorphism. If $gd = f = g'd$, then

$$\langle \mathrm{id}_A, \mathrm{id}_A \rangle \cdot f = \langle g, g' \rangle \cdot d$$

in **E**. If $\langle \mathrm{id}_A, \mathrm{id}_A \rangle$ is closed, then $\langle g, g' \rangle = \langle \mathrm{id}_A, \mathrm{id}_A \rangle t$ for a morphism t. But then $g = t = g'$, and A is separated. Conversely, we can factor

$$\langle \mathrm{id}_A, \mathrm{id}_A \rangle = \langle u, v \rangle \cdot d$$

with $\langle u, v \rangle$ closed and d dense. Then $ud = \mathrm{id}_A = vd$, and $u = v$ follows if A is separated. But then $\gamma \langle \mathrm{id}_A, \mathrm{id}_A \rangle \leqslant \langle \mathrm{id}_A, \mathrm{id}_A \rangle$, and $\langle \mathrm{id}_A, \mathrm{id}_A \rangle$ is closed.

42.3.4. *If A is separated for γ, then an object $u : X \to A$ of $\mathbf{E}/A$ is a sheaf for the topology γ/A of 41.2.(6) if and only if X is a sheaf for γ, and u is separated for γ/A if and only if X is separated for γ.*

PROOF. A monomorphism $d : v \to w$ of $\mathbf{E}/A$ is dense for γ/A iff d is dense for γ. If $f : v \to u$ in $\mathbf{E}/A$, then $f = gd$ in $\mathbf{E}/A$ iff $f = gd$ and $w = ug$ in $\mathbf{E}$. If d is dense and A separated, then $f = gd$ implies $w = ug$, and 42.3.4 follows.

42.4.   Closed partial morphisms. We say that a partial morphism $(m, f) : A \to F$ in $\mathbf{E}$ is *closed* if the strong monomorphism m is closed. If F is a sheaf, then we define the *closure* $\gamma(m, f)$ of (m, f) as the partial morphism $(\gamma m, f')$ with $f = f'd$ for the dense monomorphism d given by $m = \gamma m \cdot d$. If $\vartheta_F : F \to \tilde{F}$ represents partial morphisms with codomain a sheaf F, then we put $j_F = \mathrm{ch}_\mathrm{p} \gamma(\vartheta_F, \mathrm{id}_F) : \tilde{F} \to \tilde{F}$, using the notation of 16.4.

Proposition. *If F is a sheaf, and if $\mathrm{ch}_\mathrm{p}(m, f) = \bar{f}$ for a partial morphism (m, f) with codomain F, then $\mathrm{ch}_\mathrm{p} \gamma(m, f) = j_F \bar{f}$.*

PROOF. From the definitions and 41.5.1, we have pullback squares

$$
\begin{array}{ccccc}
\cdot & \xrightarrow{\ f\ } & F & & \\
{\scriptstyle d}\downarrow & & \downarrow{\scriptstyle d_1} & & \\
\cdot & \xrightarrow{\ f_1\ } & \cdot & \xrightarrow{\ t\ } & F \\
{\scriptstyle \gamma m}\downarrow & & \downarrow{\scriptstyle \gamma \vartheta_F} & & \downarrow{\scriptstyle \vartheta_F} \\
A & \xrightarrow{\ \bar{f}\ } & \tilde{F} & \xrightarrow{\ j_F\ } & \tilde{F}
\end{array}
$$

in **E**, with d and d_1 dense, $m = \gamma m \cdot d$ and $\gamma \vartheta_F \cdot d_1 = \vartheta_F$, and with $t d_1 = \mathrm{id}_F$. In this situation, $\gamma(m, f) = (\gamma m, t f_1)$, and the Proposition follows.

42.5. Corollary. *A partial morphism $(m, f) : A \to F$ with $\mathrm{ch}_\mathrm{p}(m, f) = \bar{f}$ is closed if and only if $j_F \bar{f} = \bar{f}$. In particular, $j_F j_F = j_F$ and $j_F \vartheta_F = \vartheta_F$.*

PROOF. The first part follows immediately from 42.4 and the definitions. For the second part, we observe that

$$j_F = \mathrm{ch_p}\gamma(\vartheta_F, \mathrm{id}_F) \quad \text{and} \quad \vartheta_F = \mathrm{ch_p}(\mathrm{id}_F, \mathrm{id}_F),$$

with $\gamma(\vartheta_F, \mathrm{id}_F)$ and $(\mathrm{id}_F, \mathrm{id}_F)$ closed.

42.6. Theorem. *If F is a sheaf for a topology γ of a quasitopos $\mathbf{E}$, then closed partial morphisms in $\mathbf{E}$ with codomain F are represented in $\mathbf{E}$ by a closed strong monomorphism $\zeta_F : F \to F^\#$, with $F^\#$ a sheaf. In particular, $\mathbf{E}$ has a classifier $\top_\gamma : 1 \to \Omega_\gamma$ for closed strong monomorphisms, with Ω_γ a sheaf.*

PROOF. With the notations of 42.4, let $e_F : F^\# \to \tilde{F}$ be an equalizer of $\mathrm{id}_{\tilde{F}}$ and j_F. Then $\vartheta_F = e_F\zeta_F$ for a morphism $\zeta_F : F \to F^\#$ by 42.5; we claim that ζ_F represents closed partial morphisms with codomain F.

In this situation, $e_F = \mathrm{ch_p}(\zeta_F, \mathrm{id}_F)$. Since $j_F e_F = e_F$, the strong monomorphism ζ_F is closed by 42.5. If $\bar{f} = \mathrm{ch_p}(m, f) : A \to \tilde{F}$, then m is closed iff $j_F\bar{f} = \bar{f}$, by 42.5, and thus iff $\bar{f} = e_F f_1$ for a unique $f_1 : A \to F^\#$. We then have a commutative diagram

$$
\begin{array}{ccccc}
\cdot & \xrightarrow{\ f\ } & F & \xrightarrow{\ \mathrm{id}_F\ } & F \\
{\scriptstyle m}\downarrow & & {\scriptstyle \zeta_F}\downarrow & & \downarrow{\scriptstyle \vartheta_F,} \\
A & \xrightarrow{\ f_1\ } & F^\# & \xrightarrow{\ e_F\ } & \tilde{F}
\end{array}
$$

with the outer rectangle and the righthand square pullbacks. Thus the lefthand square is a pullback square, with f_1 uniquely determined by (m, f) since $e_F f_1$ is uniquely determined by (m, f).

Since closed strong monomorphisms m correspond bijectively to closed partial morphisms (m, k) with codomain 1, we get in particular a classifier $\top_\gamma = \zeta_1 : 1 \to \Omega_\gamma$ for closed strong monomorphisms of $\mathbf{E}$, with $\Omega_\gamma = 1^\#$.

It remains to prove that $F^\#$ is a sheaf. If $Y \xleftarrow{d} X \xrightarrow{f} F^\#$ is a span in $\mathbf{E}$ with d a dense monomorphism, then we construct a pullback square

$$
\begin{array}{ccc}
\cdot & \xrightarrow{\ f_1\ } & F \\
{\scriptstyle m}\downarrow & & \downarrow{\scriptstyle \zeta_F} \\
X & \xrightarrow{\ f\ } & F^\#
\end{array}
$$

(1)

in $\mathbf{E}$, with m a closed strong monomorphism. We factor $dm = m_1 d_1$ in $\mathbf{E}$, with d_1 dense and m_1 closed. Then $f_1 = g_1 d_1$ in $\mathbf{E}$ for a unique morphism g_1.

Using 41.6 and what we already proved, we construct a diagram

$$(2) \quad \begin{array}{ccccc} \cdot & \xrightarrow{\ d_1\ } & \cdot & \xrightarrow{\ g_1\ } & F \\ \Big\downarrow{\scriptstyle m} & & \Big\downarrow{\scriptstyle m_1} & & \Big\downarrow{\scriptstyle \zeta_F} \\ X & \xrightarrow{\ d\ } & Y & \xrightarrow{\ g\ } & F^\# \end{array}$$

of pullback squares in $\mathbf{E}$. Then $f = gd$, since the outer rectangle in (2) is a pullback and ζ_F represents closed partial morphisms. Conversely, if $f = gd$, then we can get pullbacks (2) with $f_1 = g_1 d_1$. This determines g_1 and hence g uniquely.

43. Associated Separated Objects and Sheaves

43.1. Codense and bidense morphisms. We assume again that $\mathbf{E}$ is a quasitopos provided with a topology γ.

For a strong epimorphism $e : A \to B$ of $\mathbf{E}$, we form the kernel pair (u, v) of e, with $eu = ev$ a pullback square, and e a coequalizer of u and v. There is then a unique monomorphism d in $\mathbf{E}$ with $ud = \mathrm{id}_A = vd$. We say that e is a *codense morphism* of $\mathbf{E}$ if this monomorphism d is dense. We define a *bidense morphism* of $\mathbf{E}$ as a composition de, with e a codense strong epimorphism, and d a dense monomorphism.

43.2. Lemma. *Consider a pullback square*

$$\begin{array}{ccc} \cdot & \xrightarrow{\ e_1\ } & \cdot \\ \Big\downarrow{\scriptstyle m_1} & & \Big\downarrow{\scriptstyle m} \ , \\ \cdot & \xrightarrow{\ e\ } & \cdot \end{array}$$

with m and m_1 monomorphic and e a strong epimorphism. If m_1 is closed, then m is closed. If m_1 is dense, then m is dense.

PROOF. If we factor $m = m'd$ with d dense and m' closed, then we have pullback squares

$$\begin{array}{ccccc} \cdot & \xrightarrow{\ d'\ } & \cdot & \xrightarrow{\ m''\ } & \cdot \\ \Big\downarrow{\scriptstyle e_1} & & \Big\downarrow{\scriptstyle e'} & & \Big\downarrow{\scriptstyle e} \ , \\ \cdot & \xrightarrow{\ d\ } & \cdot & \xrightarrow{\ m'\ } & \cdot \end{array}$$

with d' dense and m'' closed and $m''d' = m_1$, and with strong epimorphisms e' and e_1. If m_1 is closed, then d' is an isomorphism, and de_1 a strong epimorphism. But then d is an isomorphism, and m is closed. If m_1 is dense, then m'' and m' are isomorphism, by the same reasoning, and m is dense.

43.3. Theorem. *Separated objects of a quasitopos* $\mathbf{E}$, *for a topology* γ *of* $\mathbf{E}$, *define a full reflective subcategory* $\mathrm{Sep}_\gamma \mathbf{E}$ *of* $\mathbf{E}$. *Reflections for* $\mathrm{Sep}_\gamma \mathbf{E}$ *are all codense morphisms* $e : A \to B$ *with* B *separated*.

We say that B is an *associated separated object* of A in $\mathbf{E}$ if there is a reflection $e : A \to B$ for separated objects in $\mathbf{E}$.

PROOF. For an object A of $\mathbf{E}$, we factor $\langle \mathrm{id}_A, \mathrm{id}_A \rangle$ as

$$\langle \mathrm{id}_A, \mathrm{id}_A \rangle \;=\; \langle u, v \rangle \circ d,$$

with $\langle u, v \rangle$ closed and d dense. A cokernel $e : A \to B$ of u and v is codense, with kernel pair (u, v); we claim that B is separated. By the definitions, we have a pullback diagram

$$
\begin{array}{ccc}
\cdot & \longrightarrow & B \\
{\scriptstyle \langle u, v \rangle} \downarrow & & \downarrow {\scriptstyle \langle \mathrm{id}_B, \mathrm{id}_B \rangle} \\
A \times A & \xrightarrow{\; e \times e \;} & B \times B
\end{array}
$$

in $\mathbf{E}$ with $\langle u, v \rangle$ closed, and with $e \times e$ a strong epimorphism by 23.2. By 43.2, $\langle \mathrm{id}_B, \mathrm{id}_B \rangle$ is closed; thus B is separated by 42.3.3. We conclude the proof by proving the following Lemma.

43.3.1. Lemma. *If* $e : A \to B$ *is codense in* $\mathbf{E}$, *then every morphism* $f : A \to C$ *in* $\mathbf{E}$ *with* C *separated has a unique factorization* $f = ge$ *in* $\mathbf{E}$.

PROOF. Let (u, v) be a kernel pair of e, with $ud = \mathrm{id}_A = vd$ for a dense monomorphism d, and with e a coequalizer of u and v. If $f : A \to C$ with C separated, then $fu = fv$, and thus f factors $f = ge$.

43.4. Proposition. *An object* A *of* $\mathbf{E}$ *is separated for* γ *if and only if there is a monomorphism* $m : A \to F$ *in* $\mathbf{E}$ *with* F *a sheaf*.

43.4.1. Corollary. *The initial object* 0 *of* $\mathbf{E}$ *is separated for every topology of* $\mathbf{E}$.

PROOF. By 42.3.3, we have a closed strong monomorphism $\langle \mathrm{id}_A, \mathrm{id}_A \rangle :$ $A \to A \times A$ if A is separated, and hence by 42.6 and its proof $\delta_A = e_1 \varphi$ for morphisms $\varphi : A \times A \to \Omega_\gamma$ and $e_1 : \Omega_\gamma \to \Omega$. By exponential adjunction, we get $s_A = e_1{}^A \hat{\varphi}$, with $\hat{\varphi} : A \to \Omega_\gamma{}^A$ monomorphic since s_A is monomorphic. By 42.6 and 42.3.2, $\Omega_\gamma{}^A$ is a sheaf.

The converse follows immediately from 42.3.1.

We note for the Corollary that the terminal object 1 of $\mathbf{E}$ is clearly a sheaf, and that the morphism $0 \to 1$ of $\mathbf{E}$ is a monomorphism.

43.5. Theorem. *For every topology γ of a quasitopos $\mathbf{E}$, sheaves for γ are the objects of a full reflective subcategory* Sh_γ *of $\mathbf{E}$. Reflections for* Sh_γ *are all bidense morphisms* $f : A \to F$ *of $\mathbf{E}$ with F a sheaf for γ. If A is separated, then reflections* $f : A \to F$ *for* $\mathrm{Sh}_\gamma \mathbf{E}$ *are dense monomorphisms.*

If there is a reflection $f : A \to F$ for $\mathrm{Sh}_\gamma \mathbf{E}$, then we call F an *associated sheaf* of A for γ.

PROOF. For an object A of $\mathbf{E}$, we have by 43.3 a codense morphism $e : A \to B$ with B separated. By 43.4, there is a monomorphism $m : B \to F$ with F a sheaf. If $m = m_1 d$, with m_1 closed and $d : B \to F_1$ dense, then F_1 is a sheaf by 42.3.1. Thus we have a bidense morphism $de : A \to F_1$ with F_1 a sheaf, and a dense monomorphism $d : A \to F_1$ with F_1 a sheaf if A is separated. Now the following Lemma concludes the proof.

43.5.1. Lemma. *If $u : A \to B$ is bidense in $\mathbf{E}$, then every morphism $f : A \to F$ in $\mathbf{E}$ with F a sheaf has a unique factorization $f = gu$ in $\mathbf{E}$. If A and B are sheaves, then u is an isomorphism.*

PROOF. We factor $u = de$ with e codense and d dense. By 43.3.1, there is a unique factorization $f = he$, and by the definition of a sheaf, there is a unique factorization $h = gd$.

If A is a sheaf, then $vu = \mathrm{id}_A$, and $uvu = u$, for a morphism v. If B also is a sheaf, then $uv = \mathrm{id}_B$ follows.

43.6. Theorem. *For every topology γ of a quasitopos $\mathbf{E}$, the reflective full subcategories* $\mathrm{Sh}_\gamma \mathbf{E}$ *and* $\mathrm{Sep}_\gamma \mathbf{E}$ *are quasitopoi. For* $\mathrm{Sh}_\gamma \mathbf{E}$ *and* $\mathrm{Sep}_\gamma \mathbf{E}$, *epimorphisms are all morphisms de with e epimorphic and d a dense monomorphism in $\mathbf{E}$, and strong monomorphisms are all morphisms m with m a closed strong monomorphism of $\mathbf{E}$.*

PROOF. If $\mathbf{A}$ is one of the two subcategories, then $\mathbf{A}$ has finite limits and colimits as a reflective full subcategory of $\mathbf{E}$, and limits are inherited from $\mathbf{E}$. Now it follows from 42.3.2 that $\mathbf{A}$ is cartesian closed, with exponential adjunction inherited from $\mathbf{E}$.

Every morphism f of $\mathbf{E}$ factors $f = mde$, with e epimorphic, d a dense monomorphism, and m a closed strong monomorphism. If f is in $\mathbf{A}$, then de and m are in $\mathbf{A}$, by 42.3.1, and de is epimorphic in $\mathbf{A}$. If f is epimorphic in $\mathbf{A}$, then m is epimorphic in $\mathbf{A}$, and hence isomorphic since m is an equalizer in $\mathbf{A}$, by 42.6 and 14.3. If f is a strong monomorphism in $\mathbf{A}$, then the epimorphism de of $\mathbf{A}$ is an isomorphism, and f is a closed strong monomorphism in $\mathbf{E}$. Conversely, if f in $\mathbf{A}$ is a closed strong monomorphism in $\mathbf{E}$, then by 41.5.5 and the definitions, f is a strong monomorphism of $\mathbf{A}$ since the epimorphisms of $\mathbf{A}$ are compositions de with e epimorphic and d a dense monomorphism of $\mathbf{E}$.

Now partial morphisms in $\mathrm{Sh}_\gamma E$ are represented by 42.6. For a separated object A, there is a monomorphism $e : A \to F$ with F a sheaf. Since closed partial morphisms in $\mathbf{E}$ with codomain F are represented, closed partial morphisms in $\mathbf{E}$ with codomain A are represented by 19.2. Thus partial morphisms in $\mathrm{Sep}_\gamma E$ are represented.

43.7. Proposition. *If S and T are the reflectors for $\mathrm{Sep}_\gamma E$ and for $\mathrm{Sh}_\gamma E$, then Se is isomorphic in $\mathbf{E}$ if e is codense for γ, and Tf is an isomorphism if f is bidense.*

PROOF. If $f : X \to Y$ is bidense, with reflections $\eta_X : X \to TX$ and $\eta_Y : Y \to TY$ for sheaves, then $\eta_Y f = Tf \cdot \eta_X$. By 43.5.1, we have $\eta_X = tf$ for a morphism t, and then $t = u\eta_Y$ for a morphism u. Since

$$Tf \circ u \circ \eta_Y = \eta_Y \qquad \text{and} \qquad u \circ Tf \circ \eta_X = \eta_X$$

in this situation, for the reflections η_X and η_Y, the morphisms u and Tf in $\mathrm{Sh}_\gamma E$ are inverse isomorphisms.

The first part of the Proposition is proved in the same way.

44. Reflectors for Sheaves and Separated Objects

44.1. We continue our study of a quasitopos $\mathbf{E}$ with a topology γ.

Proposition. *Bidense morphisms f for γ are characterized by the following two conditions.*

(i) *If $f = me$ with e a strong epimorphism and m monomorphic, then m is dense.*

(ii *If $ud = \mathrm{id} = vd$ for a kernel pair u, v of f, then the monomorphism d is dense.*

PROOF. Condition (i) is a part of the definition in 43.1. If $f = me$ with m monomorphic, then f and e have the same kernel pairs; thus (ii) completes the characterization.

44.2. Proposition. *Bidense morphisms of* **E** *form a subcategory of* **E** *which contains all isomorphisms.*

PROOF. Isomorphisms clearly are bidense. Consider now a diagram

$$
\begin{array}{ccc}
\cdot & \longrightarrow & B \\
\downarrow{\scriptstyle d_1} & & \downarrow{\scriptstyle d} \\
\cdot & \longrightarrow \cdot \longrightarrow & C \\
\downarrow{\scriptstyle \langle u_1, v_1 \rangle} & \downarrow{\scriptstyle \langle u, v \rangle} & \downarrow{\scriptstyle \langle \mathrm{id}_C, \mathrm{id}_C \rangle} \\
A \times A & \xrightarrow{f \times f} B \times B \xrightarrow{g \times g} & C \times C
\end{array}
\quad,
$$

with f and g bidense, (u, v) a kernel pair of g, and (u_1, v_1) a kernel pair of gf. If $ud = \mathrm{id}_B = vd$, then d is dense, and $(u_1 d_1, v_1 d_1)$ a kernel pair of f. If $u_1 d_1 d' = \mathrm{id}_A = v_1 d_1 d'$, with d' dense, then $d_1 d'$ is dense. Thus gf satisfies condition (ii) if f and g do.

If $f = m_1 e_1$ and $g = m_2 e_2$, with m_1, m_2 dense monomorphisms and e_1, e_2 strong epimorphisms, factor $e_2 m_1 = me$ with e strongly epimorphic and m monomorphic. Then $m_1 \leqslant e^{\leftarrow} m$; thus $e^{\leftarrow} m$ is dense. But then m is dense by 43.2, and gf satisfies 44.1.(i).

44.3. Proposition. *Pullbacks of bidense morphisms are bidense.*

PROOF. Let $t = me : A \to B$ be bidense, with m dense and e codense. For $f : B' \to B$, we construct a diagram

$$
\begin{array}{ccccccccc}
A' & \xrightarrow{d'} & \cdot & \underset{v'}{\overset{u'}{\rightrightarrows}} & A' & \xrightarrow{e'} & \cdot & \xrightarrow{m'} & B' \\
\downarrow{\scriptstyle f_1} & & \downarrow{\scriptstyle f''} & & \downarrow{\scriptstyle f_1} & & \downarrow{\scriptstyle f'} & & \downarrow{\scriptstyle f} \\
A & \xrightarrow{d} & \cdot & \underset{v}{\overset{u}{\rightrightarrows}} & A & \xrightarrow{e} & \cdot & \xrightarrow{m} & B
\end{array}
$$

in **E**, with two pullback squares at right, with (u, v) and (u', v') kernel pairs of e and e', with $ud = \mathrm{id}_B = vd$ and d dense, and with $u'd' = \mathrm{id}_{A'} = v'd'$. Since $ef_1 u' = ef_1 v'$, there is f'' with $uf'' = f_1 u'$ and $vf'' = f_1 v'$. Then $uf''d' = f_1 = udf_1$ and $vf''d' = vdf_1$; thus the lefthand square commutes. We claim that this square is a pullback square.

If $dx = f''y$, then $f_1 u'y = udx = x = vdx = f_1 v'y$. We also have $e'u'y = e'v'y$; thus $u'y = v'y$ since $ef_1 = f'e'$ is a pullback square. Now if $x = f_1 t$, $y = d't$, then $u'y = t = v'y$. Conversely, $u'd't = u'y = v'y = v'd't$ for this t. Thus $y = d't$, and $x = f_1 t$ follows. This verifies our claim.

Now m' and d' in the diagram are dense, and e' is a strong epimorphism; thus the pullback $m'e'$ of t is bidense.

44.4. Proposition. *If morphisms* $\cdot \overset{f}{\underset{g}{\rightrightarrows}} \cdot$ *of* $\mathbf{E}$ *satisfy* $sf = sg$ *for a bidense morphism* s, *then also* $ft = gt$ *for a dense monomorphism* t.

PROOF. If (u,v) is a kernel pair of s, then $f = ux$ and $g = vx$ for a morphism x. If $ud = vd = \mathrm{id}$ with d dense, and if $t = x^{\leftarrow}d$, then t is a dense monomorphism and $uxt = vxt$, and thus $ft = gt$.

44.5. Theorem. *For a topology* γ *of a quasitopos* $\mathbf{E}$, *the reflector for sheaves preserves finite limits.*

Corollary. *The reflector for the topos* $\mathbf{E}^{\mathrm{gr}}$ *of coarse objects of* $\mathbf{E}$ *preserves finite limits.*

PROOF. Let $T : \mathbf{E} \to \mathbf{E}$ be the reflector, with bidense reflections $\eta_A : A \to TA$. If $\lambda : L \to D$ is a finite limit cone in $\mathbf{E}$, we must prove that $T\lambda : TL \to TD$ is a limit cone.

If $\mu : A \to TD$ is a cone, let $\mu_i t_i = \eta_{Di} f_i$ be a pullback for each vertex Di of D, with t_i bidense by 44.3. Pullbacks of morphisms t_i give a bidense morphism t, with a commutative square $\mu_i t = \eta_{Di} f_i'$ for each vertex of D. For an arrow $u : Di \to Dj$ of D, we have

$$\eta_{Dj} f_j' = \mu_j t = Tu \cdot \mu_i t = Tu \cdot \eta_{Di} f_i' = \eta_{Dj} u f_i'.$$

Thus repeated application of 44.4 provides a bidense morphism s such that the morphisms $g_i = f_i' s$ form a cone γ for D, with $\eta D \cdot \gamma = \mu \cdot ts$. Now $\gamma = \lambda g$ for a morphism g, and then $\eta_L g = hts$ for a morphism $h : A \to TL$ since TL is a sheaf and ts bidense. Now

$$\mu \cdot ts = \eta D \cdot \lambda \cdot g = T\lambda \cdot \eta_L \cdot g = T\lambda \cdot h \cdot ts,$$

and $\mu = T\lambda \cdot h$ follows since ts is bidense and the vertices of $T\lambda$ are sheaves.

It remains to prove that if $T\lambda \cdot h = T\lambda \cdot h'$, then $h = h'$. In this situation, repeated use of 44.3 provides a bidense morphism t such that $ht = \eta_L f$ and $h't = \eta_L f'$ for morphisms f and f'. It follows that $\eta_{Di} \lambda_i f = \eta_{Di} \lambda_i f'$ for every vertex Di of D, and repeated application of 44.4 provides a bidense morphism s such that $\lambda f s = \lambda f' s$. But then $fs = f's$, and

$$hts = \eta_L f s = \eta_L f' s = h'ts$$

follows. But then $h = h'$ since ts is bidense and TL a sheaf.

The Corollary follows immediately from the fact, noted in 42.2, that $\mathbf{E}^{\mathrm{gr}}$ is the category of sheaves for the topology of 41.2.(3).

44.6. Proposition. *The reflector for separated objects preserves finite products and monomorphisms.*

PROOF. The first part of the Proposition follows immediately from 42.3.2 and 9.4. The second part of the proof of 44.5 remains valid for the reflector for separated objects. Using this for a cone consisting of a single monomorphism, we obtain the second part of the Proposition.

45. Strong Topologies and Coarse Sheaves

45.1. Definition. We say that a topology γ of the quasitopos $\mathbf{E}$ is *strong* if every closed monomorphism for γ is strong.

If $\mathbf{E}$ is a topos, then every topology of $\mathbf{E}$ is strong. Except for the discrete topology of 41.2.(2), the topologies in 41.2 are strong.

45.2. Lemma. *A topology γ of $\mathbf{E}$ is strong if and only if every epimorphic and monomorphic morphism of $\mathbf{E}$ is dense for γ.*

PROOF. If we factor $m = m_1 e$ in $\mathbf{E}$ with e epimorphic and monomorphic, and with m_1 a strong monomorphism, then $\gamma m \simeq \gamma m_1$ if e is dense, by 41.5.3, and then γm is strong. Conversely, if γm is strong, then $m_1 \leqslant \gamma m$, and e is dense by 41.5.2.

45.3. Proposition. *A strong topology of a quasitopos $\mathbf{E}$ is determined by its restriction to strong monomorphisms of $\mathbf{E}$. For every topology γ of $\mathbf{E}$, there is strong topology γ^s of $\mathbf{E}$, determined by γ up to equivalence, which is equivalent to γ for strong monomorphisms. Closed monomorphisms for γ^s are the closed strong monomorphisms for γ, and dense monomorphisms for γ^s are the products de with e monomorphic and epimorphic in $\mathbf{E}$, and d a strong monomorphism dense for γ.*

We call γ^s an *associated strong topology* of γ.

PROOF. If we factor a monomorphism m as $m = m'e$ with e monomorphic and epimorphic and m' strong, then $\gamma m' \simeq \gamma m$, by 45.2 and 41.5.3, if γ is strong. The four conditions of 41.1 are easily verified for γ thus defined if they are valid for γ restricted to strong monomorphisms; we omit the details.

For the second part of the Proposition, we factor $m = m'e$ as above, and we put $\gamma^s m = \gamma m'$. By the preceding paragraph, this defines a strong topology γ^s up to equivalence.

Now factor $m = m'e$ as above, and factor $m' = m_1 d$ with d dense and m_1 closed for γ. Then m_1 is strong by 41.1.4, and d is strong by 10.3.(2). In this factorization, de is dense for γ^s, and m_1 closed for γ^s. If m is closed for γ^s, then de is an isomorphism by 41.5.4, and if m is dense, then m_1 is an isomorphism.

45.4. Proposition. *Every topology γ of $\mathbf{E}$ induces a topology γ^{gr} of the topos $\mathbf{E}^{gr}$ by putting*

$$\gamma^{gr} m = \gamma m$$

for a monomorphism m of $\mathbf{E}^{gr}$. A topology γ and its associated strong topology γ^s induce the same topology on $\mathbf{E}^{gr}$.

PROOF. A monomorphism m of $\mathbf{E}^{gr}$ is strong in the reflective subcategory $\mathbf{E}^{gr}$ of $\mathbf{E}$, and hence strong in $\mathbf{E}$. Thus $\gamma m = \gamma^s m$ is a strong monomorphism of $\mathbf{E}$. But then the domain of γm is a coarse object of $\mathbf{E}$ by 21.5, and γm is a monomorphism of $\mathbf{E}^{gr}$. Thus the restriction γ^{gr} of γ, and of γ^s, to $\mathbf{E}^{gr}$ is defined. This is clearly a topology of $\mathbf{E}^{gr}$.

45.5. Theorem. *For strong monomorphisms, a topology γ of a quasitopos $\mathbf{E}$ is determined up to equivalence by the morphism*

$$(1) \qquad\qquad j = \mathrm{ch}(\gamma\top) : \Omega \to \Omega ,$$

with $\mathrm{ch}(\gamma m) \simeq j \circ \mathrm{ch}\, m$ for every strong monomorphism m of $\mathbf{E}$. This morphism j satisfies the following conditions:
(i) $j \circ j = j$,
(ii) $j \circ \top = \top$,
(iii) $j \circ \wedge = \wedge \circ (j \times j)$.
Conversely, every morphism $j : \Omega \to \Omega$ which satisfies these conditions determines a topology of $\mathbf{E}$ for strong monomorphisms, and for all monomorphisms if $\mathbf{E}$ is a topos, up to equivalence.

PROOF. If $f = \mathrm{ch}\, m$ for a strong momonorphism m, then $m \simeq f^{\leftarrow}\top$, and $\gamma m \simeq f^{\leftarrow}(\gamma\top)$ follows. Thus

$$(2) \qquad\qquad \mathrm{ch}(\gamma m) = j \circ \mathrm{ch}\, m ,$$

by 14.4. If we use (2) with condition (i), we get $\mathrm{ch}(\gamma\gamma m) = \mathrm{ch}(\gamma m)$, which is equivalent to $\gamma\gamma m \simeq \gamma m$. Condition (ii) states that

$$\mathrm{ch}(\gamma m) \circ m = j \circ \top \circ k_X = \top \circ k_X ,$$

for $m : X \to A$ and $k_X : X \to 1$; this says that $m \leqslant \gamma m$ if m is strong. From (iii) and 32.4, we get

$$\mathrm{ch}(\gamma a \cap \gamma b) = \mathrm{ch}(\gamma(a \cap b)) ,$$

for strong monomorphisms a and b. This is necessary by 41.7, and 41.1.1 for strong monomorphisms follows from it. Finally, naturality 41.1.3 of γ for strong monomorphisms m, and 41.1.4, are built into (2).

45.6. Proposition. *A coarse object A of $\mathbf{E}$ is separated for γ^{gr} iff A is separated for γ, and A is a sheaf for γ^{gr} iff A is a sheaf for γ.*

PROOF. Consider a span $Y \xleftarrow{d} X \xrightarrow{f} A$ with d dense, and let T be the reflector for $\mathbf{E}^{\mathrm{gr}}$. If $f_1 : TX \to A$ is adjoint to f for T, then $f = gd$ for a morphism g iff $f_1 = g_1 \cdot Td$ for the morphism $g_1 : TY \to A$ adjoint to g. Now the Proposition follows immediately from the following Lemma.

Lemma. *The reflector for $\mathbf{E}^{\mathrm{gr}}$ preserves dense monomorphisms.*

PROOF. Let T be the reflector, and let $d : X \to Y$ be a dense monomorphism of $\mathbf{E}$. The morphism Td is a strong momomorphism of $\mathbf{E}$, by 44.5 and 21.6. If we factor Td as $Td = m_1 d_1$, with $m_1 \simeq \gamma(Td)$ closed and d_1 dense, then d_1 and m_1 are in $\mathbf{E}^{\mathrm{gr}}$ by 41.1.4 and 21.5. We have $m_1 d_1 e = e'd$ for the reflections $e : X \to TX$ and $e' : Y \to TY$, and then $td = d_1 e$ and $m_1 t = e'$ for a morphism t. This morphism factors $t = ue'$, with u in $\mathbf{E}^{\mathrm{gr}}$. Now $e' = m_1 ue'$, and $m_1 u = \mathrm{id}_{TY}$ follows. Thus m_1 and u are inverse isomorphisms, and Td is dense.

45.7. Proposition. *The topologies γ and γ^s of $\mathbf{E}$ have the same separated objects.*

PROOF. If A is separated for γ^s, then A is separated for the finer topology γ. Conversely, if A is separated for γ, and if $hde = h'de$ for h, h' with codomain A, with e monomorphic and epimorphic and d dense for γ, then $hd = h'd$, and $h = h'$. Thus A is separated for γ^s.

45.8. Theorem. *For a topology γ of a quasitopos $\mathbf{E}$ and an object A of $\mathbf{E}$, the following statements are equivalent.*

(i) *A is a coarse object of $\mathrm{Sep}_\gamma \mathbf{E}$.*
(ii) *A is a coarse object of $\mathrm{Sh}_\gamma \mathbf{E}$.*
(iii) *A is a sheaf for the topology γ^s of $\mathbf{E}$.*
(iv) *A is a sheaf for the topology γ^{gr} of $\mathbf{E}^{\mathrm{gr}}$.*

Corollary. *If the topology γ is strong, and in particular if $\mathbf{E}$ is a topos, then $\mathrm{Sh}_\gamma \mathbf{E}$ is a topos.*

PROOF. By 43.5 and 21.6, a coarse object A for $\mathrm{Sep}_\gamma \mathbf{E}$ is a sheaf. Monomorphic and epimorphic morphisms of $\mathrm{Sh}_\gamma \mathbf{E}$ are monomorphic and epimorphic in $\mathrm{Sep}_\gamma \mathbf{E}$; thus A is coarse for $\mathrm{Sh}_\gamma \mathbf{E}$.

Now consider a span $Y \xleftarrow{d} X \xrightarrow{f} A$ in $\mathbf{E}$, with A coarse for $\mathrm{Sh}_\gamma \mathbf{E}$ and d dense for γ^s. If T is the reflector for $\mathrm{Sh}_\gamma \mathbf{E}$ and $d = d'e$, with e monomorphic and epimorphic and d' dense for γ, then Td' is dense for γ by 45.6, and thus epimorphic in $\mathrm{Sh}_\gamma \mathbf{E}$. It follows that Td is monomorphic and epimorphic in $\mathrm{Sh}_\gamma \mathbf{E}$. Now $f = gd$ for a morphism $g : Y \to A$ iff $f_1 = g_1 \cdot Td$ for the adjoint morphisms $f_1 : TX \to A$ and $g_1 : TY \to A$. This determines g_1 and hence g uniquely, and thus A is a sheaf for γ^s.

By 45.2, a sheaf for γ^s is a coarse object of $\mathbf{E}$, and thus a sheaf for the restriction γ^{gr} of γ to $\mathbf{E}^{\mathrm{gr}}$.

Finally, a sheaf A for γ^{gr} is a sheaf for γ, by 45.6. If $Y \xleftarrow{e} X \xrightarrow{f} A$ is a span in $\mathrm{Sep}_\gamma \mathbf{E}$ with e monomorphic and epimorphic, factor $e = de'$ by 43.6, with e' monomorphic and epimorphic in $\mathbf{E}$, and d dense for γ. Then $f = ge'$ and $g = hd$ for unique morphisms g and h; thus A is coarse for $\mathrm{Sep}_\gamma \mathbf{E}$.

The Corollary now follows immediately from 21.9 and the fact that every topology of a topos is strong.

45.9. Discussion. A topology for a topos $\mathbf{E}$ is often defined as a morphism $j : \Omega \to \Omega$ which satisfies conditions (i) – (iii) of 45.5. A topology in this sense corresponds to an equivalence class of topologies in our sense.

This approach still works for strong topologies of a quasitopos, but it fails for topologies which are not strong. It follows that a strong topology of a quasitopos is determined, up to equivalence, by its restriction to the topos $\mathbf{E}^{\mathrm{gr}}$.

By 45.4, the topology of 41.2.(3) is the finest strong topology, the associated strong topology of the discrete topology of $\mathbf{E}$, and its restriction to $\mathbf{E}^{\mathrm{gr}}$ is the discrete topology of $\mathbf{E}^{\mathrm{gr}}$.

46. Solid Quasitopoi

46.1. Definitions [75]. In a quasitopos $\mathbf{E}$ with initial object 0 and terminal object 1, factor the unique morphism $0 \to 1$ as $0 \xrightarrow{e} 0^* \xrightarrow{m} 1$, with e epimorphic and monomorphic, and with m a strong monomorphism. We say that an object A of $\mathbf{E}$ is *solid* if there is a morphism $0^* \to A$ in $\mathbf{E}$. We denote by $\mathrm{Sol}\,\mathbf{E}$ the full subcategory of $\mathbf{E}$ with solid objects of $\mathbf{E}$ as its objects, and we say that $\mathbf{E}$ is *solid* if every object of $\mathbf{E}$ is solid.

Every topos is a solid quasitopos. In a Heyting algebra, only the greatest element 1 is solid.

The topological quasitopoi of Section 31 are solid. In these quasitopoi, the empty set has only one structure. Thus the initial object is coarse for these examples, and the quasitopoi are solid.

46.2. For an object A of a quasitopos $\mathbf{E}$, we factor the unique morphism $0 \to A$ as $0 \longrightarrow 0_A^* \overset{\bar{o}_A}{\longrightarrow} A$, with $0 \longrightarrow 0_A^*$ monomorphic and epimorphic, and with $\bar{o}_A$ a strong monomorphism.

Lemma. *If $f : A \to B$ in $\mathbf{E}$, then $f^{\leftarrow} \bar{o}_B \simeq \bar{o}_A$.*

PROOF. Construct a commutative diagram

$$
\begin{array}{ccccc}
0 & \overset{e}{\longrightarrow} & \cdot & \overset{m}{\longrightarrow} & A \\
\downarrow{\scriptstyle \mathrm{id}_0} & & \downarrow & & \downarrow{\scriptstyle f} \\
0 & \longrightarrow & 0_B^* & \overset{\bar{o}_B}{\longrightarrow} & B
\end{array}
$$

in $\mathbf{E}$, with a pullback square at right. The outer rectangle is a pullback by 23.5; thus the lefthand square is a pullback. Now e is epimorphic and m a strong monomorphism; thus $m \simeq \bar{o}_A$.

46.3. For a monomorphism $m : X \to A$ of $\mathbf{E}$, we have by 23.8 and 46.2 a pullback-pushout diagram

(1)
$$
\begin{array}{ccc}
\bar{o}_X & \longrightarrow & \bar{o}_A \\
\downarrow & & \downarrow \\
m & \longrightarrow & m \cup \bar{o}_A
\end{array}
$$

in $\mathbf{E}/A$, with $m \cup \bar{o}_A$ monomorphic in $\mathbf{E}$.

Lemma. *Putting $\gamma m = m \cup \bar{o}_A$ for a monomorphism m of $\mathbf{E}$ defines a topology of $\mathbf{E}$, with $\gamma m \simeq m$ for every strong monomorphism.*

PROOF. By 23.4 and 46.2, pullbacks in $\mathbf{E}$ preserve pushouts (1); thus 41.1.3 is satisfied for γ. Conditions 41.1.1 and 41.1.2 are clearly valid. If m with codomain A is a strong monomorphism, then $\bar{o}_A \leqslant m$, and $\gamma m \simeq m$ follows. Thus 41.1.4 for γ is trivial.

46.4. Theorem. *If $\mathbf{E}$ is a quasitopos, then every object of $\mathbf{E}$ is separated for the topology of 46.3. The sheaves for this topology are the solid objects of $\mathbf{E}$, and every coarse object of $\mathbf{E}$ is solid.*

PROOF. A momomorphism $m : X \to A$ is dense for the topology of 46.3

iff the pullback square

(1)
$$\begin{array}{ccc} 0_X^* & \xrightarrow{\ e\ } & 0_A^* \\ \Big\downarrow \bar{o}_X & & \Big\downarrow \bar{o}_A \\ X & \xrightarrow{\ m\ } & A \end{array}$$

of monomorphisms is also a pushout square. If this is the case, then m is epimorphic since the morphism e with codomain 0_A^* is epimorphic. Thus every object of $\mathbf{E}$ is separated for γ.

If F is a sheaf for γ, then the morphism $0 \to F$ factors through the dense monomorphism $0 \to 0^*$; thus F is a solid object of $\mathbf{A}$. Conversely, if F is solid, then there is a morphism $g : 0_A^* \to F$. If $m : X \to A$ is dense for γ and $f : X \to F$, then $f\bar{o}_X = ge$ in (1) as there is at most one morphism from 0_X^* to F. But then f factors $f = tm$ since (1) is a pushout square, with t unique since m is epimorphic. Thus F is a sheaf for γ.

Since the morphism $0 \to 0^*$ is monomorphic and epimorphic, the morphism $0 \to A$ factors through 0^* if A is coarse; thus A is solid.

46.5. Proposition. Sol $\mathbf{E}$ *is a solid quasitopos with initial object* 0^*.

PROOF. Sol $\mathbf{E}$ is a quasitopos by 46.4, and 0^* is an initial object of Sol $\mathbf{E}$. Since the monomorphism $0^* \to 1$ of $\mathbf{E}$ is strong, and closed for the topology γ of 46.3, it is strong in Sol $\mathbf{E}$ by 46.4 and 43.6. Thus 0^* for Sol $\mathbf{E}$ is 0^* for $\mathbf{E}$, and Sol $\mathbf{E}$ is a solid quasitopos.

46.6. Proposition. *For an object* A *of* $\mathbf{E}$, *the solid objects of the quasitopos* $\mathbf{E}/A$ *are the objects* $u : X \to A$ *with* X *a solid object of* $\mathbf{E}$. *If* $\mathbf{E}$ *is solid, then* $\mathbf{E}/A$ *is solid.*

PROOF. If γ is the topology of 46.2, then the solid objects of $\mathbf{E}/A$ are the sheaves for the topology γ/A, by 46.4. Now the Proposition follows immediately from 42.3.4.

46.7. We recall that a coproduct $A \amalg B$ in a category $\mathbf{C}$ with an initial object 0 is called *disjoint* if the pushout square

$$\begin{array}{ccc} 0 & \longrightarrow & A \\ \Big\downarrow & & \Big\downarrow \\ B & \longrightarrow & A \amalg B \end{array}$$

is also a pullback square.

Theorem. *For a quasitopos* **E**, *the following statements are equivalent.*
(i) *The initial object* 0 *of* **E** *is coarse.*
(ii) **E** *is a solid quasitopos.*
(iii) *The monomorphism* $0 \to 1$ *of* **E** *is strong.*
(iv) *Every monomorphism* $0 \to A$ *of* **E** *is strong.*
(v) *Coproducts* $A \amalg B$ *in* **E** *are disjoint.*

PROOF. (i) and (ii) are equivalent by 21.5 and 21.6. (iii) is equivalent to the statement that $0 \to 0^*$ is an isomorphism; this is the case iff **E** is solid.

For every object A of **E**, we have a pullback square

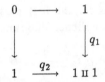

in **E**; thus $0 \to A$ is a strong monomorphism if $0 \to 1$ is one.

If (iv) is valid, then coproducts $A \amalg B$ in **E** are disjoint by 23.6.

Finally, if the pushout square

$$
\begin{array}{ccc}
0 & \longrightarrow & 1 \\
\downarrow & & \downarrow{\scriptstyle q_1} \\
1 & \xrightarrow{\ q_2\ } & 1 \amalg 1
\end{array}
$$

in **E** is a pullback square, then $0 \to 1$ is an equalizer of q_1 and q_2, and hence a strong monomorphism.

46.8. Proposition. *If* γ *is a topology of a solid quasitopos* **E**, *then* $\mathrm{Sh}_\gamma \mathbf{E}$ *is a solid quasitopos, and the quasitopos* $\mathrm{Sep}_\gamma \mathbf{E}$ *is solid if and only if the initial object* 0 *of* **E** *is a sheaf for* γ.

PROOF. If **E** is solid, then the monomorphism $0 \to 1$ of **E** is strong, and so is its closure. If $\bar{0} \to 1$ is this closure, then $0 \to \bar{0}$ is a reflection for $\mathrm{Sh}_\gamma \mathbf{E}$, by 43.5 and 42.3.1. As the reflector for sheaves preserves initial objects, $\bar{0}$ is an initial object for $\mathrm{Sh}_\gamma \mathbf{E}$. By 43.6, $\bar{0} \to 1$ is a strong monomorphism of $\mathrm{Sh}_\gamma \mathbf{E}$; thus $\mathrm{Sh}_\gamma \mathbf{E}$ is solid by 46.7.

By 43.4.1, the initial object 0 of **E** is separated for γ. Thus $\mathrm{Sep}_\gamma \mathbf{E}$ is solid, by 46.7, iff 0 is coarse in Sep_γ. By 45.8 and the preceding paragraph, this is the case iff 0 is a sheaf for γ.

46.9. Representing finite sets. For every category $\mathbf{E}$ with finite coproducts and a terminal object 1, a functor R from finite sets to $\mathbf{E}$ can be constructed as follows. For a finite set A, we let RA be a coproduct of of copies of 1, one for each element of A, with injections $\pi_i : 1 \to RA$, one for every $i \in A$. For a mapping $f : A \to B$ of finite sets, we define $Rf : RA \to RB$ by putting

$$Rf \circ \pi_i = \pi'_{f(i)},$$

for the injections π_i of RA and π'_j of RB.

The functor R clearly preserves finite coproducts and terminal objects, but it need not preserve or reflect much else. For example, if $\mathbf{E}$ is a lattice, then RA is the greatest element of $\mathbf{E}$ for every finite set A, and Ff is an identity morphism for every mapping f of finite sets. This situation changes if we specialize.

Proposition. *If $\mathbf{E}$ is a solid quasitopos, then the functor R from finite sets to $\mathbf{E}$ preserves finite limits, and R is faithful if $\mathbf{E}$ is not trivial.*

We note that the functor R need not be full, nor can we expect it to be injective for objects.

PROOF. Since R preserves terminal objects, sending a singleton to a terminal object of $\mathbf{E}$, it suffices for the first part to prove that R preserves pullbacks. For sets, we obtain a pullback diagram

$$
\begin{array}{ccc}
A \times_C B & \xrightarrow{\ q\ } & B \\
{\scriptstyle p}\downarrow & & \downarrow{\scriptstyle g} \\
A & \xrightarrow{\ f\ } & C
\end{array}
$$

by letting $A \times_C B$ be the set of all pairs (i,j) in $A \times B$ with $f(i) = g(j)$. Since pullbacks in the quasitopos $\mathbf{E}$ preserve coproducts, the pullback $RA \times_{RC} RB$ is the coproduct of all P_{ij} for pullbacks

$$
\begin{array}{ccc}
P_{ij} & \longrightarrow & 1 \\
\downarrow & & \downarrow{\scriptstyle \pi_{g(j)}} \\
1 & \xrightarrow{\ \pi_{f(i)}\ } & C
\end{array}
$$

If $\mathbf{E}$ is solid, then $P_{ij} = 1$ if $f(i) = g(j)$, and $P_{ij} = 0$ otherwise. Thus the pullback of Rf and Rg is isomorphic to $R(A \times_C B)$, as claimed.

For mappings $A \overset{f}{\underset{g}{\rightrightarrows}} B$ of finite sets, we have $Rf = Rg$ iff $\pi_{f(i)} = \pi_{g(i)}$ for all $i \in A$, for the injections of RB. If $\mathbf{E}$ is solid and non-trivial, it follows that $f(i) = g(i)$ for all $i \in A$, *i.e.* $f = g$.

47. Grothendieck Topologies

47.1. Notations. We deal in this Section with topologies and categories of sheaves for a functor category $[\mathcal{C}^{\mathrm{op}}, \mathrm{SET}]$, using the notations of Section 26, but we do not assume $\mathcal{C}$ to be small. As noted in 26.4, every monomorphism of $[\mathcal{C}^{\mathrm{op}}, \mathrm{SET}]$ is equivalent to a unique subfunctor inclusion; we use this to identify subobjects with subfunctors when dealing with characteristic morphisms or topologies. Thus if $m : R \to A$ is a subfunctor inclusion, then we denote by $\mathrm{ch}_A R$ the characteristic morphism $\mathrm{ch}\, m$, and by $\gamma_A R$, for a topology γ of $[\mathcal{C}^{\mathrm{op}}, \mathrm{SET}]$, the domain of the subfunctor inclusion equivalent to γm. The codomain subscripts could be omitted if the codomain A of the subfunctor inclusion is clearly given by the context. We note that $\mathrm{ch}_A R$ assigns to every object a of $\mathcal{C}$ a sieve σ_a on $\mathcal{C}$ at a by 26.4.(1), even if $\mathcal{C}$ is not small.

For two objects R and S of a category $[\mathcal{C}^{\mathrm{op}}, \mathrm{SET}]$, we put $R \leqslant S$ if R is a subfunctor of S, and for morphisms $\rho : A \to \Omega$ and $\sigma : A \to \Omega$ in $[\mathcal{C}^{\mathrm{op}}, \mathrm{SET}]$, with the same domain A, we put $\rho \leqslant \sigma$ if always $\rho_a(x) \subset \sigma_a(x)$, for an object a of $\mathcal{C}$ and $x \in Aa$. It follows immediately from 26.4.(1) that

$$R \leqslant S \quad \Longleftrightarrow \quad \mathrm{ch}_A R \leqslant \mathrm{ch}_A S,$$

for subobjects R and S of a presheaf $A : \mathcal{C}^{\mathrm{op}} \to \mathrm{SET}$.

Sieves have been defined in 26.4. If $\mathcal{C}$ is locally small, then we shall consider sieves at an object a of $\mathcal{C}$ as subobjects of the object $Ya = \mathcal{C}(-, a)$ of $[\mathcal{C}^{\mathrm{op}}, \mathrm{SET}]$. If $f : c \to a$ in $\mathcal{C}$ and S is a sieve on $\mathcal{C}$ at a, then we define $f^{\leftarrow} S$ to be the sieve of all morphisms u of $\mathcal{C}$ with codomain c such that $fu \in S$. If $\mathcal{C}$ is locally small, this is the inverse image of the subobject S of Ya by the morphism $\mathcal{C}(-, f) = Yf : Yc \to Ya$ of $[\mathcal{C}^{\mathrm{op}}, \mathrm{SET}]$.

47.2. Grothendieck topologies. A *Grothendieck topology* J of a category $\mathcal{C}$ assigns to every object a of $\mathcal{C}$ a collection Ja of sieves on $\mathcal{C}$ at a, subject to the following conditions.

47.2.1. If $f : c \to a$ in $\mathcal{C}$ and $S \in Ja$, then $f^{\leftarrow} S \in Jc$.

47.2.2. For every object a of $\mathcal{C}$, the *full sieve* at a, consisting of all morphisms of $\mathcal{C}$ with codomain a, is in JA.

47.2.3. If R and S are sieves on $\mathcal{C}$, at the same object a of $\mathcal{C}$, with R in Ja und $u^{\leftarrow} S$ always in Ji for $u : i \to a$ in R, then $S \in Ja$.

Grothendieck topologies of $\mathcal{C}$ can be defined for an arbitrary category $\mathcal{C}$. If $\mathcal{C}$ is small, then $f^{\leftarrow} = \Omega f$ in 47.2.1, and 47.2.1 states that J is a subfunctor of the contravariant sets of sieves functor Ω on $\mathcal{C}$.

47.3. For a small category $\mathcal{C}$, the following result establishes a bijection between Grothendieck topologies of $\mathcal{C}$ and topologies of the topos $[\mathcal{C}^{\mathrm{op}}, \mathrm{SET}]$.

Theorem. *For a small category* $\mathcal{C}$, *a subobject* J *of the sets of sieves functor* Ω *in* $[\mathcal{C}^{\mathrm{op}}, \mathrm{SET}]$ *is a Grothendieck topology of* $\mathcal{C}$ *if and only if* $J = \gamma_\Omega \top$ *for a topology* γ *of the topos* $[\mathcal{C}^{\mathrm{op}}, \mathrm{SET}]$. *In this situation,* J *determines* γ *by*

$$(1) \qquad\qquad \gamma_A S = \mu^{\leftarrow} J \qquad \text{for} \qquad \mu = \mathrm{ch}_A S,$$

for a subfunctor S *of a functor* $A : \mathcal{C}^{\mathrm{op}} \to \mathrm{SET}$. *For an object* a *of* $\mathcal{C}$, *the sieves in* Ja *are then the dense subobjects of* Ya, *and* $j_a = \gamma_{Ya}$ *for the characteristic morphism* $j : \Omega \to \Omega$ *of* J.

PROOF. As already noted, 47.2.1 states that J is a subfunctor of Ω. If $j = \mathrm{ch}_\Omega J : \Omega \to \Omega$ and $f : c \to a$ in $\mathcal{C}$, then by 26.4,

$$f \in j_a(S) \quad\Longleftrightarrow\quad f^{\leftarrow} S \in Jc.$$

Since $f \in S \iff f^{\leftarrow} S = Yc$, we can restate 47.2.2 as
  (a) $S \subset j_a(S)$,
for a sieve S at a. Now 47.2.3 becomes:
  (b) if $R \subset j_a(S)$ and $R \in Ja$, then $S \in Ja$,
for R, S in Ja. We split this into two parts:
  (b₁) if $j_a(S) \in Ja$, then $S \in Ja$;
  (b₂) if $R \subset S$ and $R \in Ja$, then $S \in Ja$,
using (a) for (b₂).
  If γ is given by (1) for a Grothendieck topology J of $\mathcal{C}$, then γ clearly satisfies 41.1.3. If $U \leqslant V$ for subobjects U, V of an object A of $[\mathcal{C}^{\mathrm{op}}, \mathrm{SET}]$, with characteristic morphisms μ and ν, then $\mu \leqslant \nu$, so that

$$x \in (\gamma_A U)a \quad\Longleftrightarrow\quad \mu_a(x) \in Ja \quad\Longrightarrow\quad \nu_a(x) \in Ja \quad\Longleftrightarrow\quad x \in (\gamma_A V)a$$

for $x \in Aa$, by (1) and (b₂) above. Thus γ satisfies 41.1.1.
  The first part of 41.1.2 follows immediately from (1) and (a) above. For the second part, we have

$$x \in (\gamma_A \gamma_A U)a \quad\Longleftrightarrow\quad j_a(\mu_a(x)) \in Ja$$
$$\Longleftrightarrow\quad \mu_a(x) \in Ja \quad\Longleftrightarrow\quad x \in (\gamma_A U)a$$

for $x \in Aa$. This proves the second part of 41.1.2.

For a sieve S at an object a of $\mathcal{C}$, regarded as a subobject of Ya with $\mathrm{ch}_{Ya}\, S = \sigma$, and for $f : c \to a$ in $\mathcal{C}$, we have

$$u \in \sigma_c(f) \quad \Longleftrightarrow \quad fu \in S$$

if fu is defined in $\mathcal{C}$. Thus $\sigma_c(f) = f^{\leftarrow} S$. But then

$$f \in \gamma_{Ya} S \quad \Longleftrightarrow \quad f^{\leftarrow} S \in Jc \quad \Longleftrightarrow \quad f \in j_a(S),$$

and $j_a = \gamma_{Ya}$, if γ is defined by (1). Since $S \in J_a \iff j_a(S) = Ya$, the sieves in Ja are the dense subobjects of Ya if γ is a topology. But then properties (a), (b$_1$), (b$_2$) in the first part of this proof are valid, and J is a Grothendieck topology of $\mathcal{C}$.

47.4. Sheaves and separated presheaves. Let $\mathcal{C}$ be a locally small category. We say that a functor $A : \mathcal{C}^{\mathrm{op}} \to \mathrm{SET}$ is a *sheaf* for a Grothendieck topology J of $\mathcal{C}$ if for an object a of $\mathcal{C}$, a sieve S in Ja, and a morphism $\xi : S \to A$ in $[\mathcal{C}^{\mathrm{op}}, \mathrm{SET}]$, there is always a unique element x in Aa such that $\xi_c(f) = (Af)(x)$ for every $f : c \to a$ in S. We say that A is a *separated presheaf* for J if for the same data, there is always at most one x in Aa such that $\xi_c(f) = (Af)(x)$ for every $f : c \to a$ in S.

Theorem. *Sheaves and separated presheaves for a Grothendieck topology of a small category $\mathcal{C}$ are the same as the sheaves and separated objects for the corresponding topology of the topos $[\mathcal{C}^{\mathrm{op}}, \mathrm{SET}]$.*

PROOF. For a sieve S at an object a of $\mathcal{C}$, and for $\xi : S \to A$ in $[\mathcal{C}^{\mathrm{op}}, \mathrm{SET}]$ and $x \in Aa$, we have $\xi_c(f) = (Af)(x)$ for every $f : c \to a$ in S iff ξ is the restriction to S of the morphism $\mu : Ya \to A$ corresponding to x by the Yoneda Lemma 3.6. Thus A is a sheaf for a Grothendieck topology J of $\mathcal{C}$ iff for every span $Ya \xleftarrow{\delta} S \xrightarrow{\xi} A$ in $[\mathcal{C}^{\mathrm{op}}, \mathrm{SET}]$, with S in Ja and the subfunctor inclusion at left, there is exactly one morphism $\mu : Ya \to A$ in $[\mathcal{C}^{\mathrm{op}}, \mathrm{SET}]$ such that $\mu\delta = \xi$. With 47.3, it follows immediately that a sheaf for the topology γ of $[\mathcal{C}^{\mathrm{op}}, \mathrm{SET}]$ corresponding to J is a sheaf for J.

For the converse, consider a span $V \xleftarrow{d} U \xrightarrow{\varphi} A$, with a dense subfunctor inclusion at left. For $x \in Va$ and the corresponding $\mu : YA \to V$, we construct a pullback square

$$
\begin{array}{ccc}
S & \xrightarrow{\;u\;} & U \\[2pt]
{\scriptstyle \delta}\Big\downarrow & & \Big\downarrow{\scriptstyle d} \\[4pt]
Ya & \xrightarrow{\;\mu\;} & V
\end{array}
$$

in $[\mathcal{C}^{\mathrm{op}}, \mathrm{SET}]$, with $S \in Ja$ and δ the subfunctor inclusion. If A is a sheaf for J, then $\varphi u = \nu\delta$ for exactly one $\nu : Ya \to A$, with $\psi\mu = \nu$ if $\psi d = \varphi$. Thus we must put

$$\psi_a(x) = \psi_a(\mu_a(\mathrm{id}_a)) = \nu_a(\mathrm{id}_a)$$

if $\psi d = \varphi$. Using this construction with pullbacks

$$
\begin{array}{ccc}
f^{\leftarrow}S & \longrightarrow & S \\
\downarrow & & \downarrow \\
Yc & \xrightarrow{\ Yf\ } & Ya
\end{array}
\qquad \text{and} \qquad
\begin{array}{ccc}
Ya & \xrightarrow{\ \mu\ } & U \\
\mathrm{id}_{Ya}\downarrow & & \downarrow d \\
Ya & \xrightarrow{\ d\mu\ } & V
\end{array}
$$

in $[\mathcal{C}^{\mathrm{op}}, \mathrm{SET}]$, we see easily that we get a morphism $\psi : V \to A$ in $[\mathcal{C}^{\mathrm{op}}, \mathrm{SET}]$ with $\psi d = \varphi$. Thus a sheaf for J is a sheaf for γ.

The proof of the Theorem for separated presheaves is almost exactly analogous to the proof for sheaves; we omit the details.

48. Canonical Topologies

48.1. Quotient sieves. A sieve at an object a of a category $\mathcal{C}$ can be viewed as a full subcategory of the slice category $\mathcal{C}/a$, defined in Section 17. If $\mathcal{C}$ is locally small, then a morphism $\mu : S \to Yx$ in $[\mathcal{C}^{\mathrm{op}}, \mathrm{SET}]$ induces a cone $\bar{\mu} : S \to x$ in $\mathcal{C}$, with $\bar{\mu}_f = \mu_c(f)$ for $f : c \to a$ in S. We say that S is a *quotient sieve* at a if the cone induced by the inclusion $S \to Ya$ is a colimit cone in $\mathcal{C}$. If every sieve $f^{\leftarrow}S$, for $f : c \to a$ in $\mathcal{C}$, is a quotient sieve at c, then we say that S is a *universal quotient sieve* at a.

Theorem. *For a Grothendieck topology J of a locally small category $\mathcal{C}$, every object Yx of $[\mathcal{C}^{\mathrm{op}}, \mathrm{SET}]$ is a sheaf for J if and only if every sieve in J is a quotient sieve.*

PROOF. For a span $Ya \xleftarrow{\ \delta\ } S \xrightarrow{\ \mu\ } Yx$ in $[\mathcal{C}^{\mathrm{op}}, \mathrm{SET}]$, for a sieve S at a and the subcategory inclusion δ at left, we clearly have $\mu = Yg \cdot \delta$ in $[\mathcal{C}^{\mathrm{op}}, \mathrm{SET}]$ iff $\bar{\mu} = g\bar{\delta}$ for the induced cones. With this observation, the Theorem follows immediately from the Corollary of 3.6 and ̇ e definitions.

48.2. The canonical topology. If we consider Grothendieck topologies of a category $\mathcal{C}$ simply as classes of sieves, then the intersection of a class of Grothendieck topologies of $\mathcal{C}$ clearly is a Grothendieck topology of $\mathcal{C}$. Thus Grothendieck topologies of $\mathcal{C}$, ordered by class inclusion, form a complete lattice. If $\mathcal{C}$ is locally small, then it follows that there is a largest Grothendieck topology for which every functor Ya, for an object a of $\mathcal{C}$, is a sheaf. This Grothendieck topology of $\mathcal{C}$ is called the *canonical Grothendieck topology* of $\mathcal{C}$. If $\mathcal{C}$ is small, then the corresponding topos topology of $[\mathcal{C}^{op}, \mathrm{SET}]$ is called the *canonical topology* of $[\mathcal{C}^{op}, \mathrm{SET}]$.

Canonical Grothendieck topologies are easily described.

48.3. Theorem. *For every category $\mathcal{C}$, the canonical Grothendieck topology of $\mathcal{C}$ consists of all universal quotient sieves in $\mathcal{C}$.*

PROOF. If a Grothendieck topology of $\mathcal{C}$ consists of quotient sieves, then these quotient sieves must be universal by 47.2.1.

Conversely, let Ja be the collection of all universal quotient sieves at a, for every object a of $\mathcal{C}$. This clearly satisfies 47.2.1. The full sieve Ya at a is a colimit cone, and $f^{\leftarrow}Ya = Yc$ for $f : c \to a$ in $\mathcal{C}$; thus 47.2.2 is valid.

Now let R be a universal quotient sieve at a, and S a sieve at a such that $u^{\leftarrow}S$ is a quotient sieve for every u in R. If $f : c \to a$ in $\mathcal{C}$, then $v^{\leftarrow}f^{\leftarrow}S$ is a quotient sieve for every v in $f^{\leftarrow}R$, i.e. $fv \in R$. Thus $f^{\leftarrow}S$ satisfies the hypothesis of 47.2.3 for $f^{\leftarrow}R$ in Jc, and it suffices to prove that S is a quotient sieve.

A cone $\mu : S \to x$ in $\mathcal{C}$ induces for each $u : i \to a$ in R a cone $\nu : u^{\leftarrow}S \to x$, with $\nu_v = \mu_{uv}$ for $v \in u^{\leftarrow}S$. Since $u^{\leftarrow}S \in Ji$, we have $\mu_{uv} = \gamma_u v$ in $\mathcal{C}$, for all uv in S and a unique $\gamma_u : i \to x$. The unique morphisms γ_u clearly form a cone $\gamma : R \to x$, and thus there is a unique $g : a \to x$ in $\mathcal{C}$ with $\gamma_u = gu$ for every $u \in R$.

If $\mu_f = hf$ for every $f \in S$, with $h : a \to x$ in $\mathcal{C}$, then in particular $huv = \mu_{uv} = \gamma_u v$ for $u \in R$ and $uv \in S$. Thus $hu = \gamma_u = gu$ for $u \in R$, and $h = g$ follows.

Now consider $f \in S$ and $u \in f^{\leftarrow}R$, with $fu \in R$. Then

$$\mu_f uv = \mu_{fuv} = \gamma_{fu} v = gfuv$$

for every v in the quotient sieve $(fu)^{\leftarrow}S$, and $\mu_f u = gfu$ follows. Since $f^{\leftarrow}R$ is a quotient sieve, we conclude that $\mu_f = gf$, for every $f \in S$, so that S is a quotient sieve.

48.4. Example. In an ordered set L, considered as a category, a sieve at an element a of L is a decreasing subset of $\downarrow a$. If $c \leqslant a$ in L and S is a sieve at a, then $f^{\leftarrow}S$, for the unique morphism $f : c \to a$, is the sieve $\downarrow c \cap S$ at c. This sieve is sometimes denoted by $S \mid c$. We note that

$$\downarrow c \cap S = \{c \cap x \mid x \in S\}$$

if L is a meet semilattice.

Colimits in an ordered set are suprema; thus a quotient sieve at $a \in L$ is a subset S of L with $\sup S = a$ in L. If L is a meet semilattice, then S is a universal quotient sieve at $a = \sup S$ iff

$$(1) \qquad\qquad \sup(\downarrow c \cap S) = c$$

for every $c \in \downarrow a$. This is the case by 24.6 if L is a complete Heyting algebra, since then the functor $\downarrow c \cap -$ on L has a right adjoint and thus preserves suprema. Thus the canonical Grothendieck topology of a Heyting algebra H has Ja consisting of all decreasing subsets S of $\downarrow a$ with $\sup S = a$ in H.

If S is a sieve on $\mathcal{C}$, at an object a of a category $\mathcal{C}$, then a morphism $\xi : S \to A$ of $[\mathcal{C}^{op}, \text{SET}]$ assigns to every $f : c \to a$ an element $\xi_c(f) = x_f$ of Ac, subject to the condition that $x_{fu} = (Au)(x_f)$ if $f \in S$ and fu is defined in $\mathcal{C}$. For the particular case that $\mathcal{C}$ is an ordered set L and S a decreasing subset of $\downarrow a$, this means that ξ assigns to each $i \in S$ an element x_i of the set A_i, with $x_i = x_j|i$ for $j \in S$ and $i \leqslant j$. If we specialize further to a complete Heyting algebra H, then it follows that the sheaves and separated presheaves defined in Sections 28 and 29 are the sheaves and separated presheaves for the canonical Grothendieck topology of H.

Chapter 5

GEOMETRIC MORPHISMS

The first two sections of this Chapter provide a proof of the basic theorem that for a left exact comonad $\mathcal{G}$ on a quasitopos $\mathbf{E}$, the category $\mathbf{E}_\mathcal{G}$ of coalgebras for $\mathcal{G}$ is again a quasitopos. This theorem was proved for topoi by LAWVERE and TIERNEY, and extended to quasitopoi in [104]. However, the construction in [104] for representing partial morphisms in $\mathbf{E}_\mathcal{G}$ is wrong. We give a correct construction for partial morphisms with cofree codomain. This provides a subobject classifier for $\mathbf{E}_\mathcal{G}$, and we follow J. PENON [83] for a proof that $\mathbf{E}_\mathcal{G}$ is locally cartesian closed.

Injective and surjective geometric morphisms are defined in Section 51, and we construct topologies from geometric morphisms in Section 52.

For a morphism f in a quasitopos, the geometric morphism $f^* \dashv f_*$ is injective if and only if f is monomorphic, and surjective if and only if f is a strong epimorphism. We show in Section 53 that injective geometric morphisms are equivalent to full embeddings $\mathrm{Sh}_\gamma \mathbf{E} \to \mathbf{E}$ of categories of sheaves, and surjective geometric morphisms to morphisms $U_\mathcal{G} \dashv F_\mathcal{G} : \mathbf{E} \to \mathbf{E}_\mathcal{G}$ for a left exact comonad $\mathcal{G}$ on a quasitopos $\mathbf{E}$, as for topoi. We also extend to quasitopoi the theorem of LAWVERE and TIERNEY that every geometric morphism can be factored, uniquely up to equivalence, as a surjective morphism followed by an injective morphism.

For quasitopoi which are not topoi, there are injective geometric morphisms $F \dashv U$ which are not surjective, but have a faithful left adjoint F. We discuss these geometric morphisms in 53.5.

49. Coalgebras for Left Exact Comonads

49.1. Definitions. We say that a comonad $\mathcal{G} = (G, \varepsilon, \psi)$ on a category $\mathbf{E}$ is *left exact* if the functor G is left exact, *i.e.* G preserves finite limits.

Throughout this Section, we consider a comonad $\mathcal{G} = (G, \varepsilon, \psi)$ on a category $\mathbf{E}$. We denote by $\mathbf{E}_\mathcal{G}$ the category of coalgebras for $\mathcal{G}$ (see 8.4), and by $U_\mathcal{G} : \mathbf{E}_\mathcal{G} \to \mathbf{E}$ the forgetful functor, with $U_\mathcal{G}(A, \alpha) = A$ for a coalgebra (A, α), and $U_\mathcal{G}f = f : A \to B$ for a homomorphism $f : (A, \alpha) \to (B, \beta)$ of coalgebras. The *cofree coalgebra* $F_\mathcal{G}A$ over an object A of $\mathbf{E}$ is defined as $F_\mathcal{G}A = (GA, \psi_A)$. This defines a functor $F_\mathcal{G} : \mathbf{E} \to \mathbf{E}_\mathcal{G}$, with $U_\mathcal{G}F_\mathcal{G} = G$.

We note that $U_{\mathcal{G}} \dashv F_{\mathcal{G}}$, with

$$g = Gf \cdot \alpha \quad \Longleftrightarrow \quad f = \varepsilon_B \cdot g$$

for morphisms $f : A \to B$ of $\mathbf{E}$ and $g : (A, \alpha) \to F_{\mathcal{G}}B$ of $\mathbf{E}_{\mathcal{G}}$. This is easily verified; we omit the proof. The unit of this adjunction consists of algebra structures $\alpha : (A, \alpha) \to F_{\mathcal{G}}A$; the counit is the counit ε of $\mathcal{G}$.

49.2. Finite limits and colimits. Since the forgetful functor $U_{\mathcal{G}}$ for coalgebras is faithful and a left adjoint, $U_{\mathcal{G}}$ reflects monomorphisms and epimorphisms, and $U_{\mathcal{G}}$ preserves epimorphisms and all colimits. We recall that $U_{\mathcal{G}}$ *creates* colimits in the following sense. If D is a diagram of coalgebras for which the diagram $U_{\mathcal{G}}D$ has a colimit in $\mathbf{E}$, then a colimit cone $\lambda : U_{\mathcal{G}}D \to A$ in $\mathbf{E}$ lifts to exactly one cone $\lambda : D \to (A, \alpha)$ in $\mathbf{E}_{\mathcal{G}}$, and this cone is a colimit cone. It follows that $U_{\mathcal{G}}$ reflects colimits. In particular, $\mathbf{E}_{\mathcal{G}}$ has finite colimits if $\mathbf{E}$ has finite colimits.

Proposition. *If $\mathcal{G}$ is left exact, then $U_{\mathcal{G}}$ creates and hence reflects finite limits.*

PROOF. Let D be a finite diagram of coalgebras, with coalgebra structures $\varphi : D \to F_{\mathcal{G}}U_{\mathcal{G}}D$. If $U_{\mathcal{G}}D$ has a limit cone $\lambda : A \to U_{\mathcal{G}}D$, then $G\lambda$ is a limit cone, and thus

$$(1) \qquad\qquad G\lambda \cdot \alpha = U_{\mathcal{G}}\varphi \cdot \lambda$$

for a unique morphism $\alpha : A \to GA$ in $\mathbf{E}$. Using the fact that λ is a limit cone, one proves easily that $\varepsilon_A \alpha = \mathrm{id}_A$. Since G preserves finite limits, $GG\lambda$ is also a limit cone, and $G\alpha \cdot \alpha = \psi_A \cdot \alpha$ follows easily. We omit the details. Now it follows from (1) that λ lifts to a cone $\lambda : (A, \alpha) \to D$ in $\mathbf{E}_{\mathcal{G}}$.

49.3. Proposition. *If $\mathcal{G}$ is left exact and $\mathbf{E}$ has finite limits, then $\mathbf{E}_{\mathcal{G}}$ has and $U_{\mathcal{G}}$ preserves finite limits, and $U_{\mathcal{G}}$ and G preserve monomorphisms.*

PROOF. To get a limit cone for a finite diagram D in $\mathbf{E}_{\mathcal{G}}$, we construct a limit cone for $U_{\mathcal{G}}D$ in $\mathbf{E}$. This creates the desired limit cone for D, and $U_{\mathcal{G}}$ preserves this limit cone. Now if m is monomorphic in $\mathbf{E}_{\mathcal{G}}$, then m has a kernel pair of identity morphisms. This is preserved by $U_{\mathcal{G}}$; thus m is monomorphic in $\mathbf{E}$. Since the right adjoint $F_{\mathcal{G}}$ preserves monomorphisms, so does $G = U_{\mathcal{G}}F_{\mathcal{G}}$ if $U_{\mathcal{G}}$ does.

49.4. Factorizations. We assume now that $\mathcal{G}$ is left exact, and we recall that the left adjoint $U_{\mathcal{G}}$ preserves strong epimorphisms.

Proposition. *The functor* $U_\mathcal{G}$ *reflects equalizers, and* $U_\mathcal{G}$ *creates and reflects factorizations* $f = me$ *with* e *epimorphic and* m *an equalizer. If* **E** *has finite limits, then* $U_\mathcal{G}$ *preserves these factorizations, and* $U_\mathcal{G}$ *reflects strong epimorphisms.*

PROOF. If $m : (A, \alpha) \to (B, \beta)$ in $\mathbf{E}_\mathcal{G}$ is an equalizer of a pair $B \overset{u}{\underset{v}{\rightrightarrows}} E$ in **E**, then $ug = vg$ in **E** for $g : (C, \gamma) \to (A, \alpha)$ in $\mathbf{E}_\mathcal{G}$ iff $u_1 g = v_1 g$ in $\mathbf{E}_\mathcal{G}$ for the adjoint morphisms $u_1 : (B, \beta) \to F_\mathcal{G} E$ of u and $v_1 : (B, \beta) \to F_\mathcal{G} E$ of v, and then g factors $g = mh$ in **E**. In this situation,

$$Gm \cdot \alpha h = \beta g = Gm \cdot Gh \cdot \gamma$$

in **E**. Since Gm is an equalizer of Gu and Gv in **E**, it follows that $h : (C, \gamma) \to (A, \alpha)$ in $\mathbf{E}_\mathcal{G}$. Thus m is an equalizer of u_1 and v_1 in $\mathbf{E}_\mathcal{G}$.

Now if $f : (A, \alpha) \to (B, \beta)$ in $\mathbf{E}_\mathcal{G}$ factors $f = me$ in **E**, with e epimorphic and m an equalizer, then $Gm \cdot Ge \cdot \alpha = \beta me$ in **E** with Gm an equalizer, hence a strong monomorphism. Thus $Gm \cdot \gamma = \beta m$ and $Ge \cdot \alpha = \gamma e$ in **E** for a unique $\gamma : C \to GC$ in **E**. It is easily verified that this is a coalgebra structure of C. Thus f factors $f = me$ in $\mathbf{E}_\mathcal{G}$, with e epimorphic in $\mathbf{E}_\mathcal{G}$, and m an equalizer in $\mathbf{E}_\mathcal{G}$ by the preceding paragraph.

If $e : (A, \alpha) \to (B, \beta)$ in $\mathbf{E}_\mathcal{G}$ with $e : A \to B$ strongly epimorphic in **E**, consider a commutative square

$$
\begin{array}{ccc}
(A, \alpha) & \overset{e}{\longrightarrow} & (B, \beta) \\
\downarrow{\scriptstyle f} & & \downarrow{\scriptstyle g} \\
(C, \gamma) & \underset{m}{\longrightarrow} & (D, \delta)
\end{array}
$$

in $\mathbf{E}_\mathcal{G}$, with m monomorphic. If **E** has finite limits, then m is monomorphic in **E**; thus $f = te$ and $g = mt$ for a unique $t : B \to C$. It is easily seen that $t : (B, \beta) \to (C, \gamma)$ in $\mathbf{E}_\mathcal{G}$; thus e is strongly epimorphic in $\mathbf{E}_\mathcal{G}$.

It follows easily from what we have already proved that $U_\mathcal{G}$ preserves epimorphism-equalizer factorizations if **E** has finite limits.

49.5. Theorem ([83]). *If* $\mathcal{G} = (G, \varepsilon, \psi)$ *is a left exact comonad on a category* **E** *with finite limits, then every slice category* $\mathbf{E}_\mathcal{G}/(A, \alpha)$ *of* $\mathbf{E}_\mathcal{G}$ *is isomorphic to the category of coalgebras for a left exact comonad* $\mathcal{G}^\alpha$ *on the slice category* $\mathbf{E}/A$ *of* **E**.

PROOF. We refer to Section 17 for slice categories.

For every object $u : X \to A$ of $\mathbf{E}/A$, we construct a pullback square

$$
\begin{array}{ccc}
\cdot & \xrightarrow{\nu_u} & GX \\
{\scriptstyle G^\alpha u}\downarrow & & \downarrow{\scriptstyle Gu} \\
A & \xrightarrow{\alpha} & GA
\end{array}
$$

in $\mathbf{E}$. This clearly defines a functor G^α on $\mathbf{E}/A$, and a natural transformation $\nu : D_A G^\alpha \to G D_A$ for the domain functor $D_A : \mathbf{E}/A \to \mathbf{E}$ of $\mathbf{E}/A$, and it is easily seen, using 17.2, that G^α preserves finite limits if G does.

For $u : X \to A$, we have a commutative diagram

$$
\begin{array}{ccccc}
\cdot & \xrightarrow{\nu_u} & GX & \xrightarrow{\varepsilon_X} & X \\
{\scriptstyle G^\alpha u}\downarrow & & \downarrow{\scriptstyle Gu} & & \downarrow{\scriptstyle u} \\
A & \xrightarrow{\alpha} & GA & \xrightarrow{\varepsilon_A} & A
\end{array}
$$

in $\mathbf{E}$, with $\varepsilon_A \alpha = \mathrm{id}_A$. Thus $\varepsilon_u^\alpha = \varepsilon_X \nu_u$ defines a morphism $\varepsilon_u^\alpha : G^\alpha u \to u$ in $\mathbf{E}/A$. One verifies easily that this defines a natural transformation $\varepsilon^\alpha : G^\alpha \to \mathrm{Id}$, with $D_A \varepsilon^\alpha = \varepsilon D_A \cdot \nu$.

As G preserves pullbacks, we have a diagram

$$
\begin{array}{ccccc}
\cdot & \xrightarrow{\nu_{G^\alpha u}} & \cdot & \xrightarrow{G\nu_u} & GGX \\
{\scriptstyle G^\alpha G^\alpha u}\downarrow & & \downarrow{\scriptstyle GG^\alpha u} & & \downarrow{\scriptstyle GGu} \\
A & \xrightarrow{\alpha} & GA & \xrightarrow{G\alpha} & GGA
\end{array}
$$

of pullback squares. Since

$$
GGu \cdot \psi_X \cdot \nu_u = \psi_A \cdot Gu \cdot \nu_u = \psi_A \cdot \alpha \cdot G^\alpha u = G\alpha \cdot \alpha \cdot G^\alpha u ,
$$

we have $G\nu_u \cdot \nu_{G^\alpha u} \cdot \psi_u^\alpha = \psi_X \cdot \nu_u$ in $\mathbf{E}$ for a unique $\psi_u^\alpha : G^\alpha u \to G^\alpha G^\alpha u$ in $\mathbf{E}_{\mathcal{G}}$. The morphisms ψ_u^α clearly define a natural transformation $\psi^\alpha : G^\alpha \to G^\alpha G^\alpha$, determined by

$$
G\nu \cdot \nu G^\alpha \cdot D_A \psi^\alpha = \psi D_A \cdot \nu .
$$

Using the fact that the morphisms ε_u^α and ψ_u^α are obtained from pullbacks, straightforward (and tedious) calculations show that we have obtained a comonad $\mathcal{G}^\alpha = (G^\alpha, \varepsilon^\alpha, \psi^\alpha)$ on $\mathbf{E}/A$.

We have $Gu \cdot \xi = \alpha u$ in $\mathbf{E}$, for a morphism $\xi : X \to GX$ of $\mathbf{E}$, iff $\xi = \nu_u \xi^\alpha$ in $\mathbf{E}$ for a unique morphism $\xi^\alpha : u \to G^\alpha u$ of $\mathbf{E}/A$. It is easily seen that then ξ is a coalgebra structure of X for $\mathcal{G}$ iff ξ^{α} is a coalgebra structure of u for $\mathcal{G}^{\alpha}$. This provides a bijection between objects $u : (X, \xi) \to (A, \alpha)$ of the slice category $\mathbf{E}_{\mathcal{G}}/(A, \alpha)$ and objects (u, ξ^{α}) of the category of coalgebras for $\mathcal{G}^{\alpha}$, and in fact an isomorphism between the two categories which preserves underlying objects and morphisms of $\mathbf{E}/A$. We omit the details.

50. Coalgebras Define a Quasitopos

50.1. We consider in this Section a left exact comonad $\mathcal{G} = (G, \varepsilon, \psi)$ on a category $\mathbf{E}$ with finite limits. We begin with a useful lemma.

Lemma. *If* $m : (A, \alpha) \to (B, \beta)$ *is a monomorphism of coalgebras, then*

$$
\begin{array}{ccc}
A & \xrightarrow{\;m\;} & B \\
\downarrow{\scriptstyle \alpha} & & \downarrow{\scriptstyle \beta} \\
GA & \xrightarrow{\;Gm\;} & GB
\end{array}
$$

is a pullback square in $\mathbf{E}$.

PROOF. If $\beta v = Gm \cdot u$, then

$$v = \varepsilon_B \cdot Gm \cdot u = m\varepsilon_A u,$$

and $v = mt$ only if $t = \varepsilon_A u$. We have $v = mt$ for this t, and

$$Gm \cdot \alpha \cdot t = \beta mt = \beta v = Gm \cdot u.$$

But then $u = \alpha t$ since Gm is monomorphic by 49.3.

50.2. Proposition. *If* $\mathbf{E}$ *is cartesian closed, then* $\mathbf{E}_{\mathcal{G}}$ *is cartesian closed.*

PROOF. For a coalgebra (A, α), we have a commutative square

$$
\begin{array}{ccc}
\mathbf{E}_{\mathcal{G}} & \xrightarrow{\;- \times (A, \alpha)\;} & \mathbf{E}_{\mathcal{G}} \\
\downarrow{\scriptstyle U_{\mathcal{G}}} & & \downarrow{\scriptstyle U_{\mathcal{G}}} \\
\mathbf{E} & \xrightarrow{\;- \times A\;} & \mathbf{E}
\end{array}
$$

of functors, since $U_{\mathcal{G}}$ creates finite products by 49.2. If $\mathbf{E}$ is cartesian closed, then the functor $- \times A$ has a right adjoint. The comonadic functor $U_{\mathcal{G}}$ with right adjoint $F_{\mathcal{G}}$ satisfies the condition of 8.10, and $\mathbf{E}_{\mathcal{G}}$ has equalizers by 49.3. Thus $- \times (A, \alpha)$ has a right adjoint by 8.10.

50.3. Corollary. *If* **E** *is locally cartesian closed, then* **E**$_\varsigma$ *is locally cartesian closed.*

PROOF. This follows immediately from 50.2 and 49.5.

50.4. Proposition. *If partial morphisms in* **E** *are represented, then partial morphisms* $(m, f) : A \to F_\varsigma B$ *in* **E**$_\varsigma$ *are represented, for every object* B *of* **E**.

PROOF. By 16.2 and 14.3, strong monomorphisms in **E** are equalizers, and thus preserved by U_ς and G. Now let $\vartheta_B : B \to \tilde{B}$ represent strong monomorphisms in **E** with codomain B. Since the monomorphism $G\vartheta_B$ is strong, we have a pullback square

$$
\begin{array}{ccc}
GB & \xrightarrow{\ \varepsilon_B\ } & B \\
\downarrow{\scriptstyle G\vartheta_B} & & \downarrow{\scriptstyle \vartheta_B} \\
G\tilde{B} & \xrightarrow{\ \rho_B\ } & \tilde{B}
\end{array}
$$

in **E**. For the adjoint homomorphism $G\rho_B \cdot \psi_{\tilde{B}}$ of ρ_B, we construct an equalizer fork

$$
\widetilde{F_\varsigma B} \ \xrightarrow{\ e_B\ } \ F_\varsigma \tilde{B} \ \underset{\text{id}}{\overset{G\rho_B \cdot \psi_{\tilde{B}}}{\rightrightarrows}} \ F_\varsigma \tilde{B}
$$

in **E**$_\varsigma$. We have

$$
G\rho_B \cdot \psi_{\tilde{B}} \cdot G\vartheta_B = G\rho_B \cdot GG\vartheta_B \cdot \psi_B = G\vartheta_B \cdot G\varepsilon_B \cdot \psi_B = G\vartheta_B
$$

in **E**. Thus $F_\varsigma \vartheta_B = e_B \cdot \zeta_B$ for a morphism $\zeta_B : F_\varsigma B \to \widetilde{F_\varsigma B}$ of **E**$_\varsigma$. We claim that ζ_B represents partial morphisms with codomain $F_\varsigma B$.

For $m : (X, \xi) \to (A, \alpha)$, with m an equalizer in **E**, we have a bijection between partial morphisms $(m, f) : A \to B$ in **E** and partial morphisms $(m, g) : (A, \alpha) \to F_\varsigma B$ in **E**$_\varsigma$, with $g = Gf \cdot \xi$ and $f = \varepsilon_B g$ for corresponding partial morphisms. For a pullback square

$$
\begin{array}{ccc}
X & \xrightarrow{\ f\ } & B \\
\downarrow{\scriptstyle m} & & \downarrow{\scriptstyle \vartheta_B} \\
A & \xrightarrow{\ \bar{f}\ } & \tilde{B}
\end{array}
$$

in $\mathbf{E}$, we have pullback squares

$$X \xrightarrow{\xi} GX \xrightarrow{Gf} GB \xrightarrow{\varepsilon_B} B$$

$$\downarrow m \qquad \downarrow Gm \qquad \downarrow G\vartheta_B \qquad \downarrow \vartheta_B$$

$$A \xrightarrow{\alpha} GA \xrightarrow{G\bar{f}} G\tilde{B} \xrightarrow{\rho_B} \tilde{B}$$

in $\mathbf{E}$, using 50.1 for the lefthand square. Thus $\rho_B \cdot G\bar{f} \cdot \alpha = \bar{f}$, and

$$G\rho_B \cdot \psi_{\tilde{B}} \cdot G\bar{f} \cdot \alpha = G\rho_B \cdot GG\bar{f} \cdot \psi_A \cdot \alpha = G\rho_B \cdot GG\bar{f} \cdot G\alpha \cdot \alpha = G\bar{f} \cdot \alpha$$

follows. Thus $G\bar{f} \cdot \alpha = e_B u$ in $\mathbf{E}_{\mathcal{G}}$ for a morphism $u : (A, \alpha) \to \widetilde{F_{\mathcal{G}}B}$. We now have a commutative diagram

$$(X, \xi) \xrightarrow{g} F_{\mathcal{G}}B \xrightarrow{\mathrm{id}} F_{\mathcal{G}}B$$

$$\downarrow m \qquad \downarrow \zeta_B \qquad \downarrow G\vartheta_B$$

$$(A, \alpha) \xrightarrow{u} \widetilde{F_{\mathcal{G}}B} \xrightarrow{e_B} F_{\mathcal{G}}\tilde{B}$$

in $\mathbf{E}_{\mathcal{G}}$, with the outer rectangle and the righthand square pullbacks, hence with the lefthand square a pullback.

It remains to show that this determines u uniquely if (m, g) is given. We have $\rho_B e_B u = \bar{f}$ for $f = \varepsilon_B g$. The morphism $(A, \alpha) \to F_{\mathcal{G}}\tilde{B}$ adjoint to $\bar{f}$ for $U_{\mathcal{G}} \dashv F_{\mathcal{G}}$ is then $\sigma_B e_B u$ with σ_B adjoint to ρ_B. But this morphism is $\sigma_B = G\rho_B \cdot \psi_{\tilde{B}}$, and thus $\sigma_B e_B = e_B$. Now $e_B u$ is adjoint to $\bar{f}$; this determines $e_B u$ and hence u uniquely.

50.5. Theorem.

For a left exact comonad $\mathcal{G}$ on a quasitopos $\mathbf{E}$, the category $\mathbf{E}_{\mathcal{G}}$ of coalgebras for $\mathcal{G}$ is a quasitopos, and $\mathbf{E}_{\mathcal{G}}$ is a topos if and only if $\mathbf{E}$ is a topos.

PROOF. By 49.2 and 49.3, $\mathbf{E}_{\mathcal{G}}$ has finite limits and finite colimits. By 49.4 and 12.5, the strong monomorphisms in $\mathbf{E}_{\mathcal{G}}$ are the equalizers. By 50.4, partial monomorphisms in $\mathbf{E}_{\mathcal{G}}$ with codomain $F_{\mathcal{G}}1$ are represented. This is the terminal object of $\mathbf{E}_{\mathcal{G}}$; thus $\mathbf{E}_{\mathcal{G}}$ has a subobject classifier. By 50.2.3, $\mathbf{E}_{\mathcal{G}}$ is locally cartesian, and thus a quasitopos by 19.3.

Since the structure of a coalgebra (B, β) provides a monomorphism $\beta : (B, \beta) \to F_{\mathcal{G}}B$ in $\mathbf{E}_{\mathcal{G}}$, we can use 19.2 to obtain a morphism representing partial morphisms of $\mathbf{E}_{\mathcal{G}}$ with codomain (B, β).

Since $U_{\mathcal{G}}$ preserves and reflects monomorphisms and equalizers, all monomorphisms are equalizers in $\mathbf{E}$ iff all monomorphisms are equalizers in $\mathbf{E}_{\mathcal{G}}$. Thus $\mathbf{E}$ is a topos iff $\mathbf{E}_{\mathcal{G}}$ is a topos.

50.6. Proposition. *For a left exact comonad $\mathcal{G}$ on a quasitopos* $\mathbf{E}$, *the quasitopos* $\mathbf{E}_{\mathcal{G}}$ *is solid if and only if* $\mathbf{E}$ *is solid.*

PROOF. Since $G1$ is a terminal object of $\mathbf{E}$, there is a (unique) isomorphism $\alpha : 1 \to G1$, and it is easily seen that $(1, \alpha)$ is a coalgebra for $\mathcal{G}$, and the terminal object of $\mathbf{E}_{\mathcal{G}}$. The initial object of $\mathbf{E}_{\mathcal{G}}$ is $(0, \beta)$ for the unique morphism $\beta : 0 \to G0$ of $\mathbf{E}$. By the last paragraph of the proof of 50.5, the unique monomorphism $(0, \beta) \to (1, \alpha)$ of $\mathbf{E}_{\mathcal{G}}$ is strong in $\mathbf{E}_{\mathcal{G}}$ iff $0 \to 1$ is a strong monomorphism of $\mathbf{E}$. Now the result follows immediately from 46.7.

51. Geometric Morphisms

51.1. Definitions. For quasitopoi $\mathbf{E}$ and $\mathbf{F}$, a *geometric morphism* $\mathbf{E} \to \mathbf{F}$ is defined as an adjunction $F \dashv U : \mathbf{E} \to \mathbf{F}$ with a left exact left adjoint, *i.e.* with F preserving finite limits. The right adjoint U is often called the *direct image functor*, and F the *inverse image functor*, of a geometric morphism $F \dashv U$. We say that $F \dashv U$ is *surjective* if the functor F reflects isomorphisms, *injective* if the functor U is full and faithful, and *essential* if F has a left adjoint.

If $F \dashv U : \mathbf{E} \to \mathbf{F}$ and $F' \dashv U' : \mathbf{F} \to \mathbf{G}$ are geometric morphisms, then $FF' \dashv U'U : \mathbf{E} \to \mathbf{G}$ clearly is a geometric morphism. Identity functors of quasitopoi clearly define geometric morphisms; thus we can consider a category of quasitopoi and geometric morphisms. Injective and surjective geometric morphisms clearly define subcategories of such a category.

51.2. Examples. Embeddings $\mathrm{Sh}_\gamma \mathbf{E} \to \mathbf{E}$, for a topology γ of a quasitopos $\mathbf{E}$ define injective geometric morphisms. We shall see that these are, up to equivalence, the only injective geometric morphisms.

For a left exact comonad $\mathcal{G}$ on a quasitopos $\mathbf{E}$, the forgetful functor $U_{\mathcal{G}} : \mathbf{E}_{\mathcal{G}} \to \mathbf{E}$, and its right adjoint $F_{\mathcal{G}}$, define a surjective geometric morphism from $\mathbf{E}_{\mathcal{G}}$ to $\mathbf{E}$. We shall see that every surjective geometric morphism is equivalent to a geometric morphism of this type.

A morphism $f : A \to B$ in a quasitopos $\mathbf{E}$ determines an essential geometric morphism $f^* \dashv f_* : \mathbf{E}/A \to \mathbf{E}/B$. This is a motivating example. Thus an arbitrary geometric morphism is often denoted $f^* \dashv f_*$, and we shall use this notation in Sections 52 and 53.

51.3. Proposition. *For a geometric morphism* $F \dashv U : \mathbf{E} \to \mathbf{F}$, *the following statements are logically equivalent.*

(i) $F \dashv U$ *is injective.*

(ii) *The counit of* $F \dashv U$ *is a natural isomorphism.*

(iii) *The functors* FU *and* $\operatorname{Id} \mathbf{E}$ *are naturally isomorphic.*

If $F \dashv U$ *is essential, with* $H \dashv F$, *then the following statements are logically equivalent to the preceding ones.*

(iv) *The functor* H *is full and faithful.*

(v) *The unit of* $H \dashv F$ *is a natural isomorphism.*

(vi) *The functors* FH *and* $\operatorname{Id} \mathbf{E}$ *are naturally isomorphic*

PROOF. (i) $\iff$ (ii) by 4.8, and (ii) $\implies$ (iii) is obvious.

If $\rho : FU \to \operatorname{Id} \mathbf{E}$ is a natural equivalence and $x : UA \to UB$, then

$$\rho_B \cdot Fx = u \cdot \rho_A = \rho_B \cdot FUu$$

for exactly one $u : A \to B$, and then $Fx = FUu$, and

$$\varepsilon_B \cdot Fx = \varepsilon_B \cdot FUu = u \cdot \varepsilon_B$$

for the counit ε of $F \dashv U$. Taking adjoint morphisms, we get $x = Uu$; thus U is full and faithful if (iii) is valid.

If $F \dashv U$ is essential, then (iii) $\iff$ (vi), since HF is left adjoint to FU. The remaining equivalences for $H \dashv F$ are dual to those already proved for $F \dashv U$.

51.4. Proposition. *If* $F \dashv U : \mathbf{E} \to \mathbf{F}$, *then* Fv *is isomorphic for every morphism* v *of* $\mathbf{F}$ *such that* UFv *is isomorphic. If* $F \dashv U$ *is a surjective geometric morphism between quasitopoi, then* F *is faithful. Conversely, if* F *is faithful and* $\mathbf{F}$ *a topos, then* F *reflects isomorphisms.*

PROOF. If UFv is isomorphic for $v : A \to B$, put $x = (UFv)^{-1}\eta_B :$ $B \to UFA$ for the unit η of $F \dashv U$, with $UFv \cdot x = \eta_B$ and $xv = \eta_A$. By adjunction, $Fv \cdot \hat{x} = \operatorname{id}_{FB}$ and $\hat{x} \cdot Fv = \operatorname{id}_{FA}$ for the adjoint $\hat{x} : FB \to FA$ of x.

If $A \xrightarrow{e} B \underset{v}{\overset{u}{\rightrightarrows}} C$ is an equalizer fork in $\mathbf{F}$, then Fe is an equalizer of Fu and Fv if F is left exact. If $Fu = Fv$, then Fe is an isomorphism. Thus e is an isomorphism, and $u = v$ follows, if $F \dashv U$ is surjective. Conversely, t in $\mathbf{F}$ is monomorphic and epimorphic if Ft is an isomorphism and F faithful. If $\mathbf{F}$ is a topos, it follows that t is an isomorphism.

51.5. Equivalence. We say that a geometric morphism $F \dashv U : \mathbf{E} \to \mathbf{F}$ is an *equivalence* if F and U are adjoint equivalences of categories. In this situation, we also have a geometric morphism $U \dashv F : \mathbf{F} \to \mathbf{E}$ which is inverse to $F \dashv U$, up to equivalence.

Every equivalence of quasitopoi is a geometric morphism, and we note the following result

Proposition. *A geometric morphism* $F \dashv U : \mathbf{E} \to \mathbf{F}$ *is an equivalence if and only if it is injective and surjective.*

PROOF. An equivalence is clearly injective and surjective. Conversely, if $F \dashv U$ is injective, with unit η and counit ε, then ε is a natural isomorphism, and so is $F\eta$ since $\varepsilon F \cdot F\eta = \mathrm{id}_F$. But then η is a natural isomorphism if $F \dashv U$ is surjective, and $F \dashv U$ is an equivalence.

51.6. Theorem. *For a morphism f of a quasitopos $\mathbf{E}$, the geometric morphism $f^* \dashv f_*$ is injective if and only if f is monomorphic, and surjective if and only if f is a strong epimorphism. The pullback functor f^* is faithful if and only if f is an epimorphism.*

PROOF. Let $f : A \to B$ in $\mathbf{E}$. The functor $f_>$ is always faithful; it is clearly full if f is monomorphic. Conversely, if $f_>$ is full and $fu = fv$, i.e. $\mathrm{id} : f_> u \to f_> v$ in $\mathbf{E}/B$, then $\mathrm{id} : u \to v$ in $\mathbf{E}/A$; thus f is monomorphic. By 51.3, this proves the first part.

If f factors $f = me$ with e strongly epimorphic and m monomorphic, then we have a pullback square

$$
\begin{array}{ccc}
A & \xrightarrow{\;e\;} & \cdot \\
\Big\downarrow{\scriptstyle \mathrm{id}_A} & & \Big\downarrow{\scriptstyle m} \\
A & \xrightarrow{\;f\;} & B
\end{array}
$$

in bfE; thus f^*m is an isomorphism for $m : m \to \mathrm{id}_A$ in $\mathbf{E}/B$. If f^* reflects isomorphisms, then m is an isomorphism, and f a strong epimorphism. We note for the converse that strong epimorphisms in $\mathbf{E}$ are coequalizers by 23.3, and preserved by pullbacks. Thus it is sufficient to prove that v must be isomorphic in a pullback square

$$
\begin{array}{ccc}
\cdot & \xrightarrow{\;e_1\;} & Y \\
\Big\downarrow{\scriptstyle \mathrm{id}} & & \Big\downarrow{\scriptstyle v} \\
\cdot & \xrightarrow{\;e\;} & X
\end{array}
$$

with e strongly epimorphic. If e is a coequalizer of a and b, then $ea = ve_1 b$, and thus $e_1 a = e_1 b$. But then $e_1 = xe$ for a morphism x of $\mathbf{E}$, and $vxe = e$

and $vx = \mathrm{id}_X$ follow. Since e_1 in the diagram also is epimorphic, we also have $xv = \mathrm{id}_Y$.

By 4.8, f^* is faithful iff every counit ε_v for the adjunction $f_> \dashv f^*$ is epimorphic. Since these counits are pullbacks of f and include f, this is the case iff f is epimorphic.

52. Topologies from Sheaves

52.1. Left exact reflectors. By 44.5, the reflector for a category of sheaves in a quasitopos $\mathbf{E}$ is left exact. We want to prove a converse.

We recall that a full reflective subcategory $\mathbf{B}$ of a category $\mathbf{E}$ is determined up to equivalence by the reflections $\eta_A : A \to TA$ for $\mathbf{B}$. These reflections define a functor $T : \mathbf{E} \to \mathbf{E}$, the reflector for $\mathbf{B}$, and a monad $\mathcal{T} = (T, \eta, \mu)$ on $\mathbf{E}$, for which ηT is a natural isomorphism and μ its inverse, and $T\eta = \eta T$. We note that the largest full subcategory of $\mathbf{E}$ for which T is a reflector has as its objects all objects A of $\mathbf{E}$ for which η_A is an isomorphism.

We generalize by replacing the left exact reflection with an arbitrary geometric morphism.

52.2. Theorem. *If $f^* \to f_* : \mathbf{E} \to \mathbf{F}$ is a geometric morphism of quasitopoi, with induced monad $\mathcal{T} = (T, \eta, \mu)$ on $\mathbf{F}$ with $T = f_* f^*$, then pullbacks*

$$
(1) \qquad
\begin{array}{ccc}
\cdot & \xrightarrow{\gamma m} & A \\
\downarrow & & \downarrow{\scriptstyle \eta_A} \\
TX & \xrightarrow{Tm} & TA
\end{array}
,
$$

one for every monomorphism $m : X \to A$ of $\mathbf{F}$, define a topology γ of $\mathbf{F}$.

PROOF. Since T is left exact, Tm and hence γm is a monomorphism for every monomorphism m, with $\gamma m' \simeq \gamma m$ if $m' \simeq m$. Since $Tm \cdot \eta_X = \eta_A \cdot m$ for $m : X \to A$, we have $m \leqslant \gamma m$.

Consider now two commutative squares

$$
\begin{array}{ccc}
\cdot & \xrightarrow{T\gamma m} & TA \\
\downarrow & & \downarrow{\scriptstyle T\eta_A} \\
TTX & \xrightarrow{TTm} & TTA
\end{array}
\qquad \text{and} \qquad
\begin{array}{ccc}
TX & \xrightarrow{Tm} & TA \\
\downarrow{\scriptstyle T\eta_X} & & \downarrow{\scriptstyle T\eta_A} \\
TTX & \xrightarrow{TTm} & TTA
\end{array}
$$

in $\mathbf{E}$. Since T is left exact, the lefthand square is a pullback square. Now if $TTm \cdot x = T\eta_A \cdot y$ in $\mathbf{E}$, then

$$y = \mu_A \cdot TTm \cdot x = Tm \cdot \mu_X \cdot x,$$

and $x = T\eta_X \cdot \mu_X \cdot x$. Thus the righthand square also is a pullback square. But then $T\gamma m \simeq Tm$, and $\gamma\gamma m \simeq \gamma m$ follows from (1).

If $m' \simeq f^{\leftarrow} m$ for $f : B \to A$, then since T is left exact, we have diagrams

$$
\begin{array}{ccccc}
\cdot & \longrightarrow & \cdot & \longrightarrow & TX \\
\downarrow{\scriptstyle \gamma m'} & & \downarrow{\scriptstyle Tm'} & & \downarrow{\scriptstyle Tm} \\
B & \xrightarrow{\ \eta_B\ } & TB & \xrightarrow{\ Tf\ } & TA
\end{array}
\quad \text{and} \quad
\begin{array}{ccccc}
\cdot & \longrightarrow & \cdot & \longrightarrow & TX \\
\downarrow{\scriptstyle f^{\leftarrow}\gamma m} & & \downarrow{\scriptstyle \gamma m} & & \downarrow{\scriptstyle Tm} \\
B & \xrightarrow{\ f\ } & A & \xrightarrow{\ \eta_A\ } & TA
\end{array}
$$

of pullback squares. As $\eta_B \cdot Tf = f \cdot \eta_A$, we have $\gamma m' \simeq f^{\leftarrow}(\gamma m)$.

Finally, the left exact functor T preserves equalizers; thus Tm and hence γm are strong monomorphisms if m is strong.

52.3. Proposition. *A monomorphism m of $\mathbf{F}$ is dense for the topology γ of 52.2 if and only if Tm is an isomorphism, and a morphism f is bidense for γ if Tf is an isomorphism.*

PROOF. If Tm is isomorphic for a monomorphism $m : X \to A$, then $\gamma m \simeq \mathrm{id}_A$ by the definition of γ. Conversely, if $\gamma m \simeq \mathrm{id}_A$ in 52.2.(1), then Tm is isomorphic since $Tm \simeq T\gamma m$ by the proof of 52.2.

Now factor $f = me$ with e a strong epimorphism and m monomorphic. Then Tf is isomorphic iff Tm and Te are isomorphisms, since Tm is monomorphic, and Tm is isomorphic iff m is dense. If (u,v) is a kernel pair of e, then (Tu,Tv) is a kernel pair of Te, and $Tu = Tv$ is an isomorphism if Te is an isomorphism. If $ud = \mathrm{id} = vd$, then Td is isomorphic if Te is an isomorphism. Thus d is dense, and e codense, if Tf is an isomorphism.

52.4. Our next result deals with a more general situation.

Proposition. *For a geometric morphism $f^* \dashv f_* : \mathbf{E} \to \mathbf{F}$ and a topology γ of $\mathbf{F}$, every object f_*A is a sheaf for γ, and f_* factors through the embedding $\mathrm{Sh}_\gamma \mathbf{F} \to \mathbf{F}$, if and only if f^*d is isomorphic in $\mathbf{E}$ for every γ-dense monomorphism d of $\mathbf{F}$.*

PROOF. Since the embedding $\mathrm{Sh}_\gamma \mathbf{F} \to \mathbf{F}$ is full and faithful, the functor f_* factors through the embedding iff every object f_*A is a sheaf for γ. For a span $Y \xleftarrow{d} X \xrightarrow{u} f_*A$ in $\mathbf{E}$, we have $u = vd$ for $v : Y \to f_*A$ in $\mathbf{E}$ iff $\hat{u} = \hat{v} \cdot f^*d$ in $\mathbf{E}$ for the adjoint morphisms. Thus every object f_*A is a sheaf for γ iff every mapping $\mathbf{E}(f^*d, A)$ with d dense for γ is bijective. This is the case iff f^*d is isomorphic in $\mathbf{E}$ for every γ-dense monomorphism d of $\mathbf{F}$.

52.5. Theorem. *If $T : \mathbf{E} \to \mathbf{E}$ is a left exact reflector on a quasitopos $\mathbf{E}$, with reflections $\eta_A : A \to TA$, then an object F of $\mathbf{E}$ is a sheaf for the topology γ of 52.2 if and only if the reflection η_F is an isomorphism.*

PROOF. We use a special case of 52.2, with $\mathbf{F}$ replaced by $\mathbf{E}$, and with μ and $T\eta = \eta T$ inverse natural isomorphisms. By 52.3 and 51.4, f^*d is isomorphic if d is dense for γ. Thus every object TA is a sheaf for γ, by 52.4, and F is a sheaf if η_F is an isomorphism. Conversely, η_F is bidense by 52.3 since $T\eta_F$ is isomorphic. If F is a sheaf, then η_F is a bidense morphism between sheaves, and thus isomorphic by 43.5.1.

52.6. We note two partial converses of 52.2.

Proposition. *For a topology γ of $\mathbf{E}$, with reflector T for sheaves, and for a monomorphism $m : X \to A$ of $\mathbf{E}$, with m strong, or A and X separated for γ, there is a pullback square*

$$
\begin{array}{ccc}
\cdot & \xrightarrow{\gamma m} & A \\
\downarrow{\scriptstyle u} & & \downarrow{\scriptstyle \eta_A} \\
TX & \xrightarrow{Tm} & TA
\end{array}
$$

(1)

in $\mathbf{E}$, with $\eta_A : A \to TA$ the reflection for $\mathrm{Sh}_\gamma \mathbf{E}$.

PROOF. If m is strong and $\mathrm{ch}(\gamma m) = \varphi : A \to \Omega$, then $j\varphi = \varphi$ for the morphism j of 45.5. Thus φ factors $\varphi = e_1 v$ for the equalizer $e_1 : \Omega_\gamma \to \Omega$ of j and id_Ω, in the notation of 42.4–42.6, and we have pullback squares

$$
\begin{array}{ccccc}
\cdot & \longrightarrow & 1 & \xrightarrow{\mathrm{id}_1} & 1 \\
\downarrow{\scriptstyle \gamma m} & & \downarrow{\scriptstyle \top_\gamma} & & \downarrow{\scriptstyle \top} \\
A & \xrightarrow{v} & \Omega_\gamma & \xrightarrow{e_1} & \Omega
\end{array}
$$

(2)

in $\mathbf{E}$. We can factor $v = v_1 \eta_A$ since Ω_γ is a sheaf, and we note that $Tm \simeq T(\gamma m)$ by 43.7. Since T preserves pullbacks and the restriction of T to sheaves is an equivalence, it follows that we have a diagram

$$
\begin{array}{ccccccc}
\cdot & \longrightarrow & TX & \longrightarrow & 1 & \xrightarrow{\mathrm{id}_1} & 1 \\
\downarrow{\scriptstyle \gamma m} & & \downarrow{\scriptstyle Tm} & & \downarrow{\scriptstyle \top_\gamma} & & \downarrow{\scriptstyle \top} \\
A & \xrightarrow{\eta_A} & TA & \xrightarrow{v_1} & \Omega_\gamma & \xrightarrow{e_1} & \Omega
\end{array}
$$

(3)

in $\mathbf{E}$, with the same outer rectangle as (1), and with pullback squares in the middle and at right. Thus (1), the lefthand square of (3), is a pullback square.

If A is separated, then so is X by 42.3.1, and η_A is a dense monomorphism by 43.5. We have $T(\gamma m) \simeq Tm$ for every monomorphism m; thus we have a commutative square (1) with u a reflection for $\mathrm{Sh}_\gamma \, \mathbf{E}$, hence a dense monomorphism. By 41.6, this square is a pullback square, since Tm is closed by 42.3.1.

52.7. Discussion. By 43.4, two topologies of $\mathbf{E}$ with the same sheaves have the same separated objects. If $\mathbf{E}$ is a topos, then topologies of $\mathbf{E}$ with the same sheaves are equivalent. By 44.6, this remains true for a quasitopos $\mathbf{E}$ if the topologies are strong, and in any case for the restrictions of the topologies to strong monomorphisms and to $\mathrm{Sep}_\gamma \, \mathbf{E}$, but we do not know whether two topologies of a quasitopos with the same sheaves always are equivalent. We have commutative squares 52.6.(1) for every topology γ of $\mathbf{E}$ with a given reflector T for sheaves; thus the topology of 52.2 is the coarsest topology with this reflector.

We note that the second part of 52.3 is a converse of the second part of 43.7. Thus if T is the reflector for sheaves and γ the topology of 52.2, then Tf is an isomorphism for a morphism f of $\mathbf{E}$ if and only if f is bidense for γ.

53. Factorization of Geometric Morphisms

53.1. Factoring through sheaves. For a geometric morphism $f^* \dashv f_* : \mathbf{E} \to \mathbf{F}$, and for a topology γ of $\mathbf{F}$, we have obtained in 52.4 a necessary and sufficient condition for f_* to factor through the embedding $\mathrm{Sh}_\gamma \, \mathbf{F} \to \mathbf{F}$. If g_* is the codomain restriction of f_* to $\mathrm{Sh}_\gamma \, \mathbf{F}$, and g^* the domain restriction of f^* to sheaves, then clearly $g^* \dashv g_*$, and g^* preserves finite limits; thus we have factored $f^* \dashv f_*$ in a category of quasitopoi and geometric morphisms. We note the following properties of this factorization.

Proposition. *If a geometric morphism $f^* \dashv f_* : \mathbf{E} \to \mathbf{F}$ factors through $\mathrm{Sh}_\gamma \, \mathbf{F}$ for a topology γ of $\mathbf{F}$, by $g^* \dashv g_*$, then $f^* \dashv f_*$ and $g^* \dashv g_*$ induce the same left exact comonad on $\mathbf{E}$, and $f^* \dashv f_*$ is injective if and only if $g^* \dashv g_*$ is injective. If γ is the topology of 52.2, then $g^* \dashv g_*$ is surjective.*

Corollary. *Every injective geometric morphism $f^* \dashv f_* : \mathbf{E} \to \mathbf{F}$ is equivalent to an embedding $\mathrm{Sh}_\gamma \, \mathbf{F} \to \mathbf{F}$.*

PROOF. The adjunctions $f^* \dashv f_*$ and $g^* \dashv g_*$ clearly have the same counit, and the same unit components $\eta_{f_* A} = \eta_{g_* A}$. Thus the induced comonads have the same functor $f^* f_* = g^* g_*$, the same counit, and the same comultiplication. Clearly f_* is full and faithful iff g_* is full and faithful.

If γ is the topology of 52.2, then a morphism v of $\mathrm{Sh}_\gamma F$ is bidense for γ if $g^* v = f^* v$ is isomorphic in $\mathbf{E}$. But then v is isomorphic by 43.5.1, and $g^* \dashv g_*$ is surjective.

We observe for the Corollary that we can factor $f^* \dashv f_*$ through $\mathrm{Sh}_\gamma \mathbf{E}$, with $g^* \dashv g_*$ injective and surjective if $f^* \dashv f_*$ is injective, by using the topology of 52.2.

53.2. Comparison functors.

Consider again a geometric morphism $f^* \dashv f_* : \mathbf{E} \to \mathbf{F}$, with induced left exact comonad $\mathcal{G} = (f^* f_*, \varepsilon, \psi)$ on $\mathbf{E}$. By the dual of 8.7, this induces a comparison functor $K : \mathbf{F} \to \mathbf{E}_\mathcal{G}$, with $KX = (f^* X, f^* \eta_X)$ for an object X, and with $U_\mathcal{G} K = f^*$.

Proposition. *The comparison functor K is the inverse image part of an injective geometric morphism $K \dashv R$. This geometric morphism is surjective, and hence an equivalence, if and only if $f^* \dashv f_*$ is surjective.*

Corollary. *Every surjective geometric morphism is equivalent to a morphism $U_\mathcal{G} \dashv F_\mathcal{G}$ for a left exact comonad on a quasitopos.*

PROOF. The comparison functor K preserves finite limits since $U_\mathcal{G} K$ preserves and $U_\mathcal{G}$ creates them. By the dual of 8.7 and its proof, K has a right adjoint R determined by equalizer forks

$$R(A, \alpha) \xrightarrow{\ \rho\ } f_* A \begin{array}{c} \xrightarrow{f_* \alpha} \\[-4pt] \xrightarrow[\eta_{f_* A}]{} \end{array} f_* f^* f_* A$$

in $\mathbf{F}$, and morphisms $\varepsilon_A \cdot f^* \rho$ define the counit of $K \dashv R$, for the unit η and the counit ε of $f^* \dashv f_*$. Since f^* preserves equalizers, and α is an equalizer of $f^* f_* \alpha$ and $f^* \eta_{f_* A} = \psi_A$, we have $f^* \rho \simeq \alpha$. Thus the counit of $K \dashv R$ is a natural isomorphism as claimed.

Since $U_\mathcal{G}$ reflects isomorphisms, K reflects isomorphisms iff $f^* = U_\mathcal{G} K$ does. The last part of the Proposition and the Corollary follow.

53.3. Proposition. *If a square*

$$\begin{array}{ccc} \mathbf{E} & \xrightarrow{\ f\ } & \mathbf{E}' \\ \downarrow{\scriptstyle a} & & \downarrow{\scriptstyle b} \\ \mathbf{F}' & \xrightarrow{\ g\ } & \mathbf{F} \end{array}$$

of quasitopoi and geometric morphisms commutes up to equivalence, with f surjective and g injective, then $a \simeq tf$ and $b \simeq gt$ for a geometric morphism $t : \mathbf{E}' \to \mathbf{F}'$, determined by the data up to equivalence.

PROOF. Replacing a by an equivalent geometric morphism if necessary, we can assume without loss of generality that g is the full embedding $\mathrm{Sh}_\gamma \mathbf{F} \to \mathbf{F}$ for a topology γ of $\mathbf{F}$. If d is monomorphic in $\mathbf{F}$ and dense for γ, then the reflection of d in $\mathrm{Sh}_\gamma \mathbf{F}$ is isomorphic, and thus $f^* b^* m$ is isomorphic in $\mathbf{E}$. It follows that $b^* d$ is isomorphic in $\mathbf{E}'$. Thus b factors $b = gt$, with t unique since g is a full embedding, by 52.4 and 53.1. Now $ga \simeq gtf$, and since g is a full embedding, $a \simeq tf$ follows.

53.4. Theorem. *Every geometric morphism between quasitopoi factors into a surjective geometric morphism followed by an injective one, and this factorization is unique up to equivalence.*

PROOF. The existence of the factorization follows immediately from 53.2 and 53.1, and its uniqueness from 53.3 and 51.5, and the discussion in Section 10.

53.5. Remark. If a geometric morphism is injective with a faithful inverse image part, then it is equivalent to an embedding $\mathrm{Sh}_\gamma \mathbf{E} \to \mathbf{E}$ for a topology γ of a quasitopos $\mathbf{E}$ by 53.1, with all reflections monomorphic in $\mathbf{E}$ by the dual of 4.8. This is the case, by 43.3 and 43.4, iff every object of $\mathbf{E}$ is separated for γ. This means, by 45.8, that every coarse object of $\mathbf{E}$ is a sheaf for γ, and every strong monomorphism closed.

If $\mathbf{E}$ is a topos, then the only topology γ for which all objects are separated is the discrete topology, but if $\mathbf{E}$ is not a topos, then all we can say is that γ must be finer that the topology of 41.2.(3).

Chapter 6

INTERNAL CATEGORIES AND DIAGRAMS

The first three sections of this chapter introduce internal categories, internal diagrams, and internal functors. These constructions can be carried out for every category $\mathbf{E}$ with finite limits. Internal natural transformations can also be defined, but we shall not discuss these. Every internal category $\mathcal{C}$ in a category $\mathbf{E}$ with finite limits induces monads $- \times_I \mathcal{C}$ and $\mathcal{C} \times_I -$ on $\mathbf{E}/I$ for $I = \mathrm{Ob}\,\mathcal{C}$. The algebras for these monads are internal functors $\mathcal{C} \to \mathbf{E}$ and $\mathcal{C}^{\mathrm{op}} \to \mathbf{E}$. These functors are called internal diagrams in $\mathbf{E}$, to distinguish them from internal functors between internal categories.

Section 57 deals with internal limits and colimits for internal diagrams and internal functors. We show that topoi and quasitopoi have internal limits and colimits for internal diagrams; thus topoi and quasitopoi are internally complete and cocomplete. We show in Section 58 that internal diagrams $\mathcal{C} \to \mathbf{E}$ and $\mathcal{C}^{\mathrm{op}} \to \mathbf{E}$ are the coalgebras for left exact comonads on $\mathbf{E}/I$ if $\mathbf{E}$ is locally cartesian closed. Thus internal diagram categories $[\mathcal{C}, \mathbf{E}]$ and $[\mathcal{C}^{\mathrm{op}}, \mathbf{E}]$ over a topos or quasitopos $\mathbf{E}$ are again topoi or quasitopoi.

We work in this chapter with a category $\mathbf{E}$ with finite limits. We recall that this implies that $\mathbf{E}$ has a canonical terminal object 1, and that canonical pullbacks are given for $\mathbf{E}$. Additional assumptions for $\mathbf{E}$ will be stated when needed.

54. Internal Categories

54.1. Internal graphs. An *internal graph* in a category $\mathbf{E}$ is given by two parallel morphisms $C \underset{t}{\overset{s}{\rightrightarrows}} I$ of $\mathbf{E}$. We call I the *vertex object* of the internal graph, C the *arrow object*, and s and t the *source morphism* and the *target morphism* of the internal graph.

A *morphism of internal graphs* $A \underset{\beta}{\overset{\alpha}{\rightrightarrows}} X$ and $C \underset{t}{\overset{s}{\rightrightarrows}} I$ in $\mathbf{E}$ is a pair of morphisms $f : A \to C$ and $u : X \to I$ of $\mathbf{E}$ such that $sf = u\alpha$ and $tf = u\beta$ in $\mathbf{E}$. Internal graphs and their morphisms clearly form a category.

54.2. Pullback functors. For an internal graph $C \underset{t}{\overset{s}{\rightrightarrows}} I$ in $\mathbf{E}$, we construct two pullback functors $C \times_I - = s_> t^*$ and $- \times_I C = t_> s^*$ on $\mathbf{E}/I$, determined by the internal graph. (See 17.5 for notations.)

179

For an object $u : X \to I$ of $\mathbf{E}/I$, we put $C \times_I u = s \cdot p_u : C \times_I X \to I$ and $u \times_I C = t \cdot q'_u : X \times_I C \to I$ for canonical pullback squares

$$
(1) \qquad
\begin{array}{ccc}
C \times_I X & \xrightarrow{q_u} & X \\
\Big\downarrow{\scriptstyle p_u} & & \Big\downarrow{\scriptstyle u} \\
C & \xrightarrow{\ t\ } & I
\end{array}
\quad \text{and} \quad
\begin{array}{ccc}
X \times_I C & \xrightarrow{q'_u} & C \\
\Big\downarrow{\scriptstyle p'_u} & & \Big\downarrow{\scriptstyle s} \\
X & \xrightarrow{\ u\ } & I
\end{array}
$$

in $\mathbf{E}$. For $f : u \to v$ in $\mathbf{E}/I$, with $v : Y \to I$ in $\mathbf{E}$, we obtain $C \times_I f : C \times_I u \to C \times_I v$ from a commutative diagram

$$
(2) \qquad
\begin{array}{ccc}
C \times_I X & \xrightarrow{q_u} & X \\
\Big\downarrow{\scriptstyle C \times_I f} & & \Big\downarrow{\scriptstyle f} \\
C \times_I Y & \xrightarrow{q_v} & Y \\
\Big\downarrow{\scriptstyle p_v} & & \Big\downarrow{\scriptstyle v} \\
C & \xrightarrow{\ t\ } & I
\end{array}
$$

in $\mathbf{E}$, with $C \times_I f = s_> t^* f$ for $t^* f : p_u \to p_v$ in $\mathbf{E}/C$. In this diagram, the lower square and the outer rectangle are canonical pullback squares; thus the upper square is a pullback square. We define $f \times_I C : u \times_I C \to v \times_I C$ by analogous pullback diagrams, with

$$
p'_v \cdot (f \times_I C) = f \cdot p'_u \qquad \text{and} \qquad q'_v \cdot (f \times_I C) = q'_u
$$

in $\mathbf{E}$. For given pullbacks (1), these data clearly define functors $C \times_I -$ and $- \times_I C$ on $\mathbf{E}/I$; we omit the trivial verifications.

54.3. Standard isomorphisms. We note first that the same canonical pullback square

$$
\begin{array}{ccc}
C \times_I C & \xrightarrow{q_s} & C \\
\Big\downarrow{\scriptstyle p_s} & & \Big\downarrow{\scriptstyle s} \\
C & \xrightarrow{\ t\ } & I
\end{array}
\quad \text{and} \quad
\begin{array}{ccc}
C \times_I C & \xrightarrow{q'_t} & C \\
\Big\downarrow{\scriptstyle p'_t} & & \Big\downarrow{\scriptstyle s} \\
C & \xrightarrow{\ t\ } & I
\end{array}
$$

in $\mathbf{E}$ defines $C \times_I s$ and $t \times_I C$. Thus $p'_t = p_s$ and $q'_t = q_s$, and there is just one object $C \times_I C$.

For the terminal object id_I of $\mathbf{I}/E$, we have isomorphisms

$$\lambda = p_{\mathrm{id}_I} : C \times_I I \to C \qquad \text{and} \qquad \rho = q'_{\mathrm{id}_I} : I \times_I C \to C,$$

with $q_{\mathrm{id}_I} = t\lambda$ and $p'_{\mathrm{id}_I} = s\rho$. It follows that

$$\lambda \cdot (C \times_I \tau_u) = p_u, \qquad \text{and} \qquad \rho \cdot (\tau_u \times_I C) = q'_u,$$

for the unique morphism $\tau_u = u : u \to \mathrm{id}_I$, for an object u of E/I.

We shall also use pullback squares

$$\begin{array}{ccc}
C \times_I (C \times_I C) & \xrightarrow{\ q_{C \times_I s}\ } & C \times_I C \\[2pt]
{\scriptstyle C \times_I p_s}\Big\downarrow & & \Big\downarrow{\scriptstyle p_s} \\[2pt]
C \times_I C & \xrightarrow{\ q_s\ } & C
\end{array}$$

and

$$\begin{array}{ccc}
(C \times_I C) \times_I C & \xrightarrow{\ q'_t \times_I C\ } & C \times_I C \\[2pt]
{\scriptstyle p'_{t \times_I C}}\Big\downarrow & & \Big\downarrow{\scriptstyle p'_t} \\[2pt]
C \times_I C & \xrightarrow{\ q'_t\ } & C
\end{array}$$

in E, with $p_s \cdot (C \times_I p_s) = p_{C \times_I s}$ and $q'_t \cdot (q'_t \times_I C) = q'_{t \times_I C}$. These are not canonical pullbacks; thus we get only

$$p'_{t \times_I C} \cdot \alpha = C \times_I p_s \qquad \text{and} \qquad (q'_t \times_I C) \cdot \alpha = q_{C \times_I s},$$

for an isomorphism $\alpha : C \times_I (C \times_I C) \to (C \times_I C) \times_I C$.

54.4. Discussion. If E is the category of sets, then we recall that an object $u : X \to I$ of a slice category E/I can be regarded as a set union of disjoint sets $X_i = u^{\leftarrow}(\{i\})$, one for each $i \in I$, or as a family $(X_i)_{i \in I}$ of sets, indexed by I. The pullback $C \times_I X$ then consists of all pairs (a, x) with $a \in C$ and $x \in X_{t(a)}$, and with $(C \times_I u)(a, x) = s(a)$. Similarly, $X \times_I C$ consists of all pairs (x, b) with $b \in C$ and $x \in X_{s(b)}$, and then $(u \times_I C)(x, b) = t(b)$. In particular, $C \times_I C$ is the set of all pairs (a, b) in $C \times C$ such that $t(a) = s(b)$. If C is a small category, these are the pairs for which the composition $b \cdot a$ is defined.

The definition of an internal category is now quite straightforward.

54.5. Internal categories. An *internal category* in E is an internal graph $C \underset{t}{\overset{s}{\rightrightarrows}} I$ with identity morphisms and a composition, and these data must satisfy the formal laws of 1.1.

We obtain identity morphisms by a morphism $\iota : I \to C$, called the *unit* of the internal category, which must satisfy the conditions

(1) $$s \cdot \iota = \mathrm{id}_I = t \cdot \iota,$$

for the source and target morphisms of the underlying internal graph.

Composition for an internal category is a morphism $\gamma : C \times_I C \to C$ of $\mathbf{E}$, which must satisfy the conditions

(2) $$s \cdot \gamma = s \cdot p_s \quad \text{and} \quad t \cdot \gamma = t \cdot q_s,$$

for the pullback square which defines $C \times_I C$.

The morphisms ι and γ must satisfy identity laws

(3) $$\gamma \cdot (C \times_I \iota) = \lambda, \qquad \gamma \cdot (\iota \times_I C) = \rho,$$

and the associative law

(4) $$\gamma \cdot (C \times_I \gamma) = \gamma \cdot (\gamma \times_I C) \cdot \alpha,$$

in terms of the functors $C \times_I -$ and $- \times_I C$ of 54.2, and for the standard isomorphisms of 54.3.

54.6. Duality. For every internal graph $C \overset{s}{\underset{t}{\rightrightarrows}} I$ in $\mathbf{E}$, we have a *dual internal graph* $C \overset{t}{\underset{s}{\rightrightarrows}} I$, obtained by interchanging domain and codomain morphisms. The induced functor $C^{\mathrm{op}} \times_I -$ of this dual graph is clearly naturally isomorphic to the induced functor $- \times_I C$ of the given graph, and the induced functor $- \times_I C^{\mathrm{op}}$ of the dual graph is naturally isomorphic to the functor $C \times_I -$. From both natural isomorphisms, we obtain the same isomorphism

$$u : C^{\mathrm{op}} \times_I C^{\mathrm{op}} \to C \times_I C$$

in $\mathbf{E}$, with $p_s \cdot u = q_t^{\mathrm{op}}$ and $q_s \cdot u = p_t^{\mathrm{op}}$ for the projections of the two pullbacks.

We obtain the *dual internal category* $\mathcal{C}^{\mathrm{op}}$ of an internal category $\mathcal{C}$ by changing the underlying graph of $\mathcal{C}$ to its dual graph. This does not change the morphism $\iota : I \to C$, but the composition $\gamma : C \times_I C \to C$ must be replaced by

$$\gamma^{\mathrm{op}} = \gamma \cdot u : C^{\mathrm{op}} \times_I C^{\mathrm{op}} \to C,$$

where u is the isomorphism of the preceding paragraph.

54.7. Special cases. An internal graph $I \overset{\mathrm{id}_I}{\underset{\mathrm{id}_I}{\rightrightarrows}} I$ admits exactly one internal category structure, with $\iota = \mathrm{id}_I$, and with $\gamma = \lambda = \rho$ for the isomorphisms of 54.3. We call this a *discrete internal category*.

An internal category $\mathcal{C}$ with vertex object a terminal object of $\mathbf{E}$ is called an *internal monoid* in $\mathbf{E}$.

We shall discuss the internalization of finite categories in 57.5.

We say that $\mathcal{C}$ is an *internal preorder* of $\mathrm{Ob}\,\mathcal{C}$ in $\mathbf{E}$ if the morphism $\langle s, t \rangle : C \to I \times I$ is a strong monomorphism of $\mathbf{E}$. This means that s and t define a relation $(s, t) : I \to I$ in $\mathbf{E}$.

A relation $(s, t) : I \to I$ always determines an internal graph $C \underset{t}{\overset{s}{\rightrightarrows}} I$. In the spirit of 15.2, we call a relation $(s, t) : I \to I$ *reflexive* or *transitive* if the induced external relation is reflexive or transitive . Thus (s, t) is reflexive iff $\mathrm{id}_I\,(s, t)\,\mathrm{id}_I$, *i.e.* iff there is a morphism $\iota : I \to C$ such that

$$s \cdot \iota = \mathrm{id}_I = t \cdot \iota,$$

and (s, t) is transitive iff $sp_s\,(s, t)\,tq_s$, *i.e.* iff

$$sp_s = s\gamma \qquad \text{and} \qquad tq_s = t\gamma$$

for a morphism $\gamma : C \times_I C \to C$ of $\mathbf{E}$. With $\langle s, t \rangle$ monomorphic, it is easily verified that then ι and γ define an internal category structure of the internal graph $C \underset{t}{\overset{s}{\rightrightarrows}} I$.

The dual category $\mathcal{C}^{\mathrm{op}}$ of an internal preorder $\mathcal{C}$ is again an internal preorder. Discrete internal categories and equivalence relations are internal preorders. An internal preorder $\mathcal{C}$ with underlying graph $C \underset{t}{\overset{s}{\rightrightarrows}} I$ is called an *internal order* if (s, t) is *antisymmetric*, i.e. if the intersection of $\langle s, t \rangle : C \to I \times I$ and $\langle t, s \rangle : C^{\mathrm{op}} \to I \times I$ is $\langle \mathrm{id}_I, \mathrm{id}_I \rangle : I \to I \times I$.

55. Internal Diagrams

55.1. Induced monads. We assume now that an internal category $\mathcal{C}$ in $\mathbf{E}$ is given, with underlying graph $C \underset{t}{\overset{s}{\rightrightarrows}} I$, and with identity morphisms and composition given by $\iota : I \to C$ and $\gamma : C \times_I C \to C$.

Proposition. *An internal category $\mathcal{C}$ in $\mathbf{E}$ induces a monad*

$$\mathcal{C} \times_I - = (C \times_I -, \eta, \mu)$$

on $\mathbf{E}/I$, with η and μ obtained from pullback squares

(1)
$$
\begin{array}{ccc}
X & \xrightarrow{\ \eta_u\ } & C \times_I X \\
\downarrow{\scriptstyle u} & & \downarrow{\scriptstyle p_u} \\
I & \xrightarrow{\ \iota\ } & C
\end{array}
\qquad \text{and} \qquad
\begin{array}{ccc}
C \times_I (C \times_I X) & \xrightarrow{\ \mu_u\ } & C \times_I X \\
\downarrow{\scriptstyle C \times_I p_u} & & \downarrow{\scriptstyle p_u} \\
C \times_I C & \xrightarrow{\ \gamma\ } & C
\end{array}
$$

in $\mathbf{E}$, with $q_u \eta_u = \mathrm{id}_X$ and $q_u \mu_u = q_u q_{C \times_I u}$, and with

$$\iota \times_I C = \eta_s \cdot \rho, \qquad \text{and} \qquad (\gamma \times_I C) \cdot \alpha = \mu_s,$$

for the isomorphisms ρ and α of 54.3.

There is also a monad $- \times_I \mathcal{C}$, obtained in the same way from the functor $- \times_I C$ as $\mathcal{C} \times_I -$ is obtained from $C \times_I -$.

PROOF. We note first that

$$(C \times_I u) \cdot \eta_u = s p_u \eta_u = s \iota u = u,$$

so that $\eta_u : u \to C \times_I u$. Similarly, $\mu_u : C \times_I (C \times_I u) \to C \times_I u$.

The morphism μ_u is determined by a commutative diagram

(2)
$$
\begin{array}{ccccc}
C \times_I (C \times_I X) & \xrightarrow{\mu_u} & C \times_I X & \xrightarrow{q_u} & X \\
\big\downarrow{\scriptstyle C \times_I p_u} & & \big\downarrow{\scriptstyle p_u} & & \big\downarrow{\scriptstyle u} \\
C \times_I C & \xrightarrow{\gamma} & C & \xrightarrow{t} & I
\end{array}
$$

in $\mathbf{E}$, with a pullback square at right, and with the same outer rectangle as the diagram

$$
\begin{array}{ccccc}
C \times_I (C \times_I X) & \xrightarrow{q_{C \times_I u}} & C \times_I X & \xrightarrow{q_u} & X \\
\big\downarrow{\scriptstyle C \times_I p_u} & & \big\downarrow{\scriptstyle p_u} & & \big\downarrow{\scriptstyle u} \\
C \times_I C & \xrightarrow{q_s} & C & \xrightarrow{t} & I
\end{array}
$$

of pullback squares. Thus the righthand square in (1) is a pullback square. Using two diagrams (2) for $f : u \to v$ in $\mathbf{E}/I$, and the definitions of $C \times_I f$ and $C \times_I (C \times_I f)$, we see that μ_u is natural in u. We see similarly, with simpler diagrams, that the lefthand square of (1) is a pullback square, and that η_u is natural in u.

From the definitions, we have a commutative diagram

$$
\begin{array}{ccccc}
C \times_I X & \xrightarrow{C \times_I \eta_u} & C \times_I (C \times_I X) & \xrightarrow{\mu_u} & C \times_I X \\
\big\downarrow{\scriptstyle C \times_I \tau_u} & & \big\downarrow{\scriptstyle C \times_I p_u} & & \big\downarrow{\scriptstyle p_u} \\
C \times_I I & \xrightarrow{C \times_I \iota} & C \times_I C & \xrightarrow{\gamma} & C
\end{array}
$$

in $\mathbf{E}$, with $\lambda \cdot (C \times_I \tau_u) = p_u$ and $\gamma \cdot (C \times_I \iota) = \lambda$. Since

$$q_u \cdot \mu_u \cdot (C \times_I \eta_u) = q_u \cdot q_{C \times_I u} \cdot (C \times_I \eta_u) = q_u \eta_u q_u = q_u,$$

we have

$$\mu_u \cdot (C \times_I \eta_u) = \mathrm{id}_u.$$

in $\mathbf{E}/I$. The other monadic laws for $\mathcal{C} \times_I -$ are obtained in the same way; we omit the details.

We observe for the last part that η_s and μ_s are obtained from essentially the same pullback diagrams as $\iota \times_I C$ and $\gamma \times_I C$.

55.2. Discussion. For sets, we have already remarked in 54.4 that an object $u : X \to I$ of SET$/I$ can be viewed as a family of sets $(X_i)_{i \in I}$, with $u(x) = i$ for $x \in X_i$, and that $X \times_I C$, for a graph $C \overset{s}{\underset{t}{\rightrightarrows}} I$, consists of all pairs (a, x) with $a \in C$, and $x \in X_j$ if $a : i \to j$ in C. Thus a morphism $\xi : C \times_I u \to u$ of SET$/I$ can be regarded as a collection of mappings $\xi_a : X_j \to X_i$, one for each $a : i \to j$ in C, with $\xi_a(x) = \xi(x, a)$ for $x \in X_j$. In the same way, a morphism $\bar{\xi} : u \times_I C \to u$ of SET$/I$ can be regarded as a collection of mappings $\bar{\xi}_a : X_i \to X_j$, one for each $a : i \to j$ in C.

For a small category $\mathcal{C}$ with underlying graph $C \overset{s}{\underset{t}{\rightrightarrows}} I$, a functor $A : \mathcal{C} \to$ SET thus can be represented as a single morphism $\xi : u \times_I C \to u$ in SET$/I$, and a contravariant functor $A : \mathcal{C}^{\mathrm{op}} \to$ SET as a morphism $\xi : C \times_I u \to u$ of SET$/I$. The conditions for A to be a functor then turn out to be exactly the conditions for the pair (u, ξ) to be an algebra for the monad $- \times_I \mathcal{C}$, or for the monad $\mathcal{C} \times_I -$ if A is contravariant. This motivates the following definitions.

55.3. Internal diagrams. For an internal category $\mathcal{C}$ in $\mathbf{E}$, we define an *internal diagram* $\mathcal{C} \to \mathbf{E}$ as an algebra for the monad $- \times_I \mathcal{C}$ on $\mathbf{E}$, and an internal diagram $\mathcal{C}^{\mathrm{op}} \to \mathbf{E}$ as an algebra for the monad $\mathcal{C} \times_I -$. In either case, a *morphism of internal diagrams* is a homomorphism of monadic algebras. An internal diagram $\mathcal{C}^{\mathrm{op}} \to \mathbf{E}$ is also called an *internal presheaf* for $\mathcal{C}$.

We denote by $[\mathcal{C}, \mathbf{E}]$ the category of internal diagrams $\mathcal{C} \to \mathbf{E}$, and by $[\mathcal{C}^{\mathrm{op}}, \mathbf{E}]$ the category of internal presheaves for $\mathcal{C}$. Duality for internal categories interchanges the two categories; thus they have analogous properties. In the following, we shall work with internal presheaves.

We denote by $U^{\mathcal{C}}$ the forgetful functor from $[\mathcal{C}, \mathbf{E}]$ or $[\mathcal{C}^{\mathrm{op}}, \mathbf{E}]$ to $\mathbf{E}/I$. This functor is faithful; thus it reflects monomorphisms and epimorphisms. By the general theory of monads, the functor $U^{\mathcal{C}}$ creates limits and has a left adjoint. Thus $U^{\mathcal{C}}$ also preserves and reflects limits.

56. Internal Functors

56.1. Definitions. For internal categories $\mathcal{C}$ and $\bar{\mathcal{C}}$ in a category $\mathbf{E}$, with vertex objects I and $\bar{I}$, and with morphism objects C and $\bar{C}$, we define an *internal functor* $f : \mathcal{C} \to \bar{\mathcal{C}}$ as a morphism $f : C \to \bar{C}$ of $\mathbf{E}$, which must satisfy the conditions

$$\bar{\iota} \cdot \bar{s} \cdot f = f \cdot \iota \cdot s, \qquad \bar{\iota} \cdot \bar{t} \cdot f = f \cdot \iota \cdot t, \qquad \bar{\gamma} \cdot (f * f) = f \cdot \gamma,$$

for the operations of $\mathcal{C}$ and $\bar{\mathcal{C}}$, with $f * f$ defined by

$$\bar{p}_{\bar{s}} \cdot (f * f) = f \cdot p_s, \qquad \bar{q}_{\bar{s}} \cdot (f * f) = f \cdot q_s,$$

for the projections of $C \times_I C$ and $\bar{C} \times_{\bar{I}} \bar{C}$.

Internal categories in $\mathbf{E}$ and their internal functors clearly form a category which we denote by $\mathrm{Cat}\,\mathbf{E}$, and morphism objects and the underlying morphisms of internal functors define a faithful functor which we denote by $\mathrm{Mor} : \mathrm{Cat}\,\mathbf{E} \to \mathbf{E}$.

We note that an internal functor $f : \mathcal{C} \to \bar{\mathcal{C}}$ determines a *dual internal functor* $f^{\mathrm{op}} : \mathcal{C}^{\mathrm{op}} \to \bar{\mathcal{C}}^{\mathrm{op}}$, with the same underlying morphism $f : C \to \bar{C}$.

56.2. Object morphisms. If $f : \mathcal{C} \to \bar{\mathcal{C}}$ is an internal functor, then

$$\bar{s} \cdot f \cdot \iota = f_0 = \bar{t} \cdot f \cdot \iota$$

defines a morphism $f_0 : I \to \bar{I}$ which satisfies the equations

$$\bar{s} \cdot f = f_0 \cdot s, \qquad \bar{\iota} \cdot f_0 = f \cdot \iota, \qquad \bar{t} \cdot f = f_0 \cdot t.$$

These equations are easily verified. We call f_0 the *object morphism* of the internal functor f.

Vertex objects and object morphisms clearly define a functor; we denote this functor by $\mathrm{Ob} : \mathrm{Cat}\,\mathbf{E} \to \mathbf{E}$. A morphism $f : \mathcal{C} \to \bar{\mathcal{C}}$ of categories and its object morphism define a morphism of the underlying graphs. Thus we also have a forgetful functor from internal categories to internal graphs.

56.3. Limits for internal categories. A diagram D in the category $\mathrm{Cat}\,\mathbf{E}$ induces diagrams $\mathrm{Mor}\,D$ of morphism objects and $\mathrm{Ob}\,D$ of vertex objects in $\mathbf{E}$. These induced diagrams create limits, by the following result.

Proposition. *If* $\mathrm{Mor}\,D$ *has a limit in* $\mathbf{E}$, *for a diagram* D *in* $\mathrm{Cat}\,\mathbf{E}$, *then* $\mathrm{Ob}\,D$ *has a limit in* $\mathbf{E}$, *and these limits create a limit of* D *in* $\mathrm{Cat}\,\mathbf{E}$.

PROOF. We denote by

$$sD : \mathrm{Mor}\,D \to \mathrm{Ob}\,D, \qquad tD : \mathrm{Mor}\,D \to \mathrm{Ob}\,D,$$
$$\iota D : \mathrm{Ob}\,D \to \mathrm{Mor}\,D, \qquad \gamma D : D * D \to \mathrm{Mor}\,D,$$

the diagram morphisms induced by the operations of the vertices of D, with $D * D$ obtained from the objects $C \times_I C$ and morphisms $f * f$ for the vertices and arrows of D. If $\operatorname{Mor} D$ has a limit $\nu : C \to \operatorname{Mor} D$, then

$$\iota D \cdot sD \cdot \nu = \nu \cdot \partial$$

for a unique morphism $\partial : C \to C$ of **E**. Let $\iota : I \to C$ be an equalizer of id_C and ∂. Since the unit ι is an equalizer of id_C and ιs for every internal category, we have

$$\nu \cdot \iota = \iota D \cdot \nu_0$$

for a unique cone $\nu_0 : I \to \operatorname{Ob} D$ in **E**, and it is easily verified that ν_0 is a limit cone in **E**.

For limit cones $\nu : C \to \operatorname{Mor} D$ and $\nu_0 : I \to \operatorname{Ob} D$, we have

$$\nu_0 \cdot s = sD \cdot \nu, \quad \nu_0 \cdot t = tD \cdot \nu, \quad \nu \cdot \iota = \iota D \cdot \nu_0,$$

for unique morphisms s, t, ι of **E**. From s and t, we can construct $C \times_I C$, and a cone $\nu * \nu : C \times_I C \to D * D$, in the obvious way, and then

$$\gamma D \cdot (\nu * \nu) = \nu \cdot \gamma$$

for a unique $\gamma : C \times_I C \to C$ in **E**. These data clearly define an internal category $\mathcal{C}$, and now ν becomes a cone $\nu : \mathcal{C} \to D$ in Cat **E**. It is easily seen that this is a limit cone in Cat **E**; we omit the details.

56.4. Composition of internal functors and internal diagrams. We can compose an internal functor $f : \bar{\mathcal{C}} \to \mathcal{C}$ with internal diagrams or presheaves on $\mathcal{C}$, by the following result.

Proposition. *For an internal functor* $f : \bar{\mathcal{C}} \to \mathcal{C}$, *with object morphism* $f_0 : \bar{I} \to I$, *there is a functor*

$$[f^{\operatorname{op}}, \mathbf{E}] : [\mathcal{C}^{\operatorname{op}}, \mathbf{E}] \to [\bar{\mathcal{C}}^{\operatorname{op}}, \mathbf{E}]$$

such that $\qquad U^{\bar{\mathcal{C}}} \circ [f^{\operatorname{op}}, \mathbf{E}] = f_0^* \circ U^{\mathcal{C}},$

for the pullback functor $f_0^* : \mathbf{E}/I \to \mathbf{E}/\bar{I}$ *and the forgetful functors* $U^{\mathcal{C}}$ *and* $U^{\bar{\mathcal{C}}}$.

PROOF. We put $f_0^* u = \bar{u} : \bar{X} \to \bar{I}$ for an object $u : X \to I$ of $\mathbf{E}/I$. We have pullback squares

$$
\begin{array}{ccccc}
\bar{C} \times_{\bar{I}} \bar{X} & \xrightarrow{\bar{q}\bar{u}} & \bar{X} & \xrightarrow{\varepsilon_u} & X \\
\downarrow{\bar{p}\bar{u}} & & \downarrow{\bar{u}} & & \downarrow{u} \\
\bar{C} & \xrightarrow{\bar{t}} & \bar{I} & \xrightarrow{f_0} & I
\end{array}
$$

in $\mathbf{E}$, with $\varepsilon_u : f_0 \bar{u} \to u$ a counit for the adjunction $(f_0)_> \dashv f_0^*$ (see 17.5). We also have pullback squares

$$
\begin{array}{ccccc}
\bar{C} \times_{\bar{I}} \bar{X} & \xrightarrow{\varphi_u} & C \times_I X & \xrightarrow{q_u} & X \\
\downarrow{\bar{p}_{\bar{u}}} & & \downarrow{p_u} & & \downarrow{u} \\
\bar{C} & \xrightarrow{f} & C & \xrightarrow{t} & I
\end{array}
$$

in $\mathbf{E}$, with the same outer rectangle. This defines morphisms

$$\varphi_u : (f_0)_> (\bar{C} \times_{\bar{I}} f_0^* u) \to C \times_I u$$

of $\mathbf{E}/I$, clearly natural in u. The adjoint morphisms

$$\psi_u : \bar{C} \times_{\bar{I}} f_0^* u \to f_0^* (C \times_I u)$$

of $\mathbf{E}/\bar{I}$ then define the adjoint natural transformation.

A bit of diagram chasing shows that

$$\psi \circ \bar{\eta} f_0^* = f_0^* \circ \eta,$$

and
$$\psi \circ \bar{\mu} f_0^* = f_0^* \circ \psi (C \times_I -) \circ (\bar{C} \times_{\bar{I}} -) \psi,$$

for the units η and $\bar{\eta}$, and the multiplications μ and $\bar{\mu}$, of the monads $C \times_I -$ and $\bar{C} \times_{\bar{I}} -$. It follows quickly that

$$[f^{op}, \mathbf{E}](u, \xi) = (f_0^* u, f_0^* \xi \cdot \psi_u)$$

defines the desired functor $[f^{op}, \mathbf{E}]$ for objects. By naturality of ψ_u, we have

$$f_0^* h : (f_0^* u, f_0^* \xi \cdot \psi_u) \to (f_0^* v, f_0^* \zeta \cdot \psi_v)$$

in $[\bar{C}^{op}, \mathbf{E}]$ if $h : (u, \xi) \to (v, \zeta)$ in $[C^{op}, \mathbf{E}]$. This defines the functor $[f^{op}, \mathbf{E}]$ for morphisms.

56.5. Associated internal functors. To an internal presehaf $(u, \xi) : C^{op} \to \mathbf{E}$ on a category $\mathbf{E}$, we assign an *associated internal functor* $A_C(u, \xi)$ in $\mathrm{Cat}\,\mathbf{E}/C$ by the following result. We only sketch the proof since we shall not use the result. Associated functors for internal diagrams $(u, \xi) : C \to \mathbf{E}$ are obtained by an analogous construction.

Proposition. *There is a full embedding* $A_C : [C^{op}, \mathbf{E}] \to \mathrm{Cat}\,\mathbf{E}/C$, *with* $\mathrm{Ob}\,A_C = U^{C \times_I -} : [C^{op}, \mathbf{E}] \to \mathbf{E}/I$, *and with* $A_{\bar{C}} \cdot [f^{op}, \mathbf{E}]$ *naturally isomorphic to* $f^* \cdot A_C$ *for an internal functor* $f : \bar{C} \to C$.

PROOF. We assign to an internal diagram (u, ξ) in $[\mathcal{C}^{\mathrm{op}}, \mathbf{E}]$ an internal category $\mathfrak{X}$ and an internal functor

$$A_{\mathcal{C}}(u, \xi) = p_u : \mathfrak{X} \to \mathcal{C}$$

as follows. The morphism object of $\mathfrak{X}$ is $C_{u,\xi} = C \times_I X$, the domain and codomain morphisms are ξ and q_u, and the unit is η_u. It follows easily that then $\mathrm{Ob}\, A_{\mathcal{C}}(u, \xi) = u$. The composition of $\mathfrak{X}$ is $\mu_u \sigma$ for the isomorphism

$$\sigma : C_{u,\xi} \times_X C_{u,\xi} \to C \times_I (C \times_I X),$$

with $\qquad \bar{p}_\xi = (C \times_I p_u) \cdot \sigma \qquad$ and $\qquad \bar{q}_\xi = q_{C \times_I u} \cdot \sigma$

for the projections of the two pullbacks. The only complication in the proof that this works is the presence of isomorphisms between canonical pullbacks.

For homomorphisms $h : (u, \xi) \to (v, \eta)$ of internal diagrams, it is easily seen that $C \times_I h : A_{\mathcal{C}}(u, \xi) \to A_{\mathcal{C}}(v, \zeta)$ in $\mathrm{Cat}\,\mathbf{E}/\mathcal{C}$, with object morphism h, and that the embedding thus defined is full.

The proof of 56.4 shows that the composition functor $[f^{\mathrm{op}}, \mathbf{E}]$, for an internal functor $f : \bar{\mathcal{C}} \to \mathcal{C}$, is equivalent to the pullback functor $f^* : \mathrm{Cat}\,\mathbf{E}/\mathcal{C} \to \mathrm{Cat}\,\mathbf{E}/\bar{\mathcal{C}}$, restricted to internal functors $A_{\mathcal{C}}(u, \xi)$ and their morphisms.

56.6. Remark. It is easily seen that an internal functor $f : \bar{\mathcal{C}} \to \mathcal{C}$, with underlying graph $T \overset{\alpha}{\underset{\beta}{\rightrightarrows}} X$ for $\bar{\mathcal{C}}$ and object morphism $u : X \to I$, is isomorphic in $\mathrm{Cat}\,\mathbf{E}/\mathcal{C}$ to an associated functor $A_{\mathcal{C}}(u, \xi)$ if and only if

$$
\begin{array}{ccc}
T & \overset{\beta}{\longrightarrow} & X \\
\downarrow{\scriptstyle f} & & \downarrow{\scriptstyle u} \\
C & \underset{t}{\longrightarrow} & I
\end{array}
$$

is a pullback square. If this is the case, then $\xi \simeq \alpha$, and $f : \bar{\mathcal{C}} \to \mathcal{C}$ is called a *discrete internal fibration*.

57. Internal Limits and Colimits

57.1. Constant diagrams. The discrete category 1 at the terminal object 1 of $\mathbf{E}$ is a terminal object in $\mathrm{Cat}\,\mathbf{E}$, with a unique internal functor $\tau_C : \mathcal{C} \to 1$ for an internal category $\mathcal{C}$, for $C = \mathrm{Mor}\,\mathcal{C}$ and $\tau_C : C \to 1$ in $\mathbf{E}$.

The monads $1 \times_1 -$ and $- \times_1 1$ on $\mathbf{E}/1$ are isomorphic to the identity monad, with diagram categories isomorphic to $\mathbf{E}/1$ and to $\mathbf{E}$. The functors

$[\tau_C, \mathbf{E}]$ and $[\tau_C{}^{\mathrm{op}}, \mathbf{E}]$ thus can be replaced by functors $\Delta_{\mathcal{C}}$ with domain $\mathbf{E}$. We say that these functors assign *constant diagrams* $\Delta_{\mathcal{C}}X$ to objects X of $\mathbf{E}$.

Replacing $\mathbf{E}/1$ by $\mathbf{E}$ also means replacing a pullback functor $(\tau_A)^*$ by a pullback functor $A^* : \mathbf{E} \to \mathbf{E}/A$, right adjoint to the domain functor D_A of $\mathbf{E}/A$. When working with internal presheaf categories $[\mathcal{C}^{\mathrm{op}}, \mathbf{E}]$, we shall use the functor A^* which assigns to an object X of $\mathbf{E}$ the projection $A \times X \to A$ of the product $A \times X$, and to a morphism f the morphism $\mathrm{id}_A \times f$. For an internal category C, with underlying graph $C \underset{t}{\overset{s}{\rightrightarrows}} I$, we shall put

$$I^*X = \pi_X : I \times X \to I \qquad \text{and} \qquad C^*X = \pi'_X : C \times X \to C.$$

With these notations, the equations

$$(1) \qquad p_{\pi_X} \cdot \sigma_X = \pi'_X, \qquad \text{and} \qquad q_{\pi_X} \cdot \sigma_X = t \times \mathrm{id}_X,$$

for the projections p_{π_X} and q_{π_X} of $C \times_I (I \times X)$, define a natural isomorphism $\sigma : C^* \to t^* I^*$, and we can describe constant diagrams as follows.

57.2. Proposition. *For an internal category* $\mathcal{C}$ *in* $\mathbf{E}$, *the constant diagram functor* $\Delta_{\mathcal{C}} : \mathbf{E} \to [\mathcal{C}^{\mathrm{op}}, \mathbf{E}]$ *is given by*

$$\Delta_{\mathcal{C}}X = (\pi_X, \psi_X), \qquad \text{with} \qquad \psi_X \cdot \sigma_X = s \times \mathrm{id}_X : C \times X \to I \times X,$$

and by

$$\Delta_{\mathcal{C}}f \cdot \sigma_X = \sigma_Y \cdot (\mathrm{id}_C \times f),$$

for objects X *and morphisms* $f : X \to Y$ *of* $\mathbf{E}$, *with* $\sigma_X : C \times X \to C \times_I (I \times X)$ *and* σ_Y *defined by* 4.1.(1).

PROOF. If $u : X \to 1$ in $\mathbf{E}$, then $\varphi_u \sigma_X = \tau_C \times \mathrm{id}_X : C \times X \to 1 \times X$ for the morphism φ_u in the proof of 56.4, and thus

$$\psi_u \sigma_X = s \times \langle u, \mathrm{id}_I \rangle : C \times X \to I \times (1 \times X)$$

in $\mathbf{E}$ for the adjoint morphism ψ_u in $\mathbf{E}/I$. Replacing u and $1 \times X$ by X, we replace ψ_u by ψ_X, with $\psi_X \cdot \sigma_X = s \times \mathrm{id}_X$, to get $\Delta_{\mathcal{C}}X$. The second part of the result now follows immediately from the second part of 56.4.

57.3. Internal colimits. We say that a pair $A \underset{g}{\overset{f}{\rightrightarrows}} B$ of morphisms in a category $\mathbf{E}$ is *reflexive* in $\mathbf{E}$ if there is a morphism $q : B \to A$ in $\mathbf{E}$ such that $fq = \mathrm{id}_B = gq$.

We recall that a colimit cone in a category $\mathbf{C}$ is defined as a universal pair for a constant diagram functor with domain $\mathbf{C}$. We define *internal colimits* in the same way, using the constant diagram functor $\Delta_{\mathcal{C}}$ for a diagram category

$[\mathcal{C}^{\mathrm{op}}, \mathbf{E}]$. We say that $\mathbf{E}$ is *internally cocomplete* if every possible internal colimit exists, *i.e.* if every constant diagram functor $\Delta_{\mathcal{C}}$ for $\mathbf{E}$ has a left adjoint.

We can also define a constant diagram functor $\Delta : \mathbf{E} \to \mathrm{Cat}\,\mathbf{E}$; this functor assigns to every object X of $\mathbf{E}$ the discrete internal category with morphism and vertex object X, and to every morphism $f : X \to Y$ of $\mathbf{E}$ the internal functor $f : \Delta X \to \Delta Y$.

57.4. Theorem. *If a category $\mathbf{E}$ with finite limits has coequalizers of reflexive pairs, then $\mathbf{E}$ is internally cocomplete, and the constant diagram functor $\Delta : \mathbf{E} \to \mathrm{Cat}\,\mathbf{E}$ has a left adjoint. For an internal category $\mathcal{C}$ in $\mathbf{E}$ with underlying graph $C \overset{s}{\underset{t}{\rightrightarrows}} C$, the internal colimit of an internal presheaf $(u, \xi) : \mathcal{C}^{\mathrm{op}} \to \mathbf{E}$ is a coequalizer*

$$(1) \qquad C \times_I X \overset{\xi}{\underset{q_u}{\rightrightarrows}} X \longrightarrow \mathrm{Colim}\,(u, \xi)$$

in $\mathbf{E}$, with X the domain of u.

PROOF. For an internal category $\mathcal{C}$ with underlying graph $C \overset{s}{\underset{t}{\rightrightarrows}} I$, the pair s, t is split by ι. Internal functors $\mathcal{C} \to \Delta Y$ are morphisms $f : C \to Y$ of $\mathbf{E}$ such that $fs = ft$; thus coequalizers of pairs s, t provide a left adjoint for the functor Δ.

For an object Y of $\mathbf{E}$, we have a bijection between morphisms $f : X \to Y$ in $\mathbf{E}$ and morphisms $\langle u, f \rangle : u \to \pi_X$ in $\mathbf{E}/I$. If we put $C \times_I \langle u, f \rangle = \sigma_Y g$ for the isomorphism σ_Y of 57.1.(1), then it is easily seen that

$$g = \langle p_u, f q_u \rangle : C \times_I X \to C \times Y .$$

We have $\langle u, f \rangle : (u, \xi) \to \Delta_{\mathcal{C}} Y$ in $[\mathcal{C}^{\mathrm{op}}, \mathbf{E}]$ iff

$$\langle u, f \rangle \cdot \xi = (s \times \mathrm{id}_Y) \cdot g ,$$

by 57.2 and the definitions. Since $sp_u = u\xi$, this is the case iff $f\xi = fq_u$, hence iff f factors through the coequalizer (1).

57.5. For internal limits, we shall need a right adjoint A_* of the pullback functor A^*. This functor exists by 18.5 if $\mathbf{E}$ is cartesian closed. Having replaced $\mathbf{E}/1$ by the isomorphic category $\mathbf{E}$, and the pullback functor $(\tau_A)^*$ by A^*, we replace B and $\mathrm{id}_{A \times B}$ in 18.5.(1) by A and the isomorphic projection $p_A : A \times 1 \to A$. With these changes, 18.5.(1) becomes a pullback square

$$(1) \qquad \begin{array}{ccc} A_* u & \longrightarrow & 1 \\ {\scriptstyle \kappa_A u}\downarrow & & \downarrow{\scriptstyle \widehat{p_A}} \\ X^A & \overset{u^A}{\longrightarrow} & A^A \end{array} ,$$

for $u : X \to A$ in $\mathbf{E}$, and the exponential adjoint of the isomorphic projection $p_A = A^*1 : A \times 1 \to A$ in $\mathbf{E}$. By 18.5, this defines the right adjoint A_* of A^* if $\mathbf{E}$ is cartesian closed.

57.6. Internal limits. We define an *internal limit* in $\mathbf{E}$ as a couniversal pair for a constant diagram functor $\Delta_\mathbb{C}$ with domain $\mathbf{C}$. Thus internal limits are dual to internal colimits. We say that $\mathbf{E}$ is *internally complete* if every possible internal colimit exists, *i.e.* if every constant diagram functor $\Delta_\mathbb{C}$ for $\mathbf{E}$ has a right adjoint.

Theorem. *A cartesian closed category $\mathbf{E}$ with pullbacks is internally complete.*

PROOF. We use the notations established earlier in this Section. By 57.2, a morphism $f : \Delta_\mathbb{C} Y \to (u, \xi)$ in $[\mathbb{C}^{\mathrm{op}}, \mathbf{E}]$ is a morphism $f : I^*Y \to u$ in $\mathbf{E}/I$ which satisfies

(1) $$f \cdot (s \times \mathrm{id}_I) = \xi \cdot g$$

for the morphism $g = (C \times_I f) \cdot \sigma_Y : C \times Y \to C \times_I x$, with σ_Y given by 57.1.(1). Thus we must have

(2) $$X^s \cdot \hat{f} = \xi^C \cdot \hat{g} : Y \to X^C$$

for the exponential adjoints of f and g. We note that $g : C^*Y \to p_u$ in $\mathbf{E}/C$; thus (2) is equivalent to

(3) $$X^s \cdot \kappa_I u \cdot \tilde{f} = \xi^C \cdot \kappa_C p_u \cdot \tilde{g},$$

for the adjoints $\tilde{f} : Y \to I_*u$ and $\tilde{g} : Y \to C_*p_u$ of f and g, by 57.5. Now let $\beta_u : I_*I^*u \to u$ be a counit for $I^* \dashv I_*$, and consider the diagram

$$
\begin{array}{ccccc}
C^*Y & \xrightarrow{\;C^*\tilde{f}\;} & C^*I_*u & & \\
\Big\downarrow{\sigma_Y} & & \Big\downarrow{\sigma_{I_*u}} & & \\
t^*I^*Y & \xrightarrow{\;t^*I^*\tilde{f}\;} & t^*I^*I_*u & \xrightarrow{\;t^*\beta_u\;} & t^*u = p_u
\end{array}
$$

in $\mathbf{E}/C$. The square commutes since σ is a natural isomorphism, and the composite of the lower arrows is $t^*f = C \times_I f$ for the adjoint f of $\tilde{f}$. Thus the composite morphism in this diagram is simply g. It follows that $\tilde{g} = h\tilde{f}$ in $\mathbf{E}$, with $h : I_*u \to C_*p_u$ adjoint to $t^*\beta_u \cdot \sigma_{I_*u}$ for $C^* \dashv C_*$. Now (3) becomes

(4) $$X^s \cdot \kappa_I u \cdot \tilde{f} = \xi^C \cdot \kappa_C p_u \cdot h \cdot \tilde{f}.$$

This shows that $f : \Delta_\mathbb{C} Y \to (u, \xi)$ if and only if $\tilde{f} : Y \to I_*u$ factors through the equalizer of $X^s \cdot \kappa_I u$ and $\xi^C \cdot \kappa_C p_u \cdot h$ in $\mathbf{E}$. But then this equalizer is the desired internal limit of the diagram (u, ξ).

58. Internal Diagrams over a Quasitopos

58.1. Ajoint monads and comonads. We say that a monad (T, η, μ) on a category $\mathbf{E}$ and a comonad (G, ε, ψ) on $\mathbf{E}$ are *adjoint* if G is right adjoint to T, and the natural transformations ε and ψ are adjoint (see 5.6) to η and μ.

Proposition. *If* $T \dashv G : \mathbf{E} \to \mathbf{E}$, *then adjoint natural transformations provide a bijection between monads* $\mathcal{T} = (T, \eta, \mu)$ *and comonads* $\mathcal{G} = (G, \varepsilon, \psi)$ *on* $\mathbf{E}$. *If* $\mathcal{T}$ *and* $\mathcal{G}$ *are adjoint, then* $U_{\mathcal{G}} K = U^{\mathcal{T}}$ *for an isomorphism* K *from* $\mathcal{T}$-*algebras to* $\mathcal{G}$-*coalgebras.*

PROOF. If $\eta \dashv \varepsilon$ and $\mu \dashv \psi$, then $f \cdot \eta_A = \varepsilon_B \cdot g$, and $f \cdot \mu_A$ and $\psi_B \cdot g$ are adjoint for $TT \dashv GG$, if $f : TA \to B$ and $g : A \to GB$ are adjoint for $T \dashv G$. If $u : TA \to GB$ is adjoint to $f\mu_A$ for $F \dashv G$, and u to $\psi_B g : A \to GGB$, then $f\mu_A \cdot \eta_{TA} = \varepsilon_B u$, and this is adjoint to $G\varepsilon_B \cdot \psi_B g$. Thus $\mu \circ \eta T \dashv G\varepsilon \circ \psi$, and $\mu \circ \eta T = \mathrm{id}_T$ iff $G\varepsilon \circ \psi = \mathrm{id}_G$. Using similar arguments for the other formal laws, we see that (T, η, μ) is a monad $\mathcal{T}$ iff (G, ε, ψ) is a comonad $\mathcal{G}$.

Now if $\alpha : TA \to A$ is adjoint to $\beta : A \to GA$, then $\alpha\eta_A = \mathrm{id}_A$ iff $\varepsilon_A \beta = \mathrm{id}_A$, and $\alpha\mu_A$ and $\alpha \cdot T\alpha$ are adjoint to $\psi_A \beta$ and $G\beta \cdot \beta$ for $TT \dashv GG$. It follows that (A, α) is a $\mathcal{T}$-algebra iff (A, β) is a $\mathcal{G}$-coalgebra. It is also easy to see that a morphism f is a homomorphism of $\mathcal{T}$-algebras iff f is a homomorphism of the adjoint $\mathcal{G}$-coalgebras.

58.2. Theorem. *If* $\mathbf{E}$ *is a locally cartesian closed category, then every category* $[\mathcal{C}, \mathbf{E}]$ *or* $[\mathcal{C}^{\mathrm{op}}, \mathbf{E}]$ *is isomorphic to the category of coalgebras for a left exact comonad on the category* $\mathbf{E}/\mathrm{Ob}\,\mathcal{C}$. *Thus if* $\mathbf{E}$ *is a topos or quasitopos, then every category* $[\mathcal{C}, \mathbf{E}]$ *or* $[\mathcal{C}^{\mathrm{op}}, \mathbf{E}]$ *is a topos or a quasitopos.*

PROOF. Let $C \overset{s}{\underset{t}{\rightrightarrows}} I$ be the underlying graph of $\mathcal{C}$. If $\mathbf{E}$ is locally cartesian closed, then the functor $C \times_I - = s_> t^*$ has a right adjoint $G = t_* s^*$ which preserves limits, hence is left exact. By 58.1, $[\mathcal{C}^{\mathrm{op}}, \mathbf{E}]$ is isomorphic to the category of coalgebras for a left exact comonad on $\mathbf{E}/I$, hence to a topos or quasitopos if $\mathbf{E}$ is a topos or quasitopos, by 50.5.

58.3. Left and right Kan extensions. Left and right adjoints of a functor $[f, \mathbf{E}]$ or $[f^{\mathrm{op}}, \mathbf{E}]$ are called *left* and *right Kan extensions* of an internal functor f or f^{op}.

Theorem. *Let* $\mathbf{E}$ *be a category with finite limits. If* $\mathbf{E}$ *has coequalizers of reflexive pairs, and pullbacks preserve these coequalizers, then every functor*

$[f^{\mathrm{op}}, \mathbf{E}]$, for an internal functor $f : \bar{\mathcal{C}} \to \mathcal{C}$, has a left adjoint. If $\mathbf{E}$ is locally cartesian closed, then every functor $[f^{\mathrm{op}}, \mathbf{E}]$ has a right adjoint.

PROOF. If $\mathrm{Ob}\, f = f_0$, then we have a commutative square

$$
\begin{array}{ccc}
[\bar{\mathcal{C}}^{\mathrm{op}}, \mathbf{E}] & \xrightarrow{\;[f^{\mathrm{op}}, \mathbf{E}]\;} & [\mathcal{C}^{\mathrm{op}}, \mathbf{E}] \\
\Big\downarrow{\scriptstyle U^{\bar{\mathcal{C}}}} & & \Big\downarrow{\scriptstyle U^{\mathcal{C}}} \\
\mathbf{E}/\bar{I} & \xrightarrow{\;f_0^*\;} & \mathbf{E}/I
\end{array}
$$

in which all functors except maybe $[f^{\mathrm{op}}, \mathbf{E}]$ have left adjoints, and $U^{\mathcal{C}}$ is monadic. If $\mathbf{E}$ has reflexive coequalizers, and the pullback functor $C \times_I -$ preserves them, then $[\mathcal{C}^{\mathrm{op}}, \mathbf{E}]$ has reflexive coequalizers by 8.8, and it follows from the dual of 8.10 that $[f^{\mathrm{op}}, \mathbf{E}]$ has a left adjoint.

If $\mathbf{E}$ is locally cartesian closed, then the functors in the diagram above have right adjoints, and $U^{\mathcal{C}}$ is comonadic. Thus $[f^{\mathrm{op}}, \mathbf{E}]$ has a right adjoint by 8.10.

58.4. Remarks. Internal limit and colimit functors are special cases of Kan extensions; thus 57.4 and 57.6 could have been obtained as special cases of 58.3. For a topos or quasitopos, right Kan extensions for internal functors $f : \bar{\mathcal{C}} \to \mathcal{C}$ are essential geometric morphisms. Functors $[f, \mathbf{E}]$ and $[f^{\mathrm{op}}, \mathbf{E}]$ are also geometric morphisms if their left Kan extensions preserve finite limits. For a topos $\mathbf{E}$, it is shown in [58] that this is the case if $\mathcal{C}$ and $\bar{\mathcal{C}}$ are filtered and f is cofinal. An extension of this result to quasitopoi would be desirable but has not yet been attempted.

58.5. Finite diagrams. If $\mathbf{E}$ is a non-trivial solid quasitopos, and in particular if $\mathbf{E}$ is a non-trivial topos, then we use the functor R of 46.9 to internalize finite categories and diagrams in $\mathbf{E}$.

For a finite category $\mathcal{C}$, we construct a coproduct I of copies of a terminal object 1 of $\mathbf{E}$, with an injection $\pi_i : 1 \to I$ for every object of $\mathcal{C}$, and a coproduct C of copies of 1 with an injection $\pi_a : 1 \to C$ for every arrow a of $\mathcal{C}$. We define $s : C \to I$ and $t : C \to I$ by $s\pi_a = \pi_i$, and $t\pi_a = \pi_j$, for $a : i \to j$ in $\mathcal{C}$. The pullback $C \times_I C$ is then a coproduct of copies of 1, one for each composable pair (a, b) in $\mathcal{C}$, with $p_s\pi_{a,b} = \pi_a$ and $q_s\pi_{a,b} = \pi_b$ for each injection $\pi_{a,b}$ of $C \times_I C$. The unit ι is given by $\iota\pi_i = \pi_{\mathrm{id}_i}$, for each object i of $\mathcal{C}$, and $\gamma\pi_{a,b} = \pi_{ba}$ defines the composition $\gamma : C \times_I C \to C$. It is easily verified that this defines an internal category in $\mathbf{E}$. We are using the letter π for all injections of coproducts, but the context should always make it clear which coproduct is meant.

For a finite external presheaf $E : \mathcal{C}^{\mathrm{op}} \to \mathbf{E}$, with an object E_i of $\mathbf{E}$ for every object i of $\mathcal{C}$, and a morphism $e_a : E_j \to E_i$ of $\mathbf{E}$ for $a : i \to j$ in $\mathcal{C}$, let X be a coproduct of the objects E_i in $\mathbf{E}$, with injections π_i. We denote by $k_i : E_i \to 1$ the unique morphism in $\mathbf{E}$, and we define $u : X \to I$ by $u\pi_i = \pi_i k_i$, for the injections of X and I. The pullback $C \times_I X$ then is a coproduct of objects E_j, with an injection $\pi_a : E_j \to C \times_I X$ for $a : i \to j$ in $\mathcal{C}$. Putting $p_u \pi_a = \pi_i$ and $q_u \pi_a = \pi_j$ for $a : i \to j$ in $\mathcal{C}$ then characterizes the morphisms p_u and q_u of 54.2. Finally, putting $\xi \pi_a = e_a \pi_j$ determines $\xi : C \times_I X \to X$ in $\mathbf{E}$, and it is easily verified that this defines an algebra (u, ξ) for the monad $\mathcal{C} \times_I -$, representing the given diagram.

This construction provides a representation functor, from the category of external presheaves on $\mathcal{C}$ to the category $[\mathcal{C}^{\mathrm{op}}, \mathbf{E}]$ of internal presheaves. The functor thus obtained is clearly faithful, but it need not be full, and there may well be internal presheaves for the finite category which are not isomorphic to a representation of an external presheaf.

Chapter 7

TOPOLOGICAL QUASITOPOI

The main theme of this Chapter is to obtain full and dense embeddings, in the sense of 7.8, of a concrete category $(\mathbf{A}, p)$ over a topos or quasitopos $\mathbf{E}$, into a topological or nearly topological quasitopos over $\mathbf{E}$.

In [9], P. ANTOINE constructed the largest finally dense extension of a concrete category $(\mathbf{A}, p)$ over sets, the topological category $\mathbf{A}^{cr}$ of p-sieves, and he showed that this category $\mathbf{A}^{cr}$ is always cartesian closed. This construction generalizes the quasi-topologies of E. SPANIER [93]. It was shown in [104] that p-sieves define a quasitopos, with a condition on $\mathbf{A}$ which was shown by J. PENON (private communication) to be unnecessary. In [83], PENON extended the construction of p-sieves to a concrete category $(\mathbf{A}, p)$ over a quasitopos. He showed that the category of p-sieves then remains a quasitopos, and he discussed quasitopoi of epimorphic p-sieves over a topos. E. DUBUC [27] reproduced results of [104] and [83] for concrete quasitopoi over sets, from a different perspective and with some interesting applications. New results for quasitopos completions over a quasitopos were obtained in [107], but a general and coherent theory of dense quasitopos completions over a quasitopos is presented in this Chapter for the first time.

We define p-sieves in Section 59 and show in Section 60 that the category $\mathbf{A}^{cr}$ of p-sieves, for a concrete category $(\mathbf{A}, p)$ over a quasitopos $\mathbf{E}$, is a topological quasitopos over $\mathbf{E}$. As categories of p-sieves are often huge, it is important to replace a category $\mathbf{A}^{cr}$ by a full subcategory of managable size, with the desirable properties of p-sieves. This is done by restricting ourselves to categories of dense p-sieves.

In Sections 61 and 62, we show that under very mild conditions for $\mathbf{A}$ and a reasonable condition for $\mathbf{E}$, dense p-sieves form a monocoreflective full subcategory $\mathbf{A}^{cd}$ of $\mathbf{A}^{cr}$ over $\mathbf{E}$, closed under finite limits in $\mathbf{A}^{cr}$. Thus $\mathbf{A}^{cd}$ is a quasitopos if $\mathbf{A}$ and $\mathbf{E}$ satisfy these conditions, and the largest dense completion of $\mathbf{A}$ over $\mathbf{E}$.

If $\mathbf{E}$ is the category of sets and $\mathbf{A}$ a concrete category with constant morphisms, then dense completions of $\mathbf{A}$ are topological categories with constant morphisms. In general, dense completions have only initial lifts for finite sources and final lifts for quotient sinks; we call such categories quasitopological. We study quotient sinks and quasitopological categories in Sections 63 and 64.

197

Categories with constant morphisms make sense only over sets; we propose concrete categories with a left exact discrete object functor as a possible replacement. These categories need not be topological; we study them in Section 65. Left exact categories over sets have constant morphisms.

After some preliminaries, dense completions are studied in Section 68, and dense quasitopos completions characterized in Section 69, subject to the conditions of Sections 61 and 62. Examples are given in Section 70. We show that these extensions are, up to equivalence, the categories of sheaves for certain topologies of $\mathbf{A}^{\mathrm{cd}}$. These topologies correspond to standard Grothendieck topologies of $\mathbf{A}$, and the largest standard Grothendieck topology of $\mathbf{A}$ provides the smallest quasitopos completion, or quasitopos hull, of $\mathbf{A}$ over $\mathbf{E}$.

59. Categories of p-Sieves

59.1. Concrete functors. We have defined a concrete category over a category $\mathbf{E}$ in 11.1. For concrete categories $(\mathbf{A}, p)$ and $(\mathbf{B}, q)$ over the same category $\mathbf{E}$, we define a *concrete functor* $F : (\mathbf{A}, p) \to (\mathbf{B}, q)$ as a functor $F : \mathbf{A} \to \mathbf{B}$ such that $qF = p$. We note that a concrete functor is always faithful.

Proposition. *A full concrete functor is a full embedding.*

PROOF. Let F be the functor as above. If $FA = FA'$ for objects A, A' of $\mathbf{A}$, then there are morphisms $u : A \to A'$ and $v : A' \to A$ in $\mathbf{A}$ with $Fu = \mathrm{id}_{FA} = Fv$. Since F is faithful, u and v are inverse isomorphisms. Since $qF = p$, we have $pA = pA'$, and u and v have the underlying morphism id_{pA} in $\mathbf{E}$. But then $A = A'$ since we assume p to be amnestic.

59.2. p-sieves. For a concrete category $(\mathbf{A}, p)$ over a quasitopos $\mathbf{E}$, we define a *p-sieve* as a pair (E, Φ), consisting of an object E of $\mathbf{E}$ and a collection Φ of pairs (A, u), with A an object of $\mathbf{A}$ and $u : pA \to E$ in $\mathbf{E}$, such that if $(A, u) \in \Phi$ and $f : A' \to A$ in $\mathbf{A}$, then $(A', uf) \in \Phi$. We shall also say in this situation that Φ is a p-sieve at E.

If the category $\mathbf{E}$ is locally small and fibres $p^{\leftarrow} E$ of p are small, then a p-sieve at an object E of $\mathbf{E}$ can be regarded as a subfunctor of a set-valued contravariant functor $\mathbf{E}(p-, E) : \mathbf{A}^{\mathrm{op}} \to \mathrm{SET}$, with ΦA for an object A the set of all pairs (A, u) in Φ with given first entry A.

If (E, Φ) is a p-sieve, then Φ may also be viewed as a category, with the pairs (A, u) in Φ as objects, and with morphisms $f : (A', u') \to (A, u)$ given by morphisms $f : A' \to A$ in $\mathbf{A}$ with $u' = uf$ in $\mathbf{E}$. From this viewpoint, putting $D_\Phi(A, u) = A$ for a pair (A, u) in Φ, and $D_\Phi f = f : A' \to A$ for

$f : (A', u') \to (A, u)$, defines a *domain functor* $D_\Phi : \Phi \to \mathbf{A}$, and the second entries of pairs (A, u) in Φ define a cone $pD_\Phi \to E$ in $\mathbf{E}$. This cone is called the *induced cone* of the p-sieve (E, Φ).

59.3. The category of p-sieves. For a concrete category $(\mathbf{A}, p)$ over $\mathbf{E}$, we define a concrete *category* $(\mathbf{A}^{cr}, p^{cr})$ *of p-sieves* as follows. Objects of $\mathbf{A}^{cr}$ are all p-sieves (E, Φ). A morphism $f : (E, \Phi) \to (F, \Psi)$ of p-sieves is a morphism $f : E \to F$ of $\mathbf{E}$ such that (A, fu) is in Ψ for every pair (A, u) in Φ. The forgetful functor p^{cr} assigns to an object (E, Φ) the object E of $\mathbf{E}$, and to $f : (E, \Phi) \to (F, \Psi)$ the morphisms $f : E \to F$ of $\mathbf{E}$.

Proposition. *For a concrete category $(\mathbf{A}, p)$ over $\mathbf{E}$, p-sieves define a topological category over $\mathbf{E}$.*

PROOF. We construct initial lifts for sources and final lifts for sinks, omitting the easy verifications that these constructions work. For objects (E_i, Φ_i) and morphisms $f_i : E \to E_i$ of $\mathbf{E}$, the initial lift (E, Φ) has Φ consisting of all pairs (A, u) with A an object of $\mathbf{A}$, $u : pA \to E$ in $\mathbf{E}$, and $(A, f_i u) \in \Phi_i$ for every i. For objects (E_i, Φ_i) and morphisms $f_i : E_i \to E$ of $\mathbf{E}$, the final lift (E, Φ) has Φ consisting of all pairs of the form $(A, f_i u)$ for some i, with $(A, u) \in \Phi_i$ for this i.

59.4. Image and preimage sieves. For $f : E \to F$ in $\mathbf{E}$ and a p-sieve Φ at E, we denote by $f_* \Phi$ the p-sieve at F consisting of all pairs (A, fu) with (A, u) in Φ. The resulting morphism $f : (E, \Phi) \to (F, f_* \Phi)$ of $\mathbf{A}^{cr}$ then is a fine morphism of $\mathbf{A}^{cr}$, as defined in 11.8, and a strong epimorphism of $\mathbf{A}^{cr}$ if f is a strong epimorphism of $\mathbf{E}$.

For $f : E \to F$ in $\mathbf{E}$ and a p-sieve Ψ at F, we denote by $f^* \Psi$ the p-sieve at E consisting of all pairs (A, u) with A an object of $\mathbf{A}$, $u : pA \to E$ in $\mathbf{E}$, and (A, fu) in Ψ. The resulting morphism $f : (E, f^* \Psi) \to (F, \Psi)$ of $\mathbf{A}^{cr}$ then is a coarse morphism of $\mathbf{A}^{cr}$, as defined in 11.8, and a strong monomorphism of $\mathbf{A}^{cr}$ if f is a strong monomorphism of $\mathbf{E}$.

59.5. The Antoine functor. For a concrete category $(\mathbf{A}, p)$ over $\mathbf{E}$, we define a concrete functor $\Upsilon : (\mathbf{A}, p) \to (\mathbf{A}^{cr}, p^{cr})$ as follows. For an object A of $\mathbf{A}$, we put $\Upsilon A = (pA, \Upsilon A)$, with ΥA consisting of all pairs (X, f) with X an object of $\mathbf{A}$ and $f : pX \to pA$ in $\mathbf{E}$ obtained from a morphism $f : X \to A$ in $\mathbf{A}$. For $f : A \to B$ in $\mathbf{A}$, we put $\Upsilon f = f : \Upsilon A \to \Upsilon B$ in $\mathbf{A}^{cr}$. This clearly defines a functor Υ as claimed which we call the *Antoine functor* of the concrete category. Regarded as a category, ΥA is isomorphic to $\mathbf{A}/A$.

The following result states that $\mathbf{A}^{cr}$ is generated by $\mathbf{A}$, or more exactly by its objects ΥA.

Theorem (ANTOINE [9]). *For a concrete category $(\mathbf{A}, p)$ over $\mathbf{E}$ and a p-sieve Φ at an object E of $\mathbf{E}$, a pair (A, u), with A an object of $\mathbf{A}$ and $u : pA \to E$ in $\mathbf{E}$, is in Φ if and only if $u : YA \to (E, \Phi)$ in $\mathbf{A}^{\mathrm{cr}}$.*

Corollaries. *The functor $Y : \mathbf{A} \to \mathbf{A}^{\mathrm{cr}}$ is a full embedding, and every p-sieve (E, Φ) is the final lift for its induced cone $pD_\Phi \to E$.*

PROOF. If $(A, u) \in \Phi$, then $(X, uf) \in \Phi$ for every pair (X, f) in ΥA; thus $u : YA \to (E, \Phi)$. Conversely, if $u : YA \to (E, \Phi)$, then $(A, u) \in \Phi$ since $(A, \mathrm{id}_{pA}) \in \Upsilon A$.

We have in particular $f : YA \to YB$ iff $(A, f) \in \Upsilon B$, hence iff $f : A \to B$ in $\mathbf{A}$. Thus Y is full and faithful, and an embedding by 59.1. The second Corollary follows immediately from the Theorem and the definitions.

59.6. Proposition. *A limit cone $\lambda : A \to D$ in $\mathbf{A}$ is preserved by the Antoine functor Y if and only if λ is preserved by the functor p.*

PROOF. If Y preserves the limit cone, then so does $p = p^{\mathrm{cr}} Y$, since p^{cr} preserves all limits. Conversely, if $p\lambda : pA \to pD$ is a limit cone in $\mathbf{E}$, then the cone $p\lambda : (pA, \Phi) \to YD$ is a limit cone in $\mathbf{A}^{\mathrm{cr}}$ if (A, Φ) is the initial lift for the vertices of YD and the arrows of $p\lambda$. Pairs (B, u) in Φ then are given by morphisms $u : pB \to pA$ in $\mathbf{E}$ with each arrow of $p\lambda \cdot u$ lifting to a morphism of $\mathbf{A}$, i.e. with $p\lambda \cdot u = p\mu$ for a cone $\mu : B \to D$ in $\mathbf{A}$, since Y is full and faithful. In this situation, $\mu = \lambda f$ for a unique $f : B \to A$ in $\mathbf{A}$, with $u = pf$ since $p\lambda$ is a limit cone. Thus $\Phi = \Upsilon A$, and Y preserves the limit cone λ.

60. p-Sieves Define a Quasitopos

60.1. Theorem. *For a concrete category $(\mathbf{A}, p)$ over a quasitopos $\mathbf{E}$, the category $\mathbf{A}^{\mathrm{cr}}$ of p-sieves is a quasitopos, and the forgetful functor p^{cr} preserves the quasitopos structure.*

PROOF. Since the functor p^{cr} is topological, $\mathbf{A}^{\mathrm{cr}}$ has finite limits and co-limits by 11.7, and strong monomorphisms in $\mathbf{A}^{\mathrm{cr}}$ are obtained from 11.8. Thus the functor p^{cr} preserves and lifts finite limits and colimits, strong monomorphisms, and related properties. We describe the quasitopos structure of $\mathbf{A}^{\mathrm{cr}}$ further in 60.2 and 60.3.

60.2. For objects (E, Φ) and (F, Ψ) of $\mathbf{A}^{\mathrm{cr}}$, we construct the object $(F, \Psi)^{(E, \Phi)}$ as a pair $(F^E, [\Phi, \Psi])$, with F^E obtained in $\mathbf{E}$, and with

$$(A, \hat{u}) \in [\Phi, \Psi] \quad \Longleftrightarrow \quad u : YA \times (E, \Phi) \to (F, \Psi)$$

for $\hat{u} : pA \to F^E$ in $\mathbf{E}$ exponentially adjoint to $u : pA \times E \to F$.

For an object (G, Γ) of $\mathbf{A}^{cr}$, we must show that if $\varphi : G \times E \to F$ and $\hat{\varphi} : G \to F^E$ are exponentially adjoint for $\mathbf{E}$, then

(1) $\qquad \varphi : (G, \Gamma) \times (E, \Phi) \to (F, \Psi) \iff \hat{\varphi} : (G, \Gamma) \to (F^E, [\Phi, \Psi])$

in $\mathbf{A}^{cr}$. By the definitions and 59.4, the righthand side of (1) is valid iff

$$\varphi \cdot (u \times \mathrm{id}_E) : \mathsf{Y}A \times (E, \Phi) \to (F, \Psi)$$

for every pair (A, u) in Γ. This is clearly true if the lefthand side of (1) is valid.

Conversely, $(G, \Gamma) \times (E, \Phi)$ is $G \times E$ with the p-sieve $\Gamma \times \Phi$ of pairs $(A, \langle u, v \rangle)$ with $(A, u) \in \Gamma$ and $(A, v) \in \Phi$. If the righthand side of (1) is valid, then

$$\varphi \cdot \langle u, v \rangle = \varphi \cdot (u \times \mathrm{id}_E) \cdot \langle \mathrm{id}_{pA}, v \rangle$$

is in ΨA for $(A, \langle u, v \rangle)$ in $\Gamma \times \Phi$, and thus $(A, \langle \mathrm{id}_{pA}, v \rangle)$ is in $\mathsf{Y}A \times \Phi$.

60.3. If $\vartheta_E : E \to \tilde{E}$ represents partial monomorphisms with codomain E in $\mathbf{E}$ and Φ is a p-sieve at E, then we define a p-sieve $\tilde{\Phi}$ at $\tilde{E}$ from pullbacks

$$
\begin{array}{ccc}
X & \xrightarrow{\;u\;} & E \\
\big\downarrow{\scriptstyle m} & & \big\downarrow{\scriptstyle \vartheta_E} \\
pA & \xrightarrow{\;\bar{u}\;} & \tilde{E}
\end{array}
$$

in $\mathbf{E}$ by putting $(A, \bar{u}) \in \tilde{\Phi}$ iff $u : (X, m^* \mathsf{Y}A) \to (E, \Phi)$ in $\mathbf{A}^{cr}$. It is easily verified that this defines a p-sieve $\tilde{\Phi}$ at $\tilde{E}$. If $\bar{u} = \vartheta_E u$ for $u : pA \to E$, and $m = \mathrm{id}_{pA}$, then $(A, \vartheta_E u) \in \tilde{\Phi}$ iff $(A, u) \in \Phi$; thus we have an embedding $\vartheta_E : (E, \Phi) \to (\tilde{E}, \tilde{\Phi})$ in $\mathbf{A}^{cr}$. We claim that this embedding represents partial morphisms with codomain (E, Φ).

A partial morphism in $\mathbf{A}^{cr}$ with codomain (E, Φ) is a span

(1) $\qquad (F, \Psi) \xleftarrow{\;m\;} (X, m^* \Psi) \xrightarrow{\;f\;} (E, \Phi)$

in $\mathbf{A}^{cr}$ with m a strong monomorphism of $\mathbf{E}$. In order to represent partial morphisms, we consider pullbacks

(2)
$$
\begin{array}{ccc}
(X_1, m_1^* \mathsf{Y}A) & \xrightarrow{\;u_1\;} & (X, m^* \Psi) & \xrightarrow{\;f\;} & (E, \Phi) \\
\big\downarrow{\scriptstyle m_1} & & \big\downarrow{\scriptstyle m} & & \big\downarrow{\scriptstyle \vartheta_E} \\
\mathsf{Y}A & \xrightarrow{\;u\;} & (F, \Psi) & \xrightarrow{\;\bar{f}\;} & (\tilde{E}, \tilde{\Phi})
\end{array}
$$

in $\mathbf{A}^{cr}$. In this diagram, $\bar{f} : F \to \bar{E}$ is determined uniquely from (1) by the requirement that the underlying morphisms of the righthand square form a pullback in $\mathbf{E}$. This square is then a pullback in $\mathbf{A}^{cr}$ if $\bar{f}$ is a morphism of $\mathbf{A}^{cr}$ as shown.

To see this, let $(A, u) \in \Psi$, and construct the lefthand square of (2) as a pullback square, with an underlying pullback square in $\mathbf{E}$. Since all arrows of (2), except possibly $\bar{f}$, represent morphisms of $\mathbf{A}^{cr}$, and the outer rectangle of (2) is a pullback square, we have $(A, \bar{f}u) \in \tilde{\Phi}$ by the definition of $\tilde{\Phi}$. Thus $\bar{f}$ is a morphism of $\mathbf{A}^{cr}$, and we are done.

60.4. Solid sieves. If $(\mathbf{A}, p)$ is a concrete category over a solid quasitopos $\mathbf{E}$, then the objects of the quasitopos Sol $\mathbf{A}^{cr}$ are called *solid p-sieves*.

By 46.4, solid sieves are the sheaves for a topology of $\mathbf{A}^{cr}$.

Proposition. *If $\mathbf{E}$ is a solid quasitopos, then the quasitopos of solid p-sieves admits all possible initial and final lifts over $\mathbf{E}$, and every p-sieve $\mathsf{Y}A$ is solid if (and only if) p reflects initial objects.*

PROOF. If $(1, \Omega)$ is a terminal object of $\mathbf{A}^{cr}$, then an initial object 0^* of Sol $\mathbf{A}^{cr}$ is an object $(0, m^*\Omega)$, for an initial object 0 and the strong monomorphism $m : 0 \to 1$ of $\mathbf{E}$. For this m, the sieve $m^*\Omega$ consists of all pairs (X, u) with X an object of $\mathbf{A}$ for which pX is an initial object of $\mathbf{E}$ and $u : pX \to 0$ in $\mathbf{E}$. A solid p-sieve (E, Φ) admits a morphism $f : (0, m^*\Omega) \to (E, \Phi)$; this means that $(X, u) \in \Phi$ for every object X of $\mathbf{A}$ with pX an initial object of $\mathbf{E}$, and the unique morphisms $u : pX \to E$ of $\mathbf{E}$.

Initial lifts for solid p-sieves, and final lifts for a non-empty sink of solid p-sieves, clearly are solid. For an empty sink with target E, the final solid p-sieve consists of all pairs (X, u) with X an object of $\mathbf{A}$ for which pX is initial in $\mathbf{E}$, and $u : pX \to E$ in $\mathbf{E}$. Thus Sol $\mathbf{A}^{cr}$ admits all initial and final lifts over $\mathbf{E}$.

A p-sieve $\mathsf{Y}A$ is solid iff the unique $u : pX \to pA$ lifts to $u : X \to A$ for every object X of $\mathbf{A}$ with pX initial in $\mathbf{E}$. This is the case for every object A of $\mathbf{E}$ iff p reflects initial objects.

60.5. Coarse objects. As for every topological category, the fibre of an object E of $\mathbf{E}$ in $\mathbf{A}^{cr}$ is a complete lattice, with p-sieves at E ordered by set inclusion since $\mathrm{id}_E : (E, \Phi) \to (E, \Psi)$ iff $\Phi \subset \Psi$. The finest p-sieve at E is the empty sieve, and the coarsest p-sieve at E is the *full p-sieve*, consisting of all pairs (A, u) with A an object of $\mathbf{E}$ and $u : pA \to E$ in $\mathbf{E}$.

Proposition. *The coarse objects of $\mathbf{A}^{cr}$ are the pairs (E, Ω_E) with E a coarse object of $\mathbf{E}$ and Ω_E the full p-sieve at E.*

PROOF. For an object (E, Φ) of $\mathbf{A}^{\mathrm{cr}}$, every monomorphism $m : (E, \Phi) \to$ (F, Ψ) in $\mathbf{A}^{\mathrm{cr}}$ is strong iff every monomorphism $m : E \to F$ in $\mathbf{E}$ is strong, i.e. E is a coarse object of $\mathbf{E}$, and always $\Phi = m^* \Psi$. For $m = \mathrm{id}_E$ and Ψ full, it follows that Φ must be full. Conversely, if Φ is full, then $\Phi \subset m^* \Psi$, which is necessary for a morphism m of $\mathbf{A}^{\mathrm{cr}}$, implies that $\Phi = m^* \Psi$; thus (E, Φ) is a coarse object of $\mathbf{A}^{\mathrm{cr}}$ if E is a coarse object of $\mathbf{E}$ and Φ the full p-sieve at E.

61. Dense p-Sieves

61.1. Conditions for A. We say that a p-sieve (E, Φ) is a *dense p-sieve* if its induced cone is a colimit cone in $\mathbf{E}$, and we denote by $(\mathbf{A}^{\mathrm{cd}}, p^{\mathrm{cd}})$ the full subcategory of $(\mathbf{A}^{\mathrm{cr}}, p^{\mathrm{cr}})$ with dense p-sieves as its objects.

For an object A of $\mathbf{A}$, the p-sieve ΥA clearly is dense. In order to prove that $\mathbf{A}^{\mathrm{cd}}$ is closed under finite limits in $\mathbf{A}^{\mathrm{cr}}$, we shall impose the following three conditions on $(\mathbf{A}, p)$.

61.1.1. For a terminal object 1 of $\mathbf{E}$ the full p-sieve Ω_1 at 1 is dense.

61.1.2. For objects A and B of $\mathbf{A}$, the p-sieve $\Upsilon A \times \Upsilon B$ at $pA \times pB$, consisting of pairs $(X, \langle u, v \rangle)$ for spans $A \xleftarrow{u} X \xrightarrow{v} B$ in $\mathbf{A}$, is dense.

61.1.3. For an object A of $\mathbf{A}$ and a strong monomorphism $m : E \to pA$ in $\mathbf{E}$, the p-sieve $m^* \Upsilon A$ at E is dense.

Conditions 61.1.1 and 61.1.2 are satisfied, by 59.6, if $\mathbf{A}$ has, and the functor p preserves, finite products. The two conditions state that finite products of objects ΥA in $\mathbf{A}^{\mathrm{cr}}$ are dense. This is clearly necessary for $\mathbf{A}^{\mathrm{cd}}$ to be closed under finite limits in $\mathbf{A}^{\mathrm{cr}}$. If m in 61.1.3 admits a cokernel pair $\Upsilon A \overset{a}{\underset{b}{\rightrightarrows}} (F, \Psi)$ in $\mathbf{A}^{\mathrm{cr}}$, then Ψ is dense by 61.2; thus 61.1.3 also is necessary.

61.2. Proposition. *For a diagram D in $\mathbf{A}^{\mathrm{cd}}$ and a colimit cone λ : $p^{\mathrm{cd}} D \to E$ in $\mathbf{E}$, the final lift for λ in $\mathbf{A}^{\mathrm{cr}}$ is a dense p-sieve.*

PROOF. Let (E, Φ) be this final lift, with Φ consisting of pairs $(A, \lambda_i u)$ with (A, u) in Φ_i, for a vertex $Di = (E_i, \Phi_i)$ of D and the corresponding arrow $\lambda_i : E_i \to E$ of λ. If $\mu : D_\Phi \to F$ is a cone, then the morphisms $\mu_{\lambda_i u}$, for (A, u) in Φ_i, form a cone with domain D_{Φ_i}. Thus there is a unique morphism $\nu_i : E_i \to F$ in $\mathbf{E}$ with $\mu_{\lambda_i u} = \nu_i u$ for every (A, u) in Φ_i. The morphisms ν_i then clearly form a cone $\nu : p^{\mathrm{cd}} D \to F$, with $\nu = g\lambda$ for a unique $g : E \to F$ in $\mathbf{E}$, and if also $\mu = g'\lambda$, then it follows easily that $g' = g$. Thus (E, Φ) is dense as claimed.

61.3. Proposition. *If* $m : E_1 \to E$ *is a strong monomorphism of* **E** *and* (E, Φ) *a dense p-sieve, then the p-sieve* $m^* \Phi$ *at* E_1 *is dense.*

PROOF. By 23.4, pullbacks

$$
\begin{array}{ccc}
(X, m_1^* \Upsilon A) & \xrightarrow{\;m_1\;} & \Upsilon A \\[2mm]
\downarrow{\scriptstyle u_1} & & \downarrow{\scriptstyle u} \\[2mm]
(E_1, m^* \Phi) & \xrightarrow{\;m\;} & (E, \Phi)
\end{array}
$$

in $\mathbf{A}^{\mathrm{cr}}$ produce a colimit cone of morphisms u_1, with domains $(X, m_1^* \Upsilon A)$ which are dense by 61.1.3. Thus the codomain $(E_1, m^* \Phi)$ of this cone is dense by 61.2.

61.4. Proposition. *For a pullback square*

$$
\begin{array}{ccc}
(P, \Pi) & \xrightarrow{\;f_1\;} & (F, \Psi) \\[2mm]
\downarrow{\scriptstyle g_1} & & \downarrow{\scriptstyle g} \\[2mm]
(E, \Phi) & \xrightarrow{\;f\;} & (G, \Gamma)
\end{array}
$$

in $\mathbf{A}^{\mathrm{cr}}$, *with* (E, Φ) *and* (F, Ψ) *dense, the p-sieve* (P, Π) *is dense.*

PROOF. For pairs (A, u) in Φ and (B, v) in Ψ, we construct diagrams

$$
\begin{array}{ccccc}
\cdot & \xrightarrow{\;u_2\;} & \cdot & \xrightarrow{\;f_2\;} & \Upsilon B \\[2mm]
\downarrow{\scriptstyle v_2} & & \downarrow{\scriptstyle v_1} & & \downarrow{\scriptstyle v} \\[2mm]
\cdot & \xrightarrow{\;u_1\;} & (P, \Pi) & \xrightarrow{\;f_1\;} & (F, \Psi) \\[2mm]
\downarrow{\scriptstyle g_2} & & \downarrow{\scriptstyle g_1} & & \downarrow{\scriptstyle g} \\[2mm]
\Upsilon A & \xrightarrow{\;u\;} & (E, \Phi) & \xrightarrow{\;f\;} & (G, \Gamma)
\end{array}
$$

of pullbacks in $\mathbf{A}^{\mathrm{cr}}$. The top lefthand vertices of these diagrams are domains of strong monomorphisms in $\mathbf{A}^{\mathrm{cr}}$ with a codomain $\Upsilon A \times \Upsilon B$. Thus these vertices are dense, by 61.1.2 and 61.3. For fixed (A, u), the morphisms $v : \Upsilon B \to (F, \Psi)$ form a colimit cone in $\mathbf{A}^{\mathrm{cr}}$, and the morphisms v_2 form a pullback of this colimit cone, hence a colimit cone. Since the domains of morphisms v_2 are dense, it follows by 61.2 that their codomain is also dense. Now all morphisms $u : \Upsilon A \to (E, \Phi)$ form a colimit cone, and so do their pullbacks u_1. Since the domains of the morphisms u_1 are dense, so is their codomain (P, Π), again by 61.2.

61.5. Proposition. *If (E, Φ) is a dense p-sieve, and Ψ a p-sieve at E which contains Φ, then (E, Ψ) is dense.*

PROOF. Restricting a cone $\lambda : pD_\Psi \to F$ to pairs (A, u) in Φ defines a cone $pD_\Phi \to F$, with a unique $f : E \to F$ such that $\lambda_u = fu$ for every pair (A, u) in Φ. For a pair (A, u) in Ψ, we have a pullback square

$$
\begin{array}{ccc}
(pA, \Phi_u) & \xrightarrow{\mathrm{id}_{pA}} & \Upsilon A \\
\downarrow{\scriptstyle u} & & \downarrow{\scriptstyle u} \\
(E, \Phi) & \xrightarrow{\mathrm{id}_E} & (E, \Psi)
\end{array}
$$

in $\mathbf{A}^{\mathrm{cr}}$, with (pA, Φ_u) a dense p-sieve by 61.4 and $\Phi_u \subset \Upsilon A$. If $(B, v) \in \Phi_u$, then $(B, uv) \in \Phi$ and $v : B \to A$ in $\mathbf{A}$. Thus $\lambda_u \cdot v = \lambda_{uv} = fuv$ for (B, v) in Φ_u. Since Φ_u is dense, $\lambda_u = fu$ follows, and Ψ is dense.

61.6. Proposition. *If $e : (E, \Phi) \to (F, \Psi)$ in $\mathbf{A}^{\mathrm{cr}}$ with e a strong epimorphism of $\mathbf{E}$ and Φ dense at E, then Ψ is dense at F.*

PROOF. By 61.5, it suffices to consider $\Psi = e_* \Phi$. The elements of Ψ are then the pairs (A, eu) with (A, u) in Φ. If $\lambda : pD_\Psi \to G$ is a cone, then putting $\lambda_{eu} = \mu_u$ defines a cone $\mu : pD_\Phi \to G$, with a unique $f : E \to G$ such that $\lambda_{eu} = \mu_u = fu$ for every pair (A, eu) in Ψ. Now construct a pullback

$$
\begin{array}{ccc}
(P, \Pi) & \xrightarrow{b} & (E, \Phi) \\
\downarrow{\scriptstyle a} & & \downarrow{\scriptstyle e} \\
(E, \Phi) & \xrightarrow{e} & (F, \Psi)
\end{array}
$$

in $\mathbf{A}^{\mathrm{cr}}$. Then e is a coequalizer of a and b in $\mathbf{E}$, and for pairs (B, v) in Π, we have

$$
fav = \lambda_{eav} = \lambda_{ebv} = fbv .
$$

Since Π is dense by 61.4, $fa = fb$ follows. But then $f = ge$ for a unique $g : F \to G$, with $\lambda_{eu} = geu$ for every pair (A, eu) in Ψ. This shows that Ψ is dense.

62. Coreflections for Dense Sieves

62.1. Theorem. *If* (A, p) *is a concrete category over a quasitopos* **E** *which satisfies the conditions of* 61.1, *then all coreflections for* $\mathbf{A}^{\mathrm{cd}}$ *in* $\mathbf{A}^{\mathrm{cr}}$ *are coarse monomorphisms in* $\mathbf{A}^{\mathrm{cr}}$, *and if* $\mathbf{A}^{\mathrm{cd}}$ *is coreflective in* $\mathbf{A}^{\mathrm{cr}}$, *then* $\mathbf{A}^{\mathrm{cd}}$ *is a quasitopos.*

PROOF. If $f : (F, \Psi) \to (E, \Phi)$ is a coreflection for $\mathbf{A}^{\mathrm{cd}}$ in $\mathbf{A}^{\mathrm{cr}}$, factor f as $F \xrightarrow{e} E' \xrightarrow{m} E$ in **E**, with e strongly epimorphic and m monomorphic. Then

$$(F, \Psi) \quad \xrightarrow{\ e\ } \quad (E', m^*\Phi) \quad \xrightarrow{\ m\ } \quad (E, \Phi)$$

in $\mathbf{A}^{\mathrm{cr}}$, with $(E', m^*\Phi)$ dense by 61.6. Thus m factors $m = fu$ in $\mathbf{A}^{\mathrm{cr}}$. But then $fue = f$, and $ue = \mathrm{id}_{E'}$ follows for the coreflection f. Now e and u are inverse isomorphisms, in **E** and in $\mathbf{A}^{\mathrm{cd}}$, since e is epimorphic. Thus f is a coarse monomorphism in $\mathbf{A}^{\mathrm{cr}}$.

If $\mathbf{A}^{\mathrm{cd}}$ is coreflective in $\mathbf{A}^{\mathrm{cr}}$, then $\mathbf{A}^{\mathrm{cd}}$ is equivalent to the category of coalgebras for an idempotent comonad in $\mathbf{A}^{\mathrm{cr}}$. Since the embedding of $\mathbf{A}^{\mathrm{cd}}$ into $\mathbf{A}^{\mathrm{cr}}$ preserves finite limits, this comonad is left exact, and it follows from 50.5 and 60.1 that $\mathbf{A}^{\mathrm{cd}}$ is a quasitopos.

62.2. Proposition. *If* $\mathbf{A}^{\mathrm{cd}}$ *is monocoreflective in* $\mathbf{A}^{\mathrm{cr}}$, *then objects* E *of* **E** *for which the full* p-*sieve at* E *is dense define a monocoreflective full subcategory of* **E** *which is closed under finite limits in* **E**. *This subcategory is a quasitopos, with objects all objects* E *of* **E** *for which there is a dense* p-*sieve at* E, *including all objects* pA *for objects* A *of* **A**.

PROOF. Let Ω_E be the full p-sieve at an object E of **E**. Assigning to an object E of **E** the object (E, Ω_E) of $\mathbf{A}^{\mathrm{cr}}$ defines a right adjoint of p^{cr}. It follows that limits of objects (E, Ω_E) in $\mathbf{A}^{\mathrm{cr}}$ are again objects (E, Ω_E). Thus if **F** is the full subcategory of **E** with objects F for which Ω_F is dense, then **F** is closed under finite limits in **E**.

If $m : (E_1, m^*\Omega_E) \to (E, \Omega_E)$ is a coreflection for $\mathbf{A}^{\mathrm{cd}}$, then the p-sieve $m^*\Omega_E$ is full as well as dense, and it follows easily that $m : E_1 \to E$ is a coreflection for **F** in **E**. Thus if $\mathbf{A}^{\mathrm{cd}}$ is coreflective in $\mathbf{A}^{\mathrm{cr}}$, then **F** is coreflective in **E**, and equivalent to the category of coalgebras for the induced idempotent comonad. Since the inclusion $\mathbf{F} \to \mathbf{E}$ preserves finite limits, this comonad is left exact, and **F** is a quasitopos by 50.5.

The last part of the Proposition follows immediately from 61.5 and 61.1.

62.3. Monosieves. We define a *monosieve* at an object E of a quasitopos **E** as a full subcategory Σ of $\mathrm{Mon}\,E$ such that for an object m of Σ, every monomorphism $m' \leqslant m$ in **E** is an object of Σ. As a subcategory of $\mathbf{E}/E$, a monosieve Σ at E has a domain functor which we denote by D_Σ. The objects of Σ then define a cone $\sigma : D_\Sigma \to E$ in **E**. We say that **E** is *monosieve-complete* if the diagram D_Σ has a colimit in **E** for every monosieve Σ in **E**, and we note the following result.

Proposition. *If the domain functor D_Σ of a monosieve Σ at an object E of a quasitopos **E** has a colimit in **E**, then the induced morphism $f :$ Colim $D_\Sigma \to E$ in **E** is monomorphic, and a supremum of Σ in $\mathrm{Mon}\,E$.*

PROOF. If $\lambda : D_\Sigma \to F$ is the colimit cone, then $f\lambda_m = m$ for every object m of Σ. We construct a pullback

$$
\begin{array}{ccc}
F \times_E F & \xrightarrow{\;\;b\;\;} & F \\
\big\downarrow a & & \big\downarrow f \\
F & \xrightarrow{\;\;f\;\;} & E
\end{array}
$$

in **E**. As $\mathbf{E}/E$ is cartesian closed, the product $f \times_E f$ in $\mathbf{E}/E$ is a colimit of products (*i.e.* pullbacks)

$$
\begin{array}{ccc}
m \cap n & \xrightarrow{\;\;b_{mn}\;\;} & n \\
\big\downarrow a_{mn} & & \big\downarrow n \\
m & \xrightarrow{\;\;m\;\;} & \mathrm{id}_E
\end{array}
$$

in $\mathbf{E}/E$, with injections $\lambda_m \times_E \lambda_n$ satisfying

$$
a \cdot (\lambda_m \times_E \lambda_n) = \lambda_m \cdot a_{mn} \qquad \text{and} \qquad b \cdot (\lambda_m \times_E \lambda_n) = \lambda_n \cdot b_{mn},
$$

for objects m and n of Σ. As $a_{mn} : m \cap n \to m$ and $b_{mn} : m \cap n \to n$ in Σ if m and n are objects of Σ, we have

$$
\lambda_m \cdot a_{mn} = \lambda_{m \cap n} = \lambda_n \cdot b_{mn}
$$

for all objects m and n of Σ, and $a = b$ follows. Thus f is monomorphic, and then clearly a supremum of Σ in $\mathrm{Mon}\,E$.

62.4. Theorem. *Let $(\mathbf{A}, p)$ be a concrete category over a quasitopos **E**, with the conditions of 61.1 satisfied. If **E** is monosieve-complete, then $\mathbf{A}^{\mathrm{cd}}$ is a monocoreflective full subcategory of $\mathbf{A}^{\mathrm{cr}}$, and a quasitopos. Conversely, if $\mathbf{A}^{\mathrm{cd}}$ is monocoreflective in $\mathbf{A}^{\mathrm{cr}}$ and every full p-sieve is dense, then **E** is monosieve-complete.*

PROOF. For an object (E, Φ) of $\mathbf{A}^{\mathrm{cr}}$, let Σ be the full subcategory of $\mathrm{Mon}\, E$ with objects all monomorphism m of $\mathbf{E}$ with codomain E for which the p-sieve $m^*\Phi$ is dense, and Σ' the monosieve at E generated by Σ. Since Σ is closed under pullbacks, by 61.4, a cone $\lambda : D_\Sigma \to F$ in $\mathbf{E}$ has a unique extension to a cone $\lambda' : D_{\Sigma'} \to F$. If we factor the induced cone of Σ' as $m_1 \cdot \sigma'$ with σ' a colimit cone and $m_1 : E_1 \to E$ monomorphic, then it follows easily that the restriction σ of σ' to D_Σ is already a colimit cone. The final lift (E_1, Φ_1) for the cone σ and the dense p-sieves $m^*\Phi$, for m in σ, then is dense by 61.2. Since $\Phi_1 \leqslant m_1^*\Phi$ by our construction, the p-sieve $m_1^*\Phi$ at E_1 is dense by 61.5.

If $f : (F, \Psi) \to (E, \Phi)$ in $\mathbf{A}^{\mathrm{cr}}$, with Ψ dense, factor $f = me$ in $\mathbf{E}$, with e a strong epimorphism and m monomorphic. Then $m \in \Sigma$ by 61.6, and it follows that f factors $f = m_1 f_1$ in $\mathbf{E}$, with $f_1 : (F, \Psi) \to (E_1, m_1^*\Phi)$ in $\mathbf{A}^{\mathrm{cd}}$. This factorization is unique, and thus $m_1 : (E_1, m_1^*\Phi) \to (E, \Phi)$ is the desired coreflection.

For the converse, let Σ be a monosieve at an object E of $\mathbf{E}$, and let Φ be the p-sieve at E consisting of all pairs (A, mu) with m in Σ and $mu : pA \to E$ defined in $\mathbf{E}$, with a coreflection $m_1 : (E_1, \Psi) \to (E, \Phi)$ for $\mathbf{A}^{\mathrm{cd}}$ with $\Psi = m_1^*\Phi$. With the full p-sieve Ω_m at E_m dense for $m : E_m \to E$ in Σ, the morphism $m : (E_m, \Omega_m) \to (E, \Phi)$ of $\mathbf{A}^{\mathrm{cr}}$ factors $m = m_1 m'$ in $\mathbf{E}$ with $m' : (E_m, \Omega_m) \to (E_1, \Psi)$ in $\mathbf{A}^{\mathrm{cd}}$. The morphisms m' thus obtained clearly form a cone $\sigma : D_\Sigma \to E_1$ in $\mathbf{E}$; we claim that this is a colimit cone.

A cone $\lambda : D_\Sigma \to F$ assigns to every object $m : E_m \to E$ of Σ a morphism $\lambda_m : E_m \to F$ of $\mathbf{E}$, with $\lambda_{mt} = \lambda_m \cdot t$ for $t : mt \to m$ in Σ. If $mu = nv$ in $\mathbf{E}$, with m, n objects of Σ, we have $\lambda_m \cdot u = \lambda_n \cdot v$ since Σ is closed under pullbacks. Thus putting $\mu_{m'u} = \lambda_m \cdot u$, for an object $m : E_1 \to E$ of Σ with $m = m_1 m'$ and (A, u) in Ω_m, defines a cone $\mu : pD_\Psi \to F$. With (E_1, Ψ) dense, this factors $\lambda_m \cdot u = fm'u$, for all pairs $(A, m'u)$ of Ψ and a unique $f : E_1 \to F$. With Ω_m dense, it follows that $\lambda_m = fm'$ for every object m of Σ, and σ is a colimit cone as claimed.

62.5. Bounded quasitopoi.

Let $\mathfrak{U}$ be a class of sets which is closed under the formation of subsets and product sets. The class of all sets and the class of finite sets are examples. We say that a set S is $\mathfrak{U}$-small if S is the codomain of a surjective mapping $f : I \to S$ with I in $\mathfrak{U}$. Following J. Penon [83], we say that a quasitopos $\mathbf{E}$ is $\mathfrak{U}$-*bounded* if the following two conditions are satisfied. (i) Coproducts $\coprod_{i \in I} E_i$ exist in $\mathbf{E}$ for every family of objects indexed by a set I in $\mathfrak{U}$. (ii) For every object E of $\mathbf{E}$, the category $\mathrm{Mon}\, E$ has a $\mathfrak{U}$-small skeleton. We say that a quasitopos $\mathbf{E}$ is *bounded* if $\mathbf{E}$ is $\mathfrak{U}$-bounded for some $\mathfrak{U}$.

The topos of finite sets is $\mathcal{U}$-bounded for $\mathcal{U}$ the class of all finite sets, but not for $\mathcal{U}$ the class of all sets, and the topos of all sets is $\mathcal{U}$-bounded for $\mathcal{U}$ the class of all sets, but not for $\mathcal{U}$ the class of finite sets. All examples in Chapter 2 are $\mathcal{U}$-bounded for one of these two classes $\mathcal{U}$.

Proposition. *Every bounded quasitopos is monosieve-complete.*

PROOF. A monosieve Σ at an object E of a $\mathcal{U}$-bounded quasitopos **E** is equivalent to a $\mathcal{U}$-small full subcategory Σ' of Σ, by condition (ii) for **E**. The $\mathcal{U}$-small diagram $D_{\Sigma'}$ then has a colimit in **E** by condition (i), and D_Σ clearly has the same colimit.

63. Quotient Sinks

63.1. Quotient cones. In most of this Section, we assume only that **E** is a category with finite limits. We say that a cone $\varphi : D \to E$ in **E**, for a diagram D over **E** and an object E of **E**, is a *quotient cone* at E if for a factorization $g\varphi = m\psi$ in **E**, with g a morphism, m a monomorphism, and ψ a cone, there is always a (unique) morphism t in **E** such that $g = mt$, and $t\varphi = \psi$. We note that it suffices to require that $g\varphi_i = m\psi_i$ for every vertex Di of the domain D of φ; the morphisms ψ_i then define a cone with domain D. Thus if D' is the discrete diagram with the same vertices as D, then a cone $\varphi : D \to E$ is a quotient cone if and only if the induced cone $\varphi' : D' \to E$, with the same arrows as φ, is a quotient cone.

Proposition. *A quotient cone is epimorphic.*

PROOF. If $a\varphi = b\varphi$ for a quotient cone φ and morphisms a and b, we must show that $a = b$. We have $\langle a, b \rangle \cdot \varphi = \langle \mathrm{id}, \mathrm{id} \rangle \cdot \psi$ for $\psi = a\varphi$, hence $\langle a, b \rangle = \langle \mathrm{id}, \mathrm{id} \rangle \cdot t$ for some t, and $a = t = b$ follows.

A morphism f of **E** induces a (co-)cone with a one-vertex domain. By the definitions and the result just proved, this (co-)cone is a quotient cone if and only if f is a strong epimorphism.

63.2. Proposition. *Every colimit cone is a quotient cone.*

PROOF. If $\varphi : D \to E$ is a colimit cone, and if $g\varphi = m\psi$, then $\psi = t\varphi$ for a morphism t. Since then $mt\varphi = g\varphi$, $mt = g$ follows.

63.3. Proposition. *A left adjoint functor preserves quotient cones.*

PROOF. Let $F \dashv G : \mathbf{F} \to \mathbf{E}$. If $g \cdot F\varphi = m\psi$ for cones φ with domain D and ψ with domain FD, then ψ has an adjoint cone $\hat{\psi}$ with domain D, and with $Gm \cdot \hat{\psi} = \hat{g} \cdot \varphi$ for the adjoint morphism $\hat{g}$ of g, with Gm monomorphic if m is monomorphic. Now if $Gm \cdot t = \hat{g}$, and t is adjoint to s, then $ms = g$, and $s \cdot F\varphi = \psi$ follows.

63.4. Proposition. *For an object A of* **E**, *the domain functor* D_A : **E**$/A \to$ **E** *preserves and reflects quotient cones.*

PROOF. The functor D_A clearly reflects quotient cones, and D_A preserves quotient cones by 63.3 since D_A has a right adjoint.

63.5. Proposition. *If the pullback functor f^*, for a morphism f of* **E**, *has a right adjoint, then pullbacks by f preserve quotient cones.*

In particular, all pullbacks in a quasitopos preserve quotient cones.

PROOF. Let $f : A \to B$ in **E**. A quotient cone φ at B induces a cone φ in **E**$/B$, at id_B, which is a quotient cone by 63.4. Pulling this cone back by f defines a cone $f^*\varphi$ in **E**$/A$, at id_A, which is a quotient cone by 63.3. The functor D_A then produces a cone $f^*\varphi$ in **E** at A, again a quotient cone by 63.4.

63.6. Quotient cone factorizations. We say that **E** has *quotient cone factorizations* if every (co-)cone $\varphi : D \to E$ factors $\varphi = m\psi$ with ψ a quotient cone and m a monomorphism of **E**. By the defining universal property of quotient cones, this factorization is unique if it exists, up to an isomorphism in the middle.

Proposition. *If* **E** *is monosieve-complete (62.3), then* **E** *has quotient cone factorizations if and only if every morphism f of* **E** *factors $f = me$ with e a strong epimorphism and m monomorphic in* **E**.

PROOF. Factorization $f = me$ are quotient cone factorizations for a cone of one morphism. Conversely, if $\varphi : D \to E$ is a cone, factor each arrow of φ as $\varphi_i = m_i e_i$, with e_i strongly epimorphic and m_i monomorphic. Let Σ be the monosieve generated by the morphisms m_i, and factor the induced cone of Σ as $m_1 \cdot \sigma$, with σ a colimit cone, and m_1 monomorphic by 62.3. Then every morphism m_i factors $m_i = m_1 m_i'$, and the cone φ factors $\varphi = m_1 \cdot \psi$ for a cone ψ.

If $g\psi = m\lambda$ for a cone λ, with m monomorphic, then for each arrow $\psi_i = m_i' e_i$ of ψ, there is a morphism r_i with $m r_i = g m_i'$ and $\lambda_i = r_i e_i$. If $m_i u = m_j v$ for arrows φ_i and φ_j of φ, then clearly $r_i u = r_j v$; thus the morphisms r_i generate a unique cone ρ with domain D_Σ. This cone factors $\rho = t\sigma$, with $mt\sigma = g\sigma$ by our construction. But then $g = mt$, and $\varphi = m_1\psi$ is the desired factorization.

63.7. Final lifts for dense p-sieves. We return now to the situation that $(\mathbf{A}, p)$ is a concrete category over a monosieve-complete quasitopos $\mathbf{E}$, with the properties of 61.1. A sink for the functor p induces a cone in $\mathbf{E}$ with a discrete domain. We say that the sink is a *quotient sink* if this cone is a quotient cone.

Proposition. *For a sink consisting of morphisms $f_i : E_i \to E$, and a dense p-sieve (E_i, Φ_i) for each f_i, the final lift (E, Φ) in $\mathbf{A}^{\mathrm{cr}}$ is dense if and only if the morphisms f_i form a quotient sink.*

PROOF. If m is a coreflection for $\mathbf{A}^{\mathrm{cd}}$ at (E, Φ), then each f_i factors $f_i = m f_i'$. If the f_i form a quotient sink, it follows that $mt = \mathrm{id}_E$ for a morphism t of $\mathbf{E}$. Then m and t are inverse isomorphisms since m is monomorphic, and thus Φ is dense.

Conversely, if $g f_i = m g_i$ for every f_i, with m monomorphic, then putting $\mu(f_i u) = g_i u$ for every pair $(A, f_i u)$ in Φ defines a cone μ with domain $p D_\Phi$. If (E, Φ) is dense, then there is a unique morphism t with $g_i u = t f_i u$ for every pair $(A, f_i u)$ in Φ, and with $mt = g$, and the f_i form a quotient sink.

63.8. Proposition. *If the full p-sieve at an object E of $\mathbf{E}$ is dense, then $\mathbf{A}^{\mathrm{cd}}$ has initial lifts for finite sources with domain E.*

PROOF. If morphisms $f_i : E \to E_i$ of $\mathbf{E}$ and objects (E_i, Φ_i) of $\mathbf{A}^{\mathrm{cd}}$ define a finite source, then we have pullbacks

$$
\begin{array}{ccc}
(E, f_i^* \Phi_i) & \xrightarrow{f_i} & (E_i, \Phi_i) \\
\Big\downarrow{\scriptstyle \mathrm{id}_E} & & \Big\downarrow{\scriptstyle \mathrm{id}_{E_i}} \\
(E, \Omega) & \xrightarrow{f_i} & (E_i, \Omega_i)
\end{array}
$$

in $\mathbf{A}^{\mathrm{cr}}$, with full p-sieves Ω and Ω_i. If the full p-sieve (E, Ω) is dense, then the p-sieves $f_i^* \Phi_i$ are dense by 61.4. Since finite intersections of dense p-sieves at E are dense, again by 61.4, the Proposition follows.

63.9. Remark. We may call a p-sieve (E, Φ) *quotient-dense* if the induced cone of Φ is a quotient cone. If $\mathbf{E}$ has quotient cone factorizations, and we replace "dense" with "quotient-dense" throughout, beginning with the conditions in 61.1, then everything in Sections 61 and 62, including 63.7 and 63.8, remains valid, but some proofs must be changed. Note that if $\mathbf{E}$ is monosieve-complete, then every quotient-dense p-sieve is dense.

64. Quasitopological Categories

64.1. Definition. Propositions 63.7 and 63.8 motivate the following definition. We say that a concrete category $(\mathbf{T}, p)$ over a category $\mathbf{E}$ is *quasitopological* over $\mathbf{E}$ if the functor p admits initial lifts for finite sources and final lifts for all quotient sinks, and $(\mathbf{T}, p)$ is transportable.

For $\mathbf{A}^{cd}$ to be quasitopological over $\mathbf{E}$, we must require that all full p-sieves are dense. By 62.2, this can be done without loss of generality if $\mathbf{A}$ satisfies the conditions of 61.1 and $\mathbf{E}$ is monosieve-complete.

If $(\mathbf{T}, p)$ is quasitopological over $\mathbf{E}$, then it is easily seen that the functor p preserves and lifts finite limits and strong monomorphisms, as for topological categories (11.7). Trivial objects of $\mathbf{T}$ provide a left inverse right adjoint functor of p, and thus p preserves colimits and strong epimorphisms. Since colimit cones are quotient cones, p also lifts colimits and strong epimorphisms. Quasitopological categories share other properties with topological categories. They deserve a fuller treatment (and maybe a better name), but an adequate theory of quasitopological categories is outside the scope of these Notes.

64.2. A topological category clearly is quasitopological, and we note the following result.

Proposition. *A quasitopological category is topological if and only if it has discrete objects.*

PROOF. The condition is clearly necessary. Conversely, if a source of morphisms $f_i : E \to pA_i$ is given, construct the sink of all morphisms $g : pB \to E$ of $\mathbf{E}$ such that $gf_i : B \to A_i$ in $\mathbf{T}$ for every f_i. Since $\mathrm{id}_E : pA \to E$ is in this sink for the discrete lift A of E in $\mathbf{T}$, the sink is a quotient sink. Its final lift is an initial lift for the given source.

64.3. We assume in the remainder of this Section that $\mathbf{E}$ is a category with finite limits and $(\mathbf{T}, p)$ a quasitopological category over $\mathbf{E}$. We also assume that canonical finite limits in $\mathbf{T}$ are obtained by lifting the canonical finite limits which are assumed to exist for $\mathbf{E}$.

Proposition. *If $\mathbf{T}$ is cartesian closed, then $\mathbf{E}$ is cartesian closed, and evaluation morphisms* $\mathrm{ev}_{\mathbf{T}}$ *and* $\mathrm{ev}_{\mathbf{E}}$ *satisfy*

$$(1) \qquad \mathrm{ev}_{\mathbf{E}}(pY, pZ) = p\,\mathrm{ev}_{\mathbf{T}}(Y, Z) \circ (\mu_{Y,Z} \times \mathrm{id}_{pY}),$$

for natural monomorphisms $\mu_{Y,Z} : p[Y, Z]_{\mathbf{T}} \to [pY, pZ]_{\mathbf{E}}$. *If Z is a trivial object of $\mathbf{T}$, then $[Y, Z]$ is a trivial object of $\mathbf{T}$, and $\mu_{Y,Z}$ an isomorphism, for every object Y of $\mathbf{T}$.*

We say that $\mathbf{T}$ is *concretely cartesian closed* over $\mathbf{E}$ if all monomorphisms $\mu_{Y,Z}$ are isomorphisms of $\mathbf{E}$.

PROOF. Trivial objects provide a full embedding $\mathbf{T} \to \mathbf{E}$, with a left adjoint p which preserves finite products. Thus $\mathbf{E}$ is cartesian closed by 9.4 if $\mathbf{T}$ is cartesian closed, with all objects $[Y, Z]$ trivial if Z is trivial.

Natural transformations $\mu_{Y,-}$ which satisfy (1) are obtained as adjoint natural transformations (5.6) of identities $p(- \times Y) = p - \times pY$. If $\mu_{Y,Z} \cdot a = \mu_{Y,Z} \cdot b$ for morphisms $a, b : E \to p[Y, Z]$, let X be a lift for the source given by a and b. Then if $\alpha : X \times Y \to Z$ and $\beta : X \times Y \to Z$ have exponential adjoints a and b for $\mathbf{T}$, the morphisms α and β from $E \times pY$ to pZ have exponential adjoints $\mu_{Y,Z}a$ and $\mu_{Y,Z}b$ for $\mathbf{E}$, and $\alpha = \beta$ follows. But then $a = b$, and $\mu_{Y,Z}$ is monomorphic as claimed. If Z is a trivial object of $\mathbf{T}$ and F is $[pY, pZ]$ with the trivial structure, then $\mathrm{ev}_{\mathbf{E}} : F \times Y \to Z$ in $\mathbf{T}$, with $\mu_{Y,Z} \cdot \psi = \mathrm{id}_{pF}$ for the exponential adjoint ψ in $\mathbf{T}$. But then $\mu_{Y,Z}$ and ψ are inverse isomorphisms.

64.4. Theorem. *If a quasitopological category $(\mathbf{T}, p)$ over a cartesian closed category $\mathbf{E}$ is cartesian closed, then product functors $- \times Y$ in $\mathbf{T}$ preserve final lifts of quotient sinks. Conversely, if $\mathbf{E}$ has quotient cone factorizations, and product functors $- \times Y$ in $\mathbf{T}$ preserve final lifts of quotient sinks, then $\mathbf{T}$ is cartesian closed.*

PROOF. If $\mathbf{T}$ is cartesian closed, let X be the final lift for a quotient sink of morphisms $f_i : X_i \to X$, and consider $g : pX \times pY \to pZ$ in $\mathbf{E}$ with $g \cdot (f_i \times \mathrm{id}_Y) : X_i \times Y \to Z$ in $\mathbf{T}$ for all f_i. If $\hat{g}$ is exponentially adjoint to g in $\mathbf{E}$, and h_i exponentially adjoint to $g \cdot (f_i \times \mathrm{id}_Y)$ in $\mathbf{T}$, it follows that $\mu_{Y,Z}h_i = \hat{g}f_i$ for the natural monomorphism $\mu_{Y,Z}$ of 64.3. Since the f_i form a quotient sink, there is $t : pX \to p[Y, Z]$ with $\hat{g} = \mu_{X,Y}t$, and $h_i = tf_i$ for all f_i. But then $t : X \to [Y, Z]$ in $\mathbf{T}$, and $g : X \times Y \to Z$ in $\mathbf{T}$ by 64.3, with exponential adjoint t for $\mathbf{T}$. Thus the morphisms $f_i \times \mathrm{id}_{pY}$ form a quotient sink with final lift $X \times Y$.

For the converse, let Y and Z be objects of $\mathbf{T}$, and consider the sink of all exponential adjoints $\hat{\varphi} : pX \to [pY, pZ]$ for morphisms $\varphi : X \times Y \to Z$ of $\mathbf{T}$. We factor $\hat{\varphi} = \mu \cdot \varphi^{\#}$, with μ monomorphic and the morphisms $\varphi^{\#}$ forming a quotient sink. If $[Y, Z]$ is the final lift for this sink in $\mathbf{T}$, then

$$\varphi = (\mathrm{ev} \cdot (\mu \times \mathrm{id}_{pY})) \circ (\varphi^{\#} \times \mathrm{id}_Y)$$

for all $\varphi : X \times Y$ in $\mathbf{T}$, with evaluation in $\mathbf{E}$. If the functor $- \times Y$ preserves the final lift $[Y, Z]$, then $\mathrm{ev} \cdot (\mu \times \mathrm{id}_{pY}) : [Y, Z] \times Y \to Z$ in $\mathbf{T}$, and it follows easily that this is evaluation for a cartesian closed structure of $\mathbf{T}$.

64.5. We recall from 16.1 that a partial morphism is a span $X \xleftarrow{m} X' \xrightarrow{f} Y$ with m a strong monomorphism. We shall denote by $\vartheta_Y : Y \to \tilde{Y}$ a morphism representing partial morphisms with codomain Y in $\mathbf{T}$, and by $\tau_E : E \to E^*$ a morphism representing partial morphisms with codomain E in $\mathbf{E}$.

Proposition. *If partial morphisms in* $\mathbf{T}$ *are represented, then partial morphisms in* $\mathbf{E}$ *are represented, with*

$$\tau_{pY} = \nu_Y \circ p\vartheta_Y$$

in $\mathbf{E}$ *for a monomorphism* $\nu_Y : p\tilde{Y} \to (pY)^*$, *for every object* Y *of* $\mathbf{E}$. *If* Y *is trivial over* $\mathbf{E}$, *then so is* $\tilde{Y}$, *and* ν_Y *is an isomorphism.*

We say that partial morphisms in $\mathbf{T}$ are *concretely represented* if all monomorphisms ν_Y are isomorphisms of $\mathbf{E}$.

PROOF. If Y is trivial and Z the trivial lift of $p\tilde{Y}$, then $\vartheta_Y : Y \to Z$ remains an embedding, and we have pullbacks

$$
\begin{array}{ccccc}
Y & \xrightarrow{\mathrm{id}_{pY}} & Y & \xrightarrow{\mathrm{id}_{pY}} & Y \\
\downarrow{\vartheta_Y} & & \downarrow{\vartheta_Y} & & \downarrow{\vartheta_Y} \\
\tilde{Y} & \xrightarrow{\mathrm{id}} & Z & \xrightarrow{u} & \tilde{Y}
\end{array}
$$

in $\mathbf{T}$, with the bottom of the outside rectangle determined by the remainder of the diagram. It follows that $u = \mathrm{id}_{pZ}$, and hence by amnesticity of p that $\tilde{Y} = Z$.

Now if Y is the trivial lift of an object E of $\mathbf{E}$, then we have a bijection between pullbacks

$$
\begin{array}{ccc}
A' & \xrightarrow{f} & E \\
\downarrow{m} & & \downarrow{\vartheta_Y} \\
A & \xrightarrow{\bar{f}} & p\tilde{Y}
\end{array}
\quad \text{and} \quad
\begin{array}{ccc}
X' & \xrightarrow{f} & Y \\
\downarrow{m} & & \downarrow{\vartheta_Y} \\
X & \xrightarrow{\bar{f}} & \tilde{Y}
\end{array}
$$

in $\mathbf{E}$ and in $\mathbf{T}$, with X and X' the trivial lifts of A and A', and with m a strong mononorphism. Thus $\bar{f}$ at left is uniquely determined by the partial morphism (m, f), and ϑ_Y at left represents partial morphisms in $\mathbf{E}$ with codomain E.

For an object Y of $\mathbf{T}$, we have a pullback

$$
\begin{array}{ccc}
pY & \xrightarrow{\ \mathrm{id}_{pY}\ } & pY \\
{\scriptstyle p\vartheta_Y}\downarrow & & \downarrow{\scriptstyle \tau_{pY}} \\
p\tilde{Y} & \xrightarrow{\ \nu_Y\ } & (pY)^*
\end{array}
$$

in $\mathbf{E}$. If Z and Z^* are the trivial lifts of pY and of $(pY)^*$, then we have pullback squares

$$
(2) \qquad
\begin{array}{ccccc}
\cdot & \xrightarrow{\ f\ } & Y & \xrightarrow{\ \mathrm{id}_{pY}\ } & Z \\
{\scriptstyle m}\downarrow & & \downarrow{\scriptstyle \vartheta_Y} & & \downarrow{\scriptstyle \tau_{pY}} \\
X & \xrightarrow{\ \bar{f}\ } & \tilde{Y} & \xrightarrow{\ \nu_Y\ } & Z^*
\end{array}
$$

in $\mathbf{Y}$ for every partial morphism (m, f) in $\mathbf{T}$ with codomain Y. If $\nu_Y a = \nu_Y b$ in $\mathbf{E}$, let X be a lift for the source given by a and b. Then we have pullback squares (2) for $\bar{f} = a$ and $\bar{f} = b$, with the same outer rectangle. But then $a = b$ since ϑ_Y represents partial morphisms; thus ν_Y is monomorphic. If Y is trivial over $\mathbf{E}$, then $Z = Y$, and we have pullback squares (2) for the partial morphism $(\tau_{pY}, \mathrm{id}_Z) : Z^* \to Z$. As τ_{pY} represents partial morphisms in $\mathbf{E}$, we have $\nu_Y \bar{f} = \mathrm{id}$ for this choice of (m, f), and ν_Y is an isomorphism.

64.6. Theorem. *If $(\mathbf{T}, p)$ is quasitopological over a category $\mathbf{E}$ with finite limits, with partial morphisms in $\mathbf{E}$ and $\mathbf{T}$ represented, then pullbacks in $\mathbf{T}$ by strong monomorphisms preserve final lifts of quotient sinks. Conversely, if partial morphisms in $\mathbf{E}$ are represented, $\mathbf{E}$ has quotient cone factorizations, and pullbacks by strong monomorphisms in $\mathbf{T}$ preserve final lifts of quotient sinks, then partial morphisms in $\mathbf{T}$ are represented.*

PROOF. If partial morphisms are represented in $\mathbf{T}$ and in $\mathbf{E}$, then we have pullbacks

$$
\begin{array}{ccc}
m_i^* Y_i & \xrightarrow{\ g f_i'\ } & Z \\
{\scriptstyle m_i}\downarrow & & \downarrow{\scriptstyle \vartheta_Z} \\
Y_i & \xrightarrow{\ h_i'\ } & \tilde{Z}
\end{array}
\qquad \text{and} \qquad
\begin{array}{ccc}
pY & \xrightarrow{\ g\ } & pZ \\
{\scriptstyle m}\downarrow & & \downarrow{\scriptstyle \tau_{pZ}} \\
pY & \xrightarrow{\ \bar{g}\ } & (pZ)^*
\end{array}
$$

in $\mathbf{T}$ and in $\mathbf{E}$, for a quotient sink of morphisms $f_i : Y_i \to Y$, pulled back by a strong monomorphism $m : m^*Y \to Y$ to morphisms $f_i' : m_i^*Y_i \to m^*Y$, and for a morphism $g : pY \to pZ$ of $\mathbf{E}$ with $gf_i' : m_i^*Y_i \to Z$ in $\mathbf{T}$ for every f_i', with $\nu_Z h_i = \bar{g}f_i$ for all f_i by 64.5. Then $\bar{g} = \nu_Z t$, and $tf_i = h_i$ for every f_i, for a morphism t of $\mathbf{E}$, with $t : Y \to \tilde{Z}$ in $\mathbf{T}$ if Y is the final lift for the f_i. Pulling t back by ϑ_Z, with $\nu_Z \vartheta_Z = \tau_{pZ}$, we get $g : m^*Y \to Z$ in $\mathbf{T}$. Thus m^*Y is the final lift for the sink of morphism f_i', which is a quotient sink by 18.4 and 63.5.

For the converse, let Y be an object of $\mathbf{T}$, and construct pullbacks

$$
\begin{array}{ccc}
\cdot & \xrightarrow{\ f\ } & pY \\
\downarrow{\scriptstyle m} & & \downarrow{\scriptstyle \tau_{pY}} \\
pX & \xrightarrow{\ \bar{f}\ } & (pY)^*
\end{array}
$$

in $\mathbf{E}$, for every partial morphism (m, f) in $\mathbf{T}$ with codomain Y. Now factor every $\bar{f} = \nu_Y f'$, with ν_Y monomorphic and the f' forming a quotient cone. We have in particular $\tau_{pY} = \nu_Y \vartheta_Y$ for the partial morphism $(\mathrm{id}_Y, \mathrm{id}_Y)$, with ϑ_Y a strong monomorphism in $\mathbf{E}$. If $\tilde{Y}$ is the final lift for the quotient sink of morphisms f', then $\vartheta_Y : Y \to \tilde{Y}$ in $\mathbf{T}$, and we have for every partial morphism $(m, f) : X \to Y$ in $\mathbf{E}$ a unique diagram of pullbacks

(1)
$$
\begin{array}{ccccc}
\cdot & \xrightarrow{\ f\ } & Y & \xrightarrow{\ \mathrm{id}_{pY}\ } & Z \\
\downarrow{\scriptstyle m} & & \downarrow{\scriptstyle \vartheta_Y} & & \downarrow{\scriptstyle \tau_{pY}} \\
X & \xrightarrow{\ f'\ } & \tilde{Y} & \xrightarrow{\ \nu_Y\ } & Z^*
\end{array}
$$

in $\mathbf{T}$, with Z and Z^* trivial. The sink of morphisms f in these diagrams includes id_Y; thus Y is its final lift. The righthand square in (1) is a pullback in $\mathbf{T}$, and $\tau_{pY} : Z \to Z^*$ a strong monomorphism. Thus $\vartheta_y : Y \to \tilde{Y}$ is a strong monomorphism of $\mathbf{T}$, representing partial morphisms with codomain Y.

64.7. Remark. It is sufficient for most of this Section that $\mathbf{E}$ is a category with finite limits, and $(\mathbf{T}, p)$ a category over $\mathbf{E}$ with lifts for finite sources and initial lifts for finite strong monosources (which are dual to finite quotient sinks). It follows easily from these conditions that p preserves epimorphisms, and from this that initial lifts of strong monosources in $\mathbf{E}$ are strong monocones (dual to quotient cones) in $\mathbf{T}$. We also need trivial objects of $\mathbf{T}$ for parts of 64.3 and 64.5. Canonical finite limits in $\mathbf{T}$ can then be obtained as initial lifts of canonical finite limits in $\mathbf{E}$.

65. Left Exact Concrete Categories

65.1. Categories with discrete objects. We say that a concrete category $(\mathbf{A}, p)$ over a category $\mathbf{E}$ *has discrete objects* if the functor p has a right inverse left adjoint $d : \mathbf{E} \to \mathbf{A}$. Objects dE of $\mathbf{A}$ are then discrete in the sense of 11.5. The left adjoint d with $pd = \mathrm{Id}\,\mathbf{E}$ is unique if it exists, since we assume p to be amnestic.

Topological categories have discrete objects, but categories with discrete objects need not be topological. Trivial objects over $\mathbf{E}$ are discrete over $\mathbf{E}^{\mathrm{op}}$.

If $\mathbf{A}$ has discrete objects, then $\mathbf{A}$ satisfies the conditions of 61.1, and every full sieve Ω_E is dense, with (dE, id_E) in Ω_E. Condition 61.1.3 is valid because (dE, id_E) is in $m^{*}YA$ for $m : E \to pA$ in $\mathbf{E}$, with $m : dE \to A$ in $\mathbf{A}$, and 61.1.2 is obtained similarly.

65.2. Proposition. *In a concrete category with discrete objects, strong monomorphisms are embeddings.*

PROOF. Let $(\mathbf{A}, p)$ be the category and d the discrete object functor. If $m : X \to Y$ is a strong monomorphism of $\mathbf{A}$ and $mu : Z \to Y$ for $u : pZ \to pX$, then the commutative square

$$
\begin{array}{ccc}
dpZ & \xrightarrow{\ \mathrm{id}_{pZ}\ } & Z \\[2pt]
\Big\downarrow{\scriptstyle u} & & \Big\downarrow{\scriptstyle mu} \\[2pt]
X & \xrightarrow{\ m\ } & Y
\end{array}
$$

in $\mathbf{A}$ has a diagonal; this can only be $u : Z \to X$.

65.3. Left exact concrete categories. We say that a concrete category $(\mathbf{A}, p)$ over a category $\mathbf{E}$ is *left exact* if $(\mathbf{A}, p)$ has discrete objects, and the discrete object functor d preserves finite limits. If $\mathbf{A}$ and $\mathbf{E}$ are quasitopoi, this means that the adjunction $d \dashv p : \mathbf{A} \to \mathbf{E}$ is a geometric morphism.

Proposition. *If $(\mathbf{A}, p)$ has a discrete object functor d, over a category $\mathbf{E}$ with finite products, then the following conditions are necessary and sufficient for d to preserve finite limits.*

(1) *If 1 is a terminal object of $\mathbf{E}$, then $d1$ is a terminal object of $\mathbf{A}$.*

(2) *For objects E and F of $\mathbf{E}$, we have $d(E \times F) = dE \times dF$.*

(3) *If $m : A \to B$ is a strong monomorphism of $\mathbf{E}$, then $m : dE \to dF$ is a strong monomorphism of $\mathbf{A}$.*

PROOF. With 65.2, this follows easily from the definitions.

65.4. Discussion. We note first that 65.3.(1) is satisfied for a concrete category $(\mathbf{A}, p)$ with discrete objects iff the functor p reflects terminal objects. If $pT = 1$, then $d1$ is a terminal object of $\mathbf{A}$ iff $\mathrm{id}_1 : d1 \to T$ has an inverse $\mathrm{id}_1 : T \to d1$, and this means $T = d1$ since p is amnestic.

If $\mathbf{E}$ is a solid quasitopos, and in particular if $\mathbf{E}$ is a topos, then p must also reflect initial objects. If 0 is an initial object of $\mathbf{E}$, then $d0$ is an initial object of $\mathbf{A}$, and the strong monomorphism $m : 0 \to 1$ in $\mathbf{E}$ lifts to a strong monomorphism $m : d0 \to d1$ in $\mathbf{A}$. If A is an object of $\mathbf{A}$ with pA initial in $\mathbf{E}$, then the isomorphism $u : 0 \to pA$ lifts to an epimorphism $u : d0 \to A$, and $m : d0 \to d1$ factors through u since $d1$ is a terminal object of $\mathbf{A}$. But then $u : d0 \to A$ is an isomorphism, and A an initial object of $\mathbf{A}$.

For concrete categories over sets, it follows immediately from this discussion that left exact categories have constant maps (11.6).

We note also that the converse of 65.2 is valid if the functor p preserves epimorphisms, but not e.g. for Hausdorff spaces.

65.5. From now on, let $(\mathbf{A}, p)$ be a concrete category with discrete objects, over a category $\mathbf{E}$ with finite limits.

If $\mathbf{A}$ and $\mathbf{E}$ are cartesian closed, then we have natural monomorphisms $\mu_{Y,Z} : [Y, Z]_{\mathbf{T}} \to [pY, pZ]_{\mathbf{E}}$, by 64.3 with 64.7.

Proposition. *If $\mathbf{A}$ and $\mathbf{E}$ are cartesian closed, then the discrete object functor d preserves products $E \times F$ if and only if every morphism $\mu_{dF,Z}$, for objects F of $\mathbf{E}$ and Z of $\mathbf{A}$, is an isomorphism of $\mathbf{E}$.*

PROOF. With the adjunction $d \dashv p$, exponential adjunction defines bijections between morphisms

$$f : dE \times dF \to Z \quad \text{and} \quad \hat{f} : E \to p[dF, Z],$$
$$f : d(E \times F) \to Z \quad \text{and} \quad f^{\#} : E \to [F, pZ],$$

with $\quad f = \mathrm{ev}_{dF,Z} \cdot (\hat{f} \times \mathrm{id}_Z) \quad \text{and} \quad f = \mathrm{ev}_{F,pX} \cdot (f^{\#} \times \mathrm{id}_{pZ}).$

If $f^{\#}$ exists, then $\hat{f}$ exists, and $\hat{f} = \mu_{dF,Z} \cdot f^{\#}$, by 64.3. If $\mu_{dF,Z}$ is isomorphic, it follows that $f : d(E \times F) \to Z$ if $f : dE \times dF \to Z$; thus d preserves finite products. Conversely, if d preserves finite products, then we get a right inverse $f^{\#}$ of $\mu_{dF,Z}$ for $f = \mathrm{ev}_{dF,Z}$, and $\mu_{dF,Z}$ is isomorphic.

65.6. If partial morphisms in $\mathbf{A}$ and in $\mathbf{E}$ are represented, by morphisms $\vartheta_Y : Y \to \hat{Y}$ and $\tau_E : E \to E^*$, then $\tau_{pY} = \nu_Y \cdot p\vartheta_Y$ for a natural monomorphism ν_Y, by 64.5 and 64.7.

Proposition. *If partial morphisms in $\mathbf{A}$ and in $\mathbf{E}$ are represented and if $d : \mathbf{E} \to \mathbf{A}$ preserves strong monomorphisms, then every morphism ν_Y :*

$p\tilde{Y} \rightarrow (pY)^*$ *is an isomorphism. Conversely, if every morphism ν_{dE} is an isomorphism, then d preserves strong monomorphisms.*

PROOF. If d preserves strong monomorphisms, then we have a bijection between pullbacks

$$
\begin{array}{ccc}
dE & \xrightarrow{f} & Y \\
\downarrow{m} & & \downarrow{\vartheta_Y} \\
dF & \xrightarrow{\bar{f}} & \tilde{Y}
\end{array}
\quad \text{and} \quad
\begin{array}{ccc}
E & \xrightarrow{f} & pY \\
\downarrow{m} & & \downarrow{\tau_{pY}} \\
F & \xrightarrow{f^*} & (pY)^*
\end{array}
$$

in **A** and **E**, with $f^* = \nu_Y \cdot \bar{f}$. For $(m, f) = (\tau_{pY}, \mathrm{id}_{pY})$, we get $f^* = \mathrm{id}_{pY}$, and then $\bar{f}$ is an inverse of ν_Y.

For the converse, let $m : E \rightarrow F$ be a strong monomorphism of **E**, and consider pullbacks

$$
\begin{array}{ccc}
X & \xrightarrow{u} & dE \\
\downarrow{m_1} & & \downarrow{\vartheta_{dE}} \\
dF & \xrightarrow{\bar{f}} & \widetilde{dE}
\end{array}
\quad \text{and} \quad
\begin{array}{ccc}
E & \xrightarrow{\mathrm{id}_E} & E \\
\downarrow{m} & & \downarrow{\tau_E} \\
F & \xrightarrow{f^*} & E^*
\end{array} ,
$$

with $\nu_{dE} \cdot \bar{f} = f^*$. As p preserves pullbacks, $u : pX \rightarrow E$ is an isomorphism, and $m_1 = mu$. Now $u^{-1} : dE \rightarrow X$ in **A**, and $u : X \rightarrow dE$ is an isomorphism of **A**. Thus $m : dE \rightarrow dF$ is a strong monomorphism in **A**.

65.7. We specialize from now on to **E** = SET.

Proposition. *For a concrete category **A** over sets with discrete objects and with constant maps, the discrete object functor d preserves strong monomorphisms, and $d\{*\}$ is a terminal object of **A** for a singleton $\{*\}$.*

PROOF. For an injective mapping $m : E \rightarrow F$ with E not empty, there is a mapping $u : F \rightarrow E$ with $mu = \mathrm{id}_E$, and with $u : dF \rightarrow dE$ in **A**. Thus $m : dE \rightarrow dF$ is a coretraction in **A**, and hence a strong monomorphism. If E is empty, consider a commutative square

$$
\begin{array}{ccc}
X & \xrightarrow{e} & Y \\
\downarrow{f} & & \downarrow{g} \\
d\emptyset & \xrightarrow{m} & dF
\end{array}
$$

in **A**, with e epimorphic. Then X must be empty, hence $X = d\emptyset$. Now e is only epimorphic if Y also is empty, hence $f = e = \text{id}_{d\emptyset}$. But then $g = m$, and $u = \text{id}_{d\emptyset}$ provides a diagonal for the square. Thus $m : dE \to dF$ is again a strong monomorphism.

For an object A of **A**, the unique mapping $pA \to pd\{*\}$ is constant and lifts to a map $A \to d\{*\}$ of **A**. Thus $d\{*\}$ is a terminal object of **A**.

65.8. Proposition. *If* **A** *is a cartesian closed concrete category over sets, with discrete objects and with its terminal object discrete, then the discrete object functor d preserves finite products.*

PROOF. For a set F, the objects $d1 \times dF$ and $d(1 \times F)$ of **A** are isomorphic to dF. Thus for $E = 1$ in the proof of 65.5, putting $\hat{f} = \mu_{dF,Z} \cdot f^{\#}$ provides a bijection between mappings $f^{\#} : 1 \to (pZ)^F$ and $\hat{f} : 1 \to p[dF, Z]$. But then $\mu_{dF,Z}$ is bijective, and d preserves products $E \times F$ by 65.5.

65.9. Discussion. If **T** is a cartesian closed concrete category over sets, with discrete objects and constant maps, then 65.7 and 65.8 show that **T** is left exact. Thus for a topological category **T** over sets with constant maps, subspaces of discrete spaces are always discrete, and finite products of discrete spaces are discrete if **T** is cartesian closed. These results remain valid for any full subcategory **T** of a topological category over sets with constant maps, provided that **T** contains all discrete spaces and has finite products.

For all familiar topological categories with constant maps, finite products of discrete spaces are discrete. This poses the problem of either proving that finite produces of discrete spaces are discrete for all topological categories over sets with constant maps, or providing a counterexample.

66. p-Topologies

66.1. Definition. For a concrete category $(\mathbf{A}, p)$ over a quasitopos **E**, we define a p-*topology* of the quasitopos $\mathbf{A}^{\text{cr}}$ as a topology γ of $\mathbf{A}^{\text{cr}}$ for which every coarse monomorphism $m : (E, m^{*}\Psi) \to (F, \Psi)$ is closed. It follows that a p-topology γ of $\mathbf{A}^{\text{cr}}$ is, up to equivalence, of the form $\gamma m = m : (E, \bar{\Phi}) \to (F, \Psi)$, for a monomorphism $m : (E, \Phi) \to (F, \Psi)$ in $\mathbf{A}^{\text{cr}}$. Thus if γ is a p-topology, then every dense monomorphism for γ is isomorphic in **E**, and thus epimorphic in $\mathbf{A}^{\text{cr}}$. It follows that every object of $\mathbf{A}^{\text{cr}}$ is separated for a p-topology of $\mathbf{A}^{\text{cr}}$.

66.2. Covering sieves. A sieve R at an object A of **A** induces a p-sieve at pA, consisting of all pairs (X, u) with $u : pX \to pA$ in **E** underlying a morphism $u : X \to A$ in R. By *abus de langage*, we denote this p-sieve also by R, and we may put $(X, u) \in R$ for a morphism $u : X \to A$ in R. We say that R is a *covering sieve* for a p-topology γ, or that R *covers* A for γ, if the inclusion $\mathrm{id}_{pA} : (pA, R) \to \Upsilon A$ is dense for γ.

Proposition. *Covering sieves for a p-topology of $\mathbf{A}^{\mathrm{cr}}$ define a Grothendieck topology of $\mathbf{A}$, and a morphism $\mathrm{id}_E : (E, \Phi) \to (E, \Phi')$ of $\mathbf{A}^{\mathrm{cr}}$ is dense for γ if and only if R covers A for every pullback*

$$
(1) \qquad
\begin{array}{ccc}
(pA, R) & \xrightarrow{\;\mathrm{id}_{pA}\;} & \Upsilon A \\
\downarrow{\scriptstyle u} & & \downarrow{\scriptstyle u} \\
(E, \Phi) & \xrightarrow{\;\mathrm{id}_E\;} & (E, \Phi')
\end{array}
$$

in $\mathbf{A}^{\mathrm{cr}}$, with $(A, u) \in \Phi'$.

We recall that Grothendieck topologies are defined in 47.2.

PROOF. The full sieve ΥA at A clearly covers A for γ.

Pulling back $\mathrm{id}_{pA} : (pA, R) \to \Upsilon A$ by $f : \Upsilon B \to \Upsilon A$ in $\mathbf{A}^{\mathrm{cr}}$, we obtain $\mathrm{id}_{pB} : (pB, f^{\leftarrow}R) \to \Upsilon B$, with $f^{\leftarrow}R$ consisting of all $v : X \to B$ in $\mathbf{A}$ such that $fv \in R$, and pullbacks of dense morphisms are dense. Thus $f^{\leftarrow}R$ covers B if R covers A.

If R covers A, and $f^{\leftarrow}S$ covers X for every $f : X \to A$ in R, let $\mathrm{id}_{pA} : (pA, \bar{S}) \to \Upsilon A$ be the closure of $\mathrm{id}_{pA} : (pA, S) \to \Upsilon A$ for γ. Pulling this back by $v : X \to A$ in R, we get $v^{\leftarrow}\bar{S} = \Upsilon X$ since $v^{\leftarrow}S$ covers X. But then $R \subset \bar{S}$, and it follows that $\mathrm{id}_{pA} : (pA, \bar{S}) \to \Upsilon A$ is dense. Since this morphism is also closed, we have $\bar{S} = \Upsilon A$, and S covers A.

If $\mathrm{id}_E : (E, \Phi) \to (E, \Phi')$ is dense for γ, then R covers A in every pullback (1) since pullbacks of dense morphisms are dense. Conversely, consider pullback diagrams

$$
(2) \qquad
\begin{array}{ccccc}
(pA, R) & \xrightarrow{\;\mathrm{id}_{pA}\;} & (pA, R_1) & \xrightarrow{\;\mathrm{id}_{pA}\;} & \Upsilon A \\
\downarrow{\scriptstyle u} & & \downarrow{\scriptstyle u} & & \downarrow{\scriptstyle u} \\
(E, \Phi) & \xrightarrow{\;\mathrm{id}_E\;} & (E, \Phi_1) & \xrightarrow{\;\mathrm{id}_E\;} & (E, \Phi')
\end{array}
$$

in $\mathbf{A}^{\mathrm{cr}}$, with the lefthand morphisms dense and the righthand morphisms closed. If R covers A in a diagram (2), then the top right arrow is dense as well as closed, and $R_1 = \Upsilon A$ follows. If this is the case for every diagram (2), then $\Phi_1 = \Phi'$, and the given morphism $\mathrm{id}_E : (E, \Phi) \to (E, \Phi')$ is dense.

66.3. We now prove a converse of 66.2, thus obtaining a bijection between Grothendieck topologies of $\mathbf{A}$ and p-topologies of $\mathbf{A}^{\mathrm{cr}}$.

Proposition. *Every Grothendieck topology of $\mathbf{A}$ consists of the covering sieves for a unique p-topology γ of $\mathbf{A}^{\mathrm{cr}}$.*

PROOF. If $\gamma m = m : (E, \bar{\Phi}) \to (F, \Psi)$, for $m : (E, \Phi) \to (F, \Psi)$, then a pair (A, u) in $\bar{\Phi}$ must have (A, mu) in Ψ. For such a pair, we construct pullback diagrams

$$
\begin{array}{ccccc}
(pA, R) & \xrightarrow{\mathrm{id}_{pA}} & (pA, \bar{R}) & \xrightarrow{\mathrm{id}_{pA}} & \Upsilon A \\
\Big\downarrow{\scriptstyle u} & & \Big\downarrow{\scriptstyle u} & & \Big\downarrow{\scriptstyle mu} \\
(E, \Phi) & \xrightarrow{\mathrm{id}_E} & (E, \bar{\Phi}) & \xrightarrow{m} & (F, \Psi)
\end{array}
$$

(1)

in $\mathbf{A}^{\mathrm{cr}}$, with the lefthand top arrow dense and the righthand top arrow closed. Now $(A, u) \in \bar{\Phi}$ iff $\bar{R} = \Upsilon A$; i.e. $(A, u) \in \bar{\Phi}$ iff $(A, mu) \in \Psi$ and R covers A, for the induced Grothendieck topology of γ. Thus the induced Grothendieck topology determines γ uniquely.

Beginning with a Grothendieck topology of $\mathbf{A}$, we use the preceding paragraph to define an operator γ for monomorphisms of $\mathbf{A}^{\mathrm{cr}}$.

We have $R = \Upsilon A$ in (1) if $(A, u) \in \Phi$; thus $m \leqslant \gamma m$. If $u : \Upsilon A \to \gamma\gamma m$, then $\bar{R}$ in (1) covers A. For $v : X \to A$ in $\bar{R}$, we have a pullback

$$
\begin{array}{ccc}
(pX, v^{\leftarrow} R) & \xrightarrow{\mathrm{id}_{pX}} & \Upsilon X \\
\Big\downarrow{\scriptstyle v} & & \Big\downarrow{\scriptstyle v} \\
(pA, R) & \xrightarrow{\mathrm{id}_{pA}} & (pA, \bar{R})
\end{array}
$$

in $\mathbf{A}^{\mathrm{cr}}$. Putting this square on top of the lefthand square of (1), we see that $(X, uv) \in \bar{\Phi}$; thus $v^{\leftarrow} R$ covers X. But then R covers A for the given Grothendieck topology, and $(A, u) \in \bar{\Phi}$. This shows that γ is idempotent.

If $m' \leqslant m$ and $(A, m'u) \in \Psi$, then we have pullbacks

$$
\begin{array}{ccccc}
(pA, R') & \xrightarrow{\mathrm{id}_{pA}} & (pA, R) & \xrightarrow{\mathrm{id}_{pA}} & \Upsilon A \\
\Big\downarrow{\scriptstyle u} & & \Big\downarrow{\scriptstyle tu} & & \Big\downarrow{\scriptstyle mtu} \\
(E', \Phi') & \xrightarrow{t} & (E, \Phi) & \xrightarrow{m} & (F, \Psi)
\end{array}
$$

in $\mathbf{A}^{\mathrm{cr}}$, with $m' = mt$. If R' covers A, then so does R; it follows that $\gamma m' \leqslant \gamma m$.

Now consider pullbacks

$$(E_1, \Phi_1) \xrightarrow{\mathrm{id}_{E_1}} (E_1, \bar{\Phi}_1) \xrightarrow{m_1} (F_1, \Psi_1)$$

$$\left\downarrow f_1 \qquad\qquad \left\downarrow f_1 \qquad\qquad \left\downarrow f$$

$$(E, \Phi) \xrightarrow{\mathrm{id}_E} (E, \bar{\Phi}) \xrightarrow{m} (F, \Psi)$$

in $\mathbf{A}^{\mathrm{cr}}$, with the same bottom row as (1). Then a pair (A, u) is in $\bar{\Phi}_1$ iff $(A, m_1 u) \in \Psi_1$, and $(A, f_1 u) \in \bar{\Phi}$. If pulling back (E_1, Φ_1) by u produces (pA, R), then $(A, f_1 u) \in \bar{\Phi}$ iff R is a sieve covering A. This shows that pullbacks preserve closures for γ.

Finally, a strong monomorphism of $\mathbf{A}^{\mathrm{cr}}$ is coarse and hence closed, so that γm is strong if m is strong. Thus γ is a topology as claimed.

It remains to show that if γ is obtained from a Grothendieck topology J of $\mathbf{A}$, then J is the Grothendieck topology of covering sieves for γ. This is also straightforward; we omit the details.

66.4. For a sieve R at an object A of $\mathbf{A}$ and a p-sieve Φ at an object E of $\mathbf{E}$, we put $R \perp \Phi$ if (A, u) always is in Φ, for $u : pA \to E$ in $\mathbf{E}$, when $(X, uh) \in \Phi$ for every pair (X, h) in R. This means, by 59.5, that every morphism $u : (pA, R) \to (E, \Phi)$ of $\mathbf{A}^{\mathrm{cr}}$ can be lifted to a morphism $u : YA \to (E, \Phi)$.

Proposition. *A p-sieve (F, Ψ) is a sheaf for a p-topology γ of $\mathbf{A}^{\mathrm{cr}}$ if and only if $R \perp \Psi$ for every covering sieve in $\mathbf{A}$, and a sieve R at an object A of $\mathbf{A}$ covers A for γ if and only $R \perp \Psi$ for every sheaf (F, Ψ) for γ.*

PROOF. If R is a covering sieve and (F, Ψ) a sheaf, then $R \perp \Psi$ by the definitions. Conversely, consider a span

$$(E, \Phi') \xleftarrow{\mathrm{id}_E} (E, \Phi) \xrightarrow{f} (F, \Psi)$$

in $\mathbf{A}^{\mathrm{cr}}$, with the lefthand arrow dense for γ. Then $fu : (pA, R) \to (F, \Psi)$ for every pullback 66.2.(1), and $(A, fu) \in \Psi$ follows if $R \perp \Psi$. But then $f : (E, \Phi') \to (F, \Psi)$, and (F, Ψ) is a sheaf, if $R \perp \Psi$ for every covering sieve.

If R is a sieve at an object A of $\mathbf{A}$ such that $R \perp \Psi$ for every sheaf (F, Ψ), then in particular $R \perp \Phi$ for an associated sheaf (pA, Φ) of (pA, R), and $\mathrm{id}_{pA} : YA \to (pA, \Phi)$ follows. Since $\mathrm{id}_{pA} : (pA, R) \to (pA, \Phi)$ is dense for γ, the morphism $\mathrm{id}_{pA} : (pA, R) \to YA$ is dense by 41.5.2, and R covers A.

66.5. Final sieves. We say that a sieve R at an object A of $\mathbf{A}$ is *final* if A is a final lift in $\mathbf{A}$ for pA and the sink of morphisms $f : X \to A$ in R, and we say that R is a *universally final sieve* if every pullback sieve $f^{\leftarrow}A$, for a morphism $f : X \rightarrow A$ of $\mathbf{A}$, is final. A Grothendieck topology J of $\mathbf{A}$ is called a *standard Grothendieck topology* if all covering sieves for J are final (and hence universally final).

Theorem. *For a concrete category* $(\mathbf{A}, p)$ *over a quasitopos* $\mathbf{E}$ *and a* p-*topology* γ *of* $\mathbf{E}$, *the following statements are logically equivalent*
(i) *Every object* ΥA *of* $\mathbf{A}^{\mathrm{cr}}$ *is a sheaf for* γ.
(ii) *Every covering sieve for* γ *is final.*
(iii) *Every covering sieve for* γ *is universally final.*

PROOF. From the definitions, a sieve R at an object A of $\mathbf{A}$ is final iff $R \perp \Upsilon B$ for every object B of $\mathbf{A}$. Thus (i) $\Longrightarrow$ (ii) by the second part of 66.4, (ii) $\Longrightarrow$ (iii) since every pullback $f^{\leftarrow}R$ of a covering sieve R is a covering sieve, and (iii) $\Longrightarrow$ (i) by the first part of 66.4.

66.6. Combined with the results already obtained, the following result describes the largest p-topology of $\mathbf{A}^{\mathrm{cr}}$ for which every object ΥA is a sheaf.

Proposition. *Universally final sieves define a Grothendieck topology of* $\mathbf{A}$.

This topology is called the *canonical Grothendieck topology* of $\mathbf{A}$.

PROOF. A sieve ΥA clearly is final, and universally final since its pullback by $f : X \rightarrow A$ in $\mathbf{A}$ is ΥX.

Pullbacks of universally final sieves are obviously universally final.

Now let R be a universally final sieve at A, and S a sieve at A for which $u^{\leftarrow}S$ is universally final for every u in R. If $f : X \rightarrow A$ in $\mathbf{A}$, then it is easily seen that $f^{\leftarrow}R$ and $f^{\leftarrow}S$ satisfy the same conditions; thus it suffices to show that S is final.

If $h : pA \rightarrow pB$ in $\mathbf{E}$ with $hv : X \rightarrow B$ in $\mathbf{A}$ for every (X, v) in S, then $huv : Y \rightarrow B$ in $\mathbf{A}$ for (X, u) in R and (Y,v) in $u^{\leftarrow}S$. Since $u^{\leftarrow}S$ is final, $hu : X \rightarrow B$ follows. But then $h : A \rightarrow B$ as R is final, and S is indeed final.

67. Properties of Dense Sieves

67.1. Solid dense sieves. If the quasitopos $\mathbf{E}$ is solid and the functor p reflects initial objects, then it is easily seen that everything in Sections 62 and 66 remains valid if we restrict ourselves to solid p-sieves. In this situation, the only dense p-sieves which are not solid are p-sieves at an initial object of $\mathbf{E}$ which are not full, and the following result is obvious.

Proposition. *If* **E** *is a solid quasitopos and* $p : \mathbf{A} \to \mathbf{E}$ *reflects initial objects, then a dense p-sieve* (E, Φ) *is solid if and only if* Φ *contains all pairs* (T, u), *for an object T of* **A** *with pT an initial object of* **E** *and the unique morphism* $u : pT \to E$ *of* **E**.

67.2. Epimorphic sieves. We say that a p-sieve Φ is *epimorphic* if the morphisms u, for pairs (A, u) in Φ, are collectively epimorphic. A dense p-sieve is epimorphic, and there is a converse for topoi.

Proposition. *If* **E** *is a monosieve-complete topos, then the dense p-sieves coincide with the epimorphic p-sieves.*

PROOF. Let $m_1 : (E_1, \Phi_1) \to (E, \Phi)$ be a coreflection for dense p-sieves. If $am_1 = bm_1$ in **E**, then $au = bu$ for every pair (A, u) in Φ. If Φ is epimorphic, then $a = b$ follows, and m_1 is epimorphic as well as monomorphic. Thus if **E** is a topos, then m_1 is an isomorphism, and Φ is dense.

67.3. Categories with constant maps. We recall that categories with constant maps have been defined in 11.6.

Proposition. *If* $(\mathbf{A}, p)$ *is a concrete category over sets with constant maps and with an object C with pC non-empty, then all conditions for* **A** *and* **E** *are satisfied, and there is a fine dense p-sieve at every set E.*

PROOF. The topos SET is bounded, hence monosieve-complete, and surjective p-sieves are dense by 67.2. If **A** has constant morphisms, then an empty object of **A** is initial, so that p reflects initial objects. A full p-sieve at a set E contains the constant mappings $pC \to E$ and thus is dense. For objects A and B of **A**, the p-sieve $\Upsilon A \times \Upsilon B$ contains all constant mappings $\langle u, v \rangle$ for constant mappings $u : C \to A$ and $v : C \to B$; thus the sieve is surjective. For an object A of **A** and a mapping $f : E \to pA$, the sieve $f^* \Upsilon A$ contains the constant mappings $u : C \to E$; thus it is surjective. If Φ is a dense p-sieve at a set E and $x \in E$, then there is a pair (A, u) in Φ with $x = u(z)$ for some z in A. But then every constant mapping $g : pB \to E$ with range $\{x\}$, for an object B of **A**, factors $g = uh$ for a constant morphism $h : B \to A$, and $(B, g) \in \Phi$. This shows that pairs (B, g) with $g : pB \to E$ a constant mapping form a fine dense p-sieve at E.

67.4. p-dense sieves in A. We say that a sieve R in **A** is p-*dense* if the p-sieve induced by R is dense.

Proposition. *p-dense sieves in* **A** *define a Grothendieck topology of* **A**.

PROOF. Every sieve ΥA is p-dense, by 61.6.

If R is a p-dense sieve at A, then the pullback $f^{\leftarrow}R$ by $f : X \to A$ in $\mathbf{A}$ is also the pullback of R by $f : \mathsf{Y}X \to \mathsf{Y}A$ in $\mathbf{A}^{cd}$, hence p-dense by 61.4.

If R is a p-dense sieve at A, and S a sieve at A such that $u^{\leftarrow}S$ is p-dense for every u in R, consider a cone $\lambda : pD_S \to E$ in $\mathbf{E}$. For (X, u) in R, the morphisms λ_{uv} with v in $u^{\leftarrow}S$ form a cone; thus there is a unique morphism $\mu_u : pX \to E$ in $\mathbf{E}$ with $\lambda_{uv} = \mu_u \cdot v$ for every v in $u^{\leftarrow}S$. Being unique, the morphisms μ_u form a cone $pD_R \to E$. Thus $\mu_u = gu$ for all u in R, for a unique $g : pA \to E$ in $\mathbf{E}$. If $\lambda_v = fv$, for $f : pA \to B$ in $\mathbf{E}$ and every $v \in S$, then it is easily seen that $\mu_u = fu$ for every u in R, and thus $f = g$. Conversely, if $(X, v) \in S$, then $(vw)^{\leftarrow}S = \mathsf{Y}Z$ for (Z, w) in $v^{\leftarrow}R$, and $\mu_{vw} = \lambda_{vw}$ follows. Thus

$$\lambda_v \cdot w = \lambda_{vw} = \mu_{vw} = gvw$$

for w in $v^{\leftarrow}R$. As $v^{\leftarrow}R$ is p-dense, $\lambda_v = gv$ follows.

67.5. Discussion. We now have three classes of sieves in $\mathbf{A}$ with similar definitions: quotient sieves (48.1), final sieves (66.5), and p-dense sieves. These classes may well be distinct.

For examples, let $(\mathbf{A}, p)$ be the category of compact Hausdorff spaces with the forgetful functor p to sets. If A is the Stone-Čech compactification of the real open interval $(0, 1)$, then it is easily seen that the sieve R at A generated by the inclusions $[\frac{1}{n}, 1 - \frac{1}{n}] \to A$, for $n = 1, 2, 3, \ldots$, is a quotient sieve for compact Hausdorff spaces. However, R is not collectively surjective and thus neither a p-dense sieve nor a final sieve. For an infinite compact Hausdorff space K, the sieve of constant maps with codomain K is p-dense, but neither a quotient sieve nor final.

We note the following relation between the three classes of sieves.

Proposition. *A p-dense sieve R in $\mathbf{A}$ is a quotient sieve if and only if R is final.*

PROOF. If $\mu : D_R \to C$ is a cone, for a p-dense sieve R at A and an object C of $\mathbf{A}$, then $\mu_u = f \cdot u$ in $\mathbf{E}$, for every u in R and a unique $f : pA \to pC$ in $\mathbf{E}$. The sieve R then is a quotient sieve iff this f lifts to $f : A \to C$ in $\mathbf{A}$ for every cone $\mu : D_R \to C$, and the same condition says that R is final.

67.6. p-topologies for $\mathbf{A}^{cd}$. We define p-topologies for $\mathbf{A}^{cd}$ in the same way as for $\mathbf{A}^{cr}$. The properties of p-topologies obtained in Section 66 then remain valid, except that covering sieves should be restricted to covering p-dense sieves.

Since every p-sieve containing a dense p-sieve is dense, the restriction of a p-topology of $\mathbf{A}^{cr}$ to $\mathbf{A}^{cd}$ is a p-topology of $\mathbf{A}^{cd}$. It is easily seen, using 67.4 and the results of Section 66, that every p-topology of $\mathbf{A}^{cd}$ is obtained in this way, from a unique Grothendieck topology of $\mathbf{A}$ which consists of p-dense sieves. An object (E, Φ) of $\mathbf{A}^{cd}$ then is a sheaf for the p-topology if and only if $R \perp \Phi$ for every p-dense covering sieve R in $\mathbf{A}$.

68. Dense Completions

68.1. Finally dense extensions. We consider in this Section a concrete functor $G : (\mathbf{A}, p) \to (\mathbf{B}, q)$ over a quasitopos $\mathbf{E}$. We assume that $\mathbf{A}$ satisfies the conditions of 61.1, and that $\mathbf{E}$ is monosieve-complete, with full p-sieves dense (see 64.1). We associate with G a concrete functor $G^\# : \mathbf{B} \to \mathbf{A}^{cr}$ over $\mathbf{E}$, by putting $G^\# B = (qB, \Gamma B)$, with ΓB consisting of all pairs (A, u) with A an object of $\mathbf{A}$ and $u : GA \to B$ in $\mathbf{B}$. It is easily seen that this defines a functor. We say that G, and by *abus de langage* also $(\mathbf{B}, q)$, is a *finally dense extension* of $(\mathbf{A}, p)$ if G is full and finally dense, i.e. every object B of $\mathbf{B}$ is a final lift for the induced cone of ΓB.

Proposition. *A concrete functor $G : (\mathbf{A}, p) \to (\mathbf{B}, q)$ is full if and only if $G^\# G = \mathsf{Y}$, and finally dense if and only if the functor $G^\#$ is full.*

Proof. For $f : pX \to pA$, the pair (X, f) is in $\mathsf{Y}A$ iff $f : X \to A$ in $\mathbf{A}$, and in ΓGA iff $f : GX \to GA$ in $\mathbf{B}$. Thus $G^\# G = \mathsf{Y}$ iff G is full.

For $f : qB \to qC$ in $\mathbf{E}$, we have $f : G^\# B \to G^\# C$ in $\mathbf{A}^{cr}$ iff always $fu : GA \to C$ in $\mathbf{B}$ for $u : GA \to B$ in $\mathbf{B}$. Thus G is finally dense iff $G^\#$ is full.

68.2. Dense completions. If $G : (\mathbf{A}, p) \to (\mathbf{B}, q)$ is a finally dense full embedding, then $\mathbf{B}$ is equivalent to the full subcategory of $\mathbf{A}^{cr}$ with all p-sieves $(E, h^*\Gamma B)$ as objects, for isomorphisms $h : E \to GB$ of $\mathbf{E}$. This category is transportable. Thus we can and usually shall assume, without loss of generality, that $(\mathbf{B}, q)$ is transportable.

We say that $G : (\mathbf{A}, p) \to (\mathbf{B}, q)$, and by *abus de langage* also $(\mathbf{B}, q)$ or $\mathbf{B}$, is a *dense completion* of $(\mathbf{A}, p)$ if the functor G is full and dense and $(\mathbf{B}, q)$ transportable, and for every dense p-sieve (E, Φ), there is in $\mathbf{B}$ a final lift for the sink at E consisting of all morphisms $u : qGA \to E$ of $\mathbf{E}$ with $(A, u) \in \Phi$.

We recall that dense functors have been defined in 7.8. For an object B of $\mathbf{B}$, the p-sieve ΓB regarded as a category is the category $G \downarrow B$ of 7.8. We note the following result.

68.3. Proposition. *If the functor G is finally dense and every p-sieve $G^\# B$ is dense, then G is dense. Conversely, if G is a dense completion, then G is finally dense and every p-sieve $G^\# B$ is dense.*

PROOF. For the first part, consider a cone $\lambda : GD_{\Gamma B} \to C$ in $\mathbf{B}$. If ΓB is dense, then $\lambda_u = fu$ in $\mathbf{A}^{\mathrm{cd}}$, for all (A, u) in ΓB and a unique morphism $f : G^\# B \to G^\# C$. If G is finally dense, then $f : B \to C$ in $\mathbf{B}$ by 68.1; thus G is dense.

If G is dense, then G clearly is finally dense. By 62.4, the induced cone of ΓB factors $m_1 \cdot \varphi$, for a monomorphism $m_1 : E_1 \to qB$ of $\mathbf{E}$ and the induced cone φ of a dense p-sieve (E_1, Φ), with $D_\Phi = D_{\Gamma B}$. If φ has a final lift C in $\mathbf{B}$, then φ lifts to $\varphi : GD_{\Gamma B} \to C$, and $m_1 : C \to B$ in $\mathbf{B}$. As $m_1\varphi : GD_{\Gamma B} \to B$ is a colimit cone, we have $\varphi = tm_1\varphi$ for a unique $t : C \to B$. Then also $m_1\varphi = m_1 tm_1\varphi$, and $m_1 t = \mathrm{id}_B$ follows. Since m_1 is monomorphic, it follows that m_1 and t are inverse isomorphisms, and thus $G^\# B$ is dense.

68.4. Concrete adjoints. For concrete functors $F : (\mathbf{B}, q) \to (\mathbf{A}, p)$ and $G : (\mathbf{A}, p) \to (\mathbf{B}, q)$ over a category $\mathbf{E}$, we say that F is a *concrete left adjoint* of G, and G a *concrete right adjoint* of F, if a morphism $f : qB \to pA$ of $\mathbf{E}$ lifts to a morphism $f : FB \to A$ in $\mathbf{A}$ if and only if f lifts to a morphism $f : B \to GA$ of $\mathbf{B}$.

We omit the straightforward proof of the following result.

Proposition. *If F is a concrete left adjoint of G, then $F \dashv G$, with units and counits of the form $\mathrm{id}_{qB} : B \to GFB$ and $\mathrm{id}_{pA} : FGA \to B$. Conversely, if $B \leqslant GFB$ and $FGA \leqslant A$ in the fibres of qB and pA, for all objects A of $\mathbf{A}$ and B of $\mathbf{B}$, then F is a concrete left adjoint of G.*

68.5. We return now to dense completions over a quasitopos $\mathbf{E}$.

Proposition. *If $G : (\mathbf{A}, p) \to (\mathbf{B}, q)$ is a finally dense full embedding, with $\mathbf{B}$ transportable and every p-sieve $G^\# B$ dense, then G is a dense completion if and only if the functor $G^\#$ has a concrete left adjoint H. This left adjoint is obtained by assigning to every dense p-sieve (E, Φ) the final lift in $\mathbf{B}$ for the induced cone of Φ.*

PROOF. If H denotes the desired left adjoint, then for a p-sieve (E, Φ) and $f : E \to qB$ in $\mathbf{E}$, we want $f : H(E, \Phi) \to B$ in $\mathbf{B}$ iff $f : (E, \Phi) \to G^\# B$ in $\mathbf{A}^{\mathrm{cd}}$, hence iff $fu : GA \to B$ in $\mathbf{B}$ for every pair (A, u) in Φ. We get this iff $H(E, \Phi)$ is the final lift in $\mathbf{B}$ for the induced cone of Φ.

68.6. Lemma. *If* $G : (\mathbf{A}, p) \to (\mathbf{B}, q)$ *is a dense completion, and if* H *is the concrete left adjoint of* $G^{\#}$, *then* $HG^{\#} = \mathrm{Id}\, \mathbf{B}$.

PROOF. This follows immediately from the construction of H in 68.5, and the hypothesis that every object B of $\mathbf{B}$ is the final lift for the induced cone of ΓB.

68.7. Proposition. *If* $G : (\mathbf{A}, p) \to (\mathbf{B}, q)$ *is full and dense, with* $\mathbf{B}$ *transportable, then* G *is a dense completion if and only if* $(\mathbf{B}, q)$ *is quasitopological over* $\mathbf{E}$.

PROOF. With 63.2, the "if" part follows immediately from the definitions.

For objects B_i of $\mathbf{B}$ and morphisms $f_i : qB_i \to E$ in $\mathbf{E}$, the final p-sieve (E, Φ) for the morphisms f_i and the p-sieves ΓB_i, is dense if the f_i form a quotient sink, by 63.7. If (E, Φ) has a final lift $C = H(E, \Phi)$ in $\mathbf{B}$, then $g : C \to B$ in $\mathbf{B}$, for a morphism $g : E \to qB$ of $\mathbf{E}$, iff $g : (E, \Phi) \to G^{\#} B$ in $\mathbf{A}^{\mathrm{cd}}$, hence iff $g f_i : G^{\#} B_i \to G^{\#} B$ for every f_i, and hence iff $g f_i : B_i \to B$ in $\mathbf{B}$ for every f_i. Thus C is the desired final lift in $\mathbf{B}$ for the given quotient sink.

Now consider a finite source of objects B_i of $\mathbf{B}$ and morphisms $f_i : E \to qB_i$ of $\mathbf{E}$. With the full p-sieve at E dense, the source of morphisms f_i and objects $G^{\#} B_i$ has an initial lift (E, Φ) in $\mathbf{A}^{\mathrm{cd}}$, by 63.8. For the final lift $C = H(E, \Phi)$ of the induced cone of Φ, we have $f_i : C \to B_i$ in $\mathbf{B}$ for every f_i, by 68.5. If $f_i g : B \to B_i$ in $\mathbf{B}$, for every f_i, for $g : qB \to E$ in $\mathbf{E}$, then $g : G^{\#} B \to (E, \Phi)$ in $\mathbf{A}^{\mathrm{cd}}$ by our construction. With 68.6, it follows that $g : B \to C$ in $\mathbf{B}$. Thus C is the desired initial lift.

68.8. Proposition. *If* $G : (\mathbf{A}, p) \to (\mathbf{B}, q)$ *and* $G_1 : (\mathbf{A}, p) \to (\mathbf{B}_1, q_1)$ *are dense completions, and if* $G_1 = JG$ *for a concrete full embedding* $J : (\mathbf{B}, q) \to (\mathbf{B}_1, q_1)$, *then* $G_1^{\#} J = G^{\#}$, *and* J *has a concrete left adjoint.*

PROOF. For $u : pA \to qB$, and for objects A of $\mathbf{A}$ and B of $\mathbf{B}$, we have

$$(A, u) \in G_1^{\#} JB \iff u : G_1 A \to JB \iff u : GA \to B$$

by the definitions and assumptions, and $G_1^{\#} J = G^{\#}$ follows. Now if H is the concrete left adjoint of $G^{\#}$, then

$$f : C \to JB \iff f : G_1^{\#} C \to G^{\#} B \iff f : HG_1^{\#} C \to B$$

for $f : q_1 C \to qB$ in $\mathbf{E}$ and objects B of $\mathbf{B}$ and C of $\mathbf{B}_1$. Thus $HG_1^{\#} \dashv J$.

69. Quasitopos Completions and Quasitopos Hulls

69.1. We begin with a more general situation.

Theorem. *If* $(\mathbf{T}, p)$ *is a quasitopological category over a quasitopos* $\mathbf{E}$ *with quotient cone factorizations, then* $\mathbf{T}$ *is a quasitopos if and only if pullbacks in* $\mathbf{T}$ *preserve final lifts for quotient sinks.*

PROOF. This follows from 64.6 and 64.4 in the same way as 18.6 follows from 18.4 and 18.5.

69.2. Quasitopos completions. We return now to the assumptions of 68.1. We say that a dense completion $G : (\mathbf{A}, p) \to (\mathbf{B}, q)$ of $(\mathbf{A}, p)$, and by *abus de langage* also $(\mathbf{B}, q)$ or $\mathbf{B}$, is a *quasitopos completion* of $(\mathbf{A}, p)$ if $\mathbf{B}$ is a quasitopos.

Theorem. *If* $\mathbf{E}$ *is a monosieve-complete quasitopos with full* p-*sieves dense, and* $(\mathbf{A}, p)$ *a concrete category over* $\mathbf{E}$ *which satisfies the conditions of 61.1, then the following statements are equivalent for a dense completion* $G : (\mathbf{A}, p) \to (\mathbf{B}, q)$ *of* $(\mathbf{A}, p)$.

    (1) $\mathbf{B}$ *is a quasitopos.*
    (2) *Pullbacks in* $\mathbf{B}$ *preserve final lifts for quotient sinks.*
    (3) *Pullbacks in* $\mathbf{B}$ *preserve final lifts for dense* p-*sieves.*
    (4) $\mathbf{B}$ *is isomorphic to the category of sheaves for a* p-*topology of* $\mathbf{A}^{\mathrm{cd}}$, *with all covering sieves in* $\mathbf{A}$ *final and* p-*dense.*

PROOF. (4) $\Longrightarrow$ (1) trivially, (1) $\Longrightarrow$ (2) by 69.1, and (2) $\Longrightarrow$ (3) with 63.2.

Now let JA, for an object A of $\mathbf{A}$, be the class of universally final and p-dense sieves R in $\mathbf{A}$ with codomain A, such that GA is a final lift in $\mathbf{B}$ for the sink of morphisms $u : GX \to GA$ with $u : X \to A$ in R. The full sieve ΥA at A is in JA since since the functor G is dense. A pullback $f^{\leftarrow}R$ of R in JA by $f : C \to A$ in $\mathbf{A}$ is universally final, p-dense by 67.4, and thus in JC if $\mathbf{B}$ satisfies (3). If R is in JA, and $u^{\leftarrow}S$ in JX for every $u : X \to R$ in $\mathbf{A}$, then S is universally final and p-dense by 66.6 and 67.4, and one sees as in the proof of 66.6 that GA is a final lift in $\mathbf{B}$ for the sink of morphisms $v : GX \to GA$ with $v : X \to A$ in S. Thus the classes JA define a Grothendieck topology J of $\mathbf{A}$ if (3) is valid.

Let γ be the p-topology of $\mathbf{A}^{\mathrm{cd}}$ corresponding to J. For R in JA, a morphism $u : (pA, R) \to G^{\#}B$ in $\mathbf{A}^{\mathrm{cd}}$ lifts to $u : \Upsilon A \to G^{\#}B$, by the definition of J. Thus $R \perp G^{\#}B$, and $G^{\#}B$ is a sheaf for γ by 66.4. Conversely, let

(E, Φ) be an object of $\mathbf{A}^{cd}$, with (C, id_E) a final lift in $\mathbf{B}$ for the sink of morphisms $u : pA \to E$ of $\mathbf{E}$ with $(A, u) \in \Phi$. A pullback

$$
(2) \qquad
\begin{array}{ccc}
(pA, R) & \xrightarrow{\ \mathrm{id}_{pA}\ } & YA \\
\big\downarrow{\scriptstyle u} & & \big\downarrow{\scriptstyle u} \\
(E, \Phi) & \xrightarrow{\ \mathrm{id}_E\ } & G^\# C
\end{array}
$$

in $\mathbf{A}^{cd}$ preserves the final lift C in $\mathbf{B}$ if (3) is valid. Thus $R \in JA$ for a pullback (2), and $\mathrm{id}_E : (E, \Phi) \to G^\# C$ is dense for γ. If (E, Φ) is a sheaf for γ, then this is an isomorphism. Thus $(E, \Phi) = G^\# C$ by amnesticity of p^{cd}, and (4) follows from (3).

69.3. Proposition. *If $G : (\mathbf{A}, p) \to (\mathbf{B}, q)$ and $G_1 : (\mathbf{A}, p) \to (\mathbf{B}_1, q_1)$ are quasitopos completions, and if $G_1 = JG$ for a concrete full embedding $J : (\mathbf{B}, q) \to (\mathbf{B} - 1, q_1)$, then J is the right adjoint part of a geometric morphism.*

PROOF. By the proof of 68.8, the functor J has a left adjoint $HG_1^\#$, with $G_1^\#$ a right adjoint, and $H \dashv G^\#$ an associated sheaf functor by 69.2. Thus $HG_1^\#$ preserves finite limits.

69.4. Quasitopos hulls. Under the assumptions of 68.1, quasitopos completions of $\mathbf{A}$ are determined, up to isomorphism, by p-topologies of $\mathbf{A}^{cd}$ for which all objects YA of $\mathbf{A}^{cd}$ are sheaves. These p-topologies are determined by their p-dense covering sieves, which must form a Grothendieck topology of $\mathbf{A}$, consisting of p-dense sieves in $\mathbf{A}$ which are universally final, or equivalently p-dense universal quotient sieves. We obtain the smallest quasitopos completion of $\mathbf{A}$, again up to isomorphism, as the category of sheaves for the largest Grothendieck topology of $\mathbf{A}$ which satisfies these conditions. We call this quasitopos completion the *quasitopos hull* of $(\mathbf{A}, p)$, or of $\mathbf{A}$, over $\mathbf{E}$.

Theorem. *If $(\mathbf{A}, p)$ is a concrete category over a monosieve-complete quasitopos, satisfying the conditions of 61.1 and with every full p-sieve dense, then $\mathbf{A}$ has a quasitopos hull over $\mathbf{E}$. The objects of this hull are all objects (E, Φ) of $\mathbf{A}^{cd}$ which satisfy $R \perp \Phi$ (66.4) for every p-dense and universally final sieve R in $\mathbf{A}$, and the universally final p-dense sieves in $\mathbf{A}$ are the p-dense final sieves which remain final for the quasitopos hull.*

PROOF. With 66.6 and 67.4, this follows immediately from the preceding discussion and 66.4.

69.5.   The following result generalizes Theorem 1 in [7].

Theorem. *Let* $G : (\mathbf{A}, p) \to (\mathbf{B}, q)$ *be a quasitopos completion over a monosieve-complete quasitopos* $\mathbf{E}$, *and let* $\mathbf{C}$ *be the full subcategory of* $\mathbf{B}$ *with initial lifts for sources with codomains of the form* $[GA, \widetilde{GB}]$ *as its objects, for objects* A, B *of* $\mathbf{A}$. *If* $(\mathbf{A}, p)$ *satisfies the conditions of 64.7, then* $\mathbf{C}$ *is the quasitopos hull of* $\mathbf{A}$.

This result was proved in [7] with the hypotheses that $(\mathbf{B}, q)$ is topological over $\mathbf{E}$, and that partial morphisms in $\mathbf{B}$ are concretely represented, but the proof in [7] can be generalized easily. Similar results have been obtained for cartesian closed dense extensions, and for dense extensions with partial morphisms represented.

69.6.   Comments.   H. HERRLICH has observed in [46] that topological quasitopoi over sets are injective objects in a category with concrete categories as its objects, and concrete functors preserving finite products and subspaces, and hence finite limits, as its morphisms. In this framework, a topological quasitopos hull becomes an injective hull. This is an attractive result, in particular since HERRLICH has similar results for topological categories over sets, for cartesian closed topological categories over sets, and for topological categories over sets with partial morphisms represented, but time and space prevent me from including HERRLICH's results in these Notes.

Some authors consider a concrete category $(\mathbf{A}, p)$ over a locally small category $\mathbf{E}$ to be a legitimate category only if the fibres $p^\leftarrow E$, for objects E of $\mathbf{E}$ are small (*i.e.* not proper classes). For these authors, quasitopos hulls and quasitopos completions do not always exist, since the quasitopos hull of a category $\mathbf{A}$ with small fibres need not have small fibres. An example for this is given in [7]. Known quasitopos hulls of topological categories with small fibres are mostly categories with small fibres, as we shall see in Section 70.

70. Examples

70.1.   Remarks.   All examples in this Section are concrete categories $(\mathbf{A}, p)$ over sets, with constant maps. We consider only solid dense sieves; these are the sieves (E, Φ) containing all constant mappings, *i.e.* all pairs (A, u) with A an object of $\mathbf{A}$ and $u : pA \to E$ constant.

The categories of uniform spaces and of supertopological spaces (70.9), and their quasitopos hulls, are also concrete categories over B-sets and bornological sets (31.6), but we shall not discuss this aspect of the theory.

70.2.   Pseudotopological spaces. In [24], B. Day and M. Kelly gave two characterizations of quotient maps in TOP which are preserved by pullbacks in TOP. Both characterizations are easily generalized to sieves. We present one of them.

Proposition. *A sieve R in* TOP*, at a topological space X, is universally final if and only if for every pair (x, φ), consisting of a point x of X and an ultrafilter φ on X converging to x in X, there is in R a map $f : Y \rightarrow X$ such that $x = f(y)$, and $\varphi = (\mathsf{U}f)(\psi)$, for a point y of Y and an ultrafilter ψ on Y, with ψ converging to y in Y.*

Thus a sieve R in TOP *is universally final iff R remains a final sieve for pseudotopological spaces, and the category* PSTOP *of pseudotopological spaces* (31.5) *is a quasitopos hull of* TOP.

Here U is the ultrafilter functor, with $S \in (\mathsf{U}f)(\psi)$ for a subset S of X iff $f^{\leftarrow}(S) \in \psi$.

PROOF. The described sieves are clearly those which remain final in the quasitopos PSTOP. Thus the quasitopos hull of TOP includes PSTOP, and it must be PSTOP since PSTOP is a quasitopos.

Sieves which remain final in the quasitopos PSTOP are universally final. Conversely, assume that X has the final structure in TOP for morphisms $f_i : X_i \rightarrow X$, and that φ converges to x in X, but is not an image of ψ converging to y in some X_i, with $f_i(y) = x$. Let X_0 be the underlying set of X, with $X_0 \setminus \{x\}$ discrete and $\varphi \vee [\{x\}]$ the neighborhood filter of x. Pulling back f_i by id $: X_0 \rightarrow X$ results in a discrete space on the underlying set of X_i; thus the pullback sieve of the final sieve is not final.

70.3.   Comment. The other characterization of universal quotient sinks in TOP shows that the quasitopos hull of TOP is also the category of *solid convergence spaces* of M. Schroder [91]. This led to the identification of solid convergence spaces with pseudotopological spaces ([104, 105]), a prime exhibit in the small gallery of theorems discovered only by advanced categorical methods.

70.4.   Preordered sets. We recall that a preordered set is a set with a reflexive and transitive relation. The category of preordered sets, with order preserving mappings, is topological over sets, and cartesian closed. For ordered sets S and T, the "function space" $[S, T]$ is the set of all order preserving maps $f : S \rightarrow T$, with $f \leqslant g$ iff $f(y) \leqslant g(y)$ for all $y \in S$. Order preserving mappings $\varphi : R \times S \rightarrow T$ and $\hat{\varphi} : R \rightarrow [S, T]$ are exponentially adjoint if $\hat{\varphi}(x)(y) = \varphi(x, y)$ for all pairs $(x, y) \in R \times S$.

The category of order preserving maps is not a quasitopos; a coarse one-point extension of a preorder is reflexive, but not transitive.

70.5. Sets with relations. Objects are now pairs (S, σ) consisting of a set S and a binary relation σ on S, and morphisms $f : (S, \sigma) \to (T, \tau)$ are mappings $f : S \to T$ such that always $f(x)\tau f(y)$ if $x\sigma y$. This category is a topological quasitopos over sets.

Partial morphisms are represented by coarse one-point extensions $\vartheta :$ $(S, \sigma) \to (\tilde{S}, \tilde{\sigma})$, with σ the restriction of $\tilde{\sigma}$ to S, and $x\tilde{\sigma}y$ for every pair with x or y the added point.

For objects (S, σ) and (T, τ), the exponential object is the set T^S of all mappings $f : S \to T$, with the relation $[\sigma, \tau]$ obtained by putting $f[\sigma, \tau]g$ iff $f(x)\tau g(y)$ for every pair (x, y) with $x\sigma y$.

We omit the easy verifications that these constructions work properly.

70.6. Sets with reflexive relations. We restrict the preceding example to the full subcategory of sets with a reflexive relation. In this example, partial morphisms are represented in the same way, but the cartesian closed structure must be changed.

For pairs (S, σ) and (T, τ) consisting of a set and a relation on the set, we have $f[\sigma, \tau]f$ for a function $f : S \to T$ only if $f : (S, \sigma) \to (T, \tau)$. Thus if we consider sets with a reflexive relation, then exponential objects must be changed, by restricting the relation $[\sigma, \tau]$ of the preceding example to the set of morphism $f : (S, \sigma) \to (T, \tau)$. It follows that sets with reflexive relations also define a topological quasitopos over sets.

For collectively surjective mappings $f_i : S_i \to S$ and a preorder relation σ_i on each set S_i, the final lift (S, σ) in the category of sets with relations has $x\sigma y$ iff $x = f_i(x')$ and $y = f_i(y')$ for some f_i, with $x'\sigma_i y'$. The relation σ obtained in this way is reflexive, but it need not be transitive. If it is, then it is easily seen that the sink is universally final, for preordered sets as well as for sets and relations. In any case, the final lift for the given sink is the transitive closure of σ. If σ is not transitive, choose x, y, z in S so that $x\sigma y$ and $y\sigma z$, but not $x\sigma z$. Then pulling f_i back by the inclusion $\{x, z\} \to S$, results in morphisms f_i' with the final lift the identity relation on $\{x, z\}$, and the final sink is not universal in the category of preordered sets.

A final lift for sets with reflexive relations remains final for sets with relations, but the inclusion functor from preordered sets to sets with relations is not dense, and not even finally dense. If a relation σ on a set A satisfies $x\sigma y$ for distinct elements x, y of S, but $x\sigma x$ and $y\sigma y$ are not both satisfied, then σ cannot be the final lift for mappings $f_i : S_i \to S$ and reflexive relations σ_i on the sets S_i.

70.7. Quasitopologies. The category of *quasitopologies* of E. SPANIER [93] is a full subcategory of $\mathbf{K}^{\mathrm{cd}}$, for the category $\mathbf{K}$ of compact Hausdorff spaces. Its objects are p-sieves (E, Φ), for the forgetful functor p from compact Hausdorff spaces to sets, which meet the following conditions.

(1) If X is a compact Hausdorff space and $u : pX \to E$ a constant mapping, then $(X, u) \in \Phi$. (This means that (E, Φ) must be p-dense.)

(2) If $(X, ug) \in \Phi$, for a surjective map $g : X \to Y$ of compact Hausdorff spaces and a mapping $u : pY \to E$, then $(Y, u) \in \Phi$.

(3) For the injections $X \xrightarrow{h} X \amalg Y \xleftarrow{k} Y$ of a coproduct of compact Hausdorff spaces, a pair $(X \amalg Y, f)$ always is in Φ if (X, fh) and (Y, fk) are in Φ.

It follows from (2) and (3) that (X, f) always is in Φ, for a quasitopology (E, Φ), if every pair (fu_i, X_i) is in Φ for a finite family of collectively surjective maps $u_i : X_i \to X$ in $\mathbf{K}$.

In terms of 66.4, quasitopologies are the p-sieves (E, Φ) which satisfy $R \perp \Phi$ for every finitely generated collectively surjective sieve R in the concrete category $(\mathbf{K}, p)$ of compact Hausdorff spaces. Using convergence of ultrafilters, one sees easily that a collectively surjective finite sink of maps $u_i : X_i \to X$ of compact Hausdorff spaces always has X as its final lift in $\mathbf{K}$. Pullbacks preserve collectively surjective sinks; thus these final lifts are universally final. If we define a covering sieve in $\mathbf{K}$ as a sieve with a finite collectively surjective subsink, then it follows that covering sieves are universally final and form a Grothendieck topology of $\mathbf{K}$. The sheaves for the resulting p-topology of $\mathbf{K}^{\mathrm{cd}}$ are the quasitopologies; thus SPANIER's quasitopologies define a quasitopos.

70.8. Uniform and semiuniform spaces. We call a subset S of a set $E \times E$ symmetric if always $(x, x) \in S$, $(y, y) \in S$ and $(y, x) \in S$ for $(x, y) \in S$, and we denote by E_{sym} the set of all symmetric subsets of $E \times E$, ordered by set inclusion. We denote by Δ_E the set of all (x, x) for $x \in E$. For a mapping $f : E \to F$ and a subset $\mathcal{D}$ of E_{sym}, we denote by $f^{\leftarrow}\mathcal{D}$ the set of inverse images $(f \times f)^{\leftarrow}(D)$ for $D \in \mathcal{D}$. Symmetric entourages of a uniform space X with underlying set $|X|$ form a filter in $|X|_{\mathrm{sym}}$ which we denote by $\mathcal{V}X$.

We define a *semiuniform space* A as a pair $(|A|, \mathcal{V}A)$, consisting of a set $|A|$ and a filter $\mathcal{V}A$ in $|A|_{\mathrm{sym}}$, with $\Delta_{|A|}$ a subset of every set in $\mathcal{V}A$. A map $f : A \to B$ of semiuniform spaces is a mapping $f : |A| \to |B|$ such that $f^{\leftarrow}V \in \mathcal{V}A$ for every $V \in \mathcal{V}B$. Semiuniform spaces and their maps form a topological category SUNIF over sets, with a taut embedding (11.10) UNIF $\to$ SUNIF for the category UNIF of uniform spaces. J. ADÁMEK and J. REITERMAN have shown in [5] and [6] that every semiuniform space is a quotient space of a uniform space.

J. ADÁMEK and J. REITERMAN have constructed in [6] the quasitopos hull of UNIF and of SUNIF. Their work shows that a sieve R at a semiuniform space A covers A for the canonical Grothendieck topology of SUNIF iff for every dual filter $\mathcal{D}$ in E_{sym} which satisfies $\mathcal{D} \cap \mathcal{V}A = \emptyset$, there is a map $u : X \to A$ in R such that $u^\leftarrow \mathcal{D} \cap \mathcal{V}X = \emptyset$. If E is a set and $\mathcal{D}$ a dual filter in E_{sym} which covers E, i.e. every set $\{(x, x)\}$ with $x \in E$ is in $\mathcal{D}$, then the pairs (A, u) with A a semiuniform space and $u : |A| \to E$ a mapping such that $u^\leftarrow \mathcal{D} \cap \mathcal{V}A \neq \emptyset$, form a dense sieve $\Phi_{\mathcal{D}}$ at E. The sieves Φ at E with (E, Φ) an object of the quasitopos hull of SUNIF are the intersections of sieves $\Phi_{\mathcal{D}}$; it follows that the quasitopos hull of SUNIF has small fibres over sets.

The quasitopos hull of UNIF is isomorphic to the quasitopos hull of SUNIF; we obtain it and the canonical topology of UNIF by restricting all sieves to pairs (X, u) with X a uniform space.

Partial morphisms in SUNIF are represented by coarse one-point extensions, and the embedding of SUNIF into its quasitopos hull preserves these representations. For seminuniform spaces A and B, the underlying set of the function space $[A, B]$ is the set $\mathrm{SUNIF}(A, B)$ of continuous maps $f : A \to B$. For symmetric entourages U for A and V for B, let $M_{U,V}$ denote the set of all $f : A \to B$ with $U \subset (f \times f)^\leftarrow(V)$, and $[U, V]$ the set of all pairs (f, g) in $M_{U,V} \times M_{U,V}$ such that $U \subset (f \times g)^\leftarrow(V)$. For fixed V, the sets $[U, V]$ form a dual filter basis in $\||[A, B]\||_{\mathrm{sym}}$; we denote by $[V]$ the dual filter with this basis. The function space is then $(\||[A, B]\||, \Phi)$, with Φ the intersection of all sieves $\Phi_{[V]}$.

70.9. Other examples. The quasitopos hull of D. DOITCHINOV's category of supertopological spaces (see e.g. [100]) has been obtained in [108]. H. HERRLICH and H. EHRIG [47] have constructed a topos of *projection spaces* and a quasitopos of *separated projection spaces*. E. and R. LOWEN [69, 70] have contributed *approach spaces* and related quasitopos hulls. Examples of quasitopoi over sets have also been given in Section 31. J. PENON [83] has given some additional examples of quasitopos completions, but relatively few quasitopos hulls are known.

Chapter 8

QUASITOPOI OF FUZZY SETS

Fuzzy sets were introduced in 1965 by L.A. ZADEH [109] as a tool for dealing with incomplete or inexact information. ZADEH's basic idea was to replace the "crisp" True or False interpretation of set membership by a "fuzzy" degree of membership, measured by a real number in the interval $[0, 1]$. Thus a fuzzy set (A, ε) is a set A with a characteristic function $\varepsilon : A \to [0, 1]$, with the unit interval of real numbers as codomain. If ε takes only the values 0 and 1, then the fuzzy set is called crisp.

In [37], J.A. GOGUEN proposed to generalize fuzziness by interpreting set membership in a complete Heyting algebra L, thus allowing e.g. measuring the degree $\varepsilon(x)$ of set membership by a family of real numbers instead of a single real number in the interval $[0, 1]$. We shall use this generalization throughout the present Chapter. GOGUEN [38, 39] also defined a category of L-fuzzy sets and showed that this category is cartesian closed if L is a complete Heyting algebra. We show that L-fuzzy sets then form in fact a topological quasitopos over sets. We discuss L-fuzzy sets in Section 71.

For an L-fuzzy set (A, ε) and $x \in A$, it is natural to interpret the membership degree $\varepsilon(x)$ as a truth value of the statement $x \in A$. This calls for a "fuzzy logic" with L as its lattice of truth values, and with first-order logic interpreted in terms of the complete lattice structure of L. This fuzzy logic is not the internal logic for L-fuzzy sets which is the "crisp" classical logic. Moreover, powersets in the category of L-fuzzy sets are always crisp.

Various remedies have been proposed for these apparent defects. M. BARR [11] and others have argued that it is unreasonable to expect fuzzy powersets and other fuzzy set-theoretical constructs for fuzzy sets if set membership is made fuzzy, but equality remains crisp. The question also has been raised whether morphisms of fuzzy sets, with membership degrees in a complete Heyting algebra H, should be crisp mappings or fuzzy functional relations.

If equality and morphisms remain crisp, then the category of J. GOGUEN is an appropriate category of H-fuzzy sets. If equality is made fuzzy, and fuzzy conjunction is interpreted by meets in H, then H-fuzzy sets are replaced by the H-valued sets of D. HIGGS [49]; these sets have also been called totally fuzzy sets. With H-valued functional relations as morphisms, H-valued sets for a complete Heyting algebra H form a category Set H, presented by D. HIGGS [49] and also studied by M. FOURMAN and D. SCOTT [32]. This category is a topos, equivalent to the category $\operatorname{Sh} H$ of H-valued sheaves, discussed in Section 28 of these Notes.

If equality is fuzzy, but morphisms are crisp mappings, then an appropriate category of totally fuzzy sets is the category $\operatorname{Mod}^0 H$ of H-valued models, introduced by G.P. MONRO [75], and in a more convenient form by D. PONASSE [87]. A closely related category was studied in detail by J. and J.-L. COULON [18, 19, 20, 21]. The category $\operatorname{Mod}^0 H$ is a quasitopos, equivalent to the category $\operatorname{Sep} H$ of separated presheaves over H discussed in Section 29 of these Notes. We study H-valued sets, and categories $\operatorname{Set} H$, $\operatorname{Mod}^0 H$ and $\operatorname{Mod} H$, in Sections 72–78. The basic constructions for H-valued sets are simpler than the same constructions for sheaves or separated presheaves; we present these constructions in some detail.

In [28], M. EYTAN introduced a category $\operatorname{Fuz} H$ of H-valued fuzzy sets and functional relations, with crisp equality. He claimed that this category is a topos for a complete Heyting algebra H. PITTS [84] pointed out that this is the case only if H is a complete Boolean algebra. If H is irreducible in the sense of 71.7, then $\operatorname{Fuz} H$ is a solid quasitopos, with crisp mappings. Whether EYTAN's category $\operatorname{Fuz} H$ is cartesian closed or has partial morphisms represented for other complete Heyting algebras H has not been investigated and seems doubtful. We discuss $\operatorname{Fuz} H$ and related categories in Section 79.

First order fuzzy logic is presented in Section 80, in a form suitable for all reasonable categories of fuzzy sets, but without details. As already pointed out by J. GOGUEN [37], an L-valued fuzzy first-order logic can be obtained by considering a complete lattice-ordered monoid L of fuzzy truth values, with a $\bigvee$-distributive binary operation and neutral element $\top$. Conjunction and implication can then be interpreted by the monoid operation and its right adjoint. The resulting fuzzy logic then has laws which may be quite different from the usual ones; we discuss this briefly. Categories of L-valued fuzzy sets resulting from this approach have been investigated by U. HÖHLE, L.N. STOUT and others; see *e.g.* [16, 41, 56, 57, 95, 97, 98].

The foundations of the theory of fuzzy sets have been the subject of much debate and sometimes heated controversy, and this debate is still going on. Some questions being debated are as follows. Should equality be fuzzy or crisp? Should mappings between fuzzy sets be fuzzy (*i.e.* functional relations) or crisp? What are the appropriate propositional connectives and quantifiers for fuzzy logic? What are appropriate categories of fuzzy sets? The present Chapter discusses connections of these questions with quasitopos theory and exhibits some of the evidence on which answers should be based.

71. The Category of L-fuzzy Sets

71.1. Fuzzy sets. For a complete lattice L, we define an *L-fuzzy set* (A, α) as a pair consisting of a set A and a mapping $\alpha : A \to L$.

For $x \in A$, the value $\alpha(x)$ is usually interpreted as a degree or extent of membership of x in A, so that α is a generalized characteristic function. If α takes only the values 0 and 1, or $\perp$ and $\top$, then (A, α) is called a *crisp subset* of A.

In many applications, L is the real unit interval $[0, 1]$ with the usual order; L-fuzzy sets are then the fuzzy sets of L.A. ZADEH [109]. There is no mathematical reason for considering only this lattice L, and as J. GOGUEN [37] and others have pointed out, there are many applications of fuzzy sets in which membership degrees could be e.g. lists of numbers in the real unit interval, with $L = [0, 1]^n$ for some natural number n.

Two very special cases are the following. If L is a singleton, then L-fuzzy sets are just ordinary sets. If $L = \{0, 1\}$, then $\alpha : A \to L$ is an ordinary characteristic function, and thus an L-fuzzy set is a pair (A, S) with A a set and S a subset of A.

71.2. Morphisms of fuzzy sets. We define a *morphism* $f : (A, \alpha) \to (B, \beta)$ of L-fuzzy sets as a mapping $f : A \to B$ of the underlying sets such that $\alpha(x) \leqslant \beta(f(x))$ for every $x \in A$.

This obviously defines a category of L-fuzzy sets, with composition of mappings as its composition. GOGUEN denoted this category by $\mathbf{Set}(L)$; we denote it by $\mathrm{Fzs}\, L$.

If L is a singleton, then $\mathrm{Fzs}\, L$ is isomorphic to SET, and if $L = \{0, 1\}$, then $\mathrm{Fzs}\, L$ is isomorphic to the topos of pairs of sets, discussed in 31.7.

71.3. Proposition. *For a complete lattice L, the category $\mathrm{Fzs}\, L$ of L-fuzzy sets is topological over sets, and partial morphisms in $\mathrm{Fzs}\, L$ are represented by one-point extensions.*

PROOF. For a source of mappings $f : A \to A_i$, and of L-fuzzy set structures α_i on the sets A_i, putting $\alpha(x) = \bigwedge_i \alpha_i(f_i(x))$ for $x \in A$ clearly defines an initial lift; thus $\mathrm{Fzs}\, L$ is topological over sets.

In particular, embeddings $m : (A, \alpha) \to (B, \beta)$ are injective mappings m, with $\beta(m(x)) = \alpha(x)$ for all $x \in A$; it follows easily that partial morphisms $(A, \alpha) \xleftarrow{m} \cdot \xrightarrow{f} (B, \beta)$, with m an embedding, are represented by a coarsest one-point extension $(\tilde{B}, \tilde{\beta})$, with $\tilde{\beta}(\{x\}) = \beta(x)$ for $x \in B$, and $\tilde{\beta}(\emptyset) = \top$ for the added point.

71.4. Proposition. *For a complete lattice L, the category* Fzs L *of L-fuzzy sets is cartesian closed if and only if L is a complete Heyting algebra. If this is the case, then the function set $[(B, \beta), (C, \gamma)]$ is the set $(C^B, [\beta, \gamma])$, with C^B consisting of all mappings $h : B \to C$, and*

$$[\beta, \gamma](h) = \bigwedge_{y \in B} (\beta(y) \to \gamma(h(y)))$$

as degree of membership.

Proof. A morphism $\varphi : (A, \alpha) \times (B, \beta) \to (C, \gamma)$ must satisfy

$$\alpha(x) \wedge \beta(y) \leqslant \gamma(\varphi(x, y))$$

for every pair (x, y) in $A \times B$. If L is a complete Heyting algebra, then this is the case iff always

$$\alpha(x) \leqslant \beta(y) \to \gamma(\hat{\varphi}(x)(y)),$$

and hence iff $\alpha(x) \leqslant [\beta, \gamma](\hat{\varphi}(x))$ for $x \in A$, for $\hat{\varphi}$ exponentially adjoint to φ in SET. Thus Fzs L is cartesian closed.

Conversely, let $S = \{s\}$ be a singleton, and denote by (S, a), for $a \in L$, the L-fuzzy set (S, α) with $\alpha(s) = a$. A diagram of morphisms $\mathrm{id}_S : (S, a_i \wedge a_j) \to (S, a_i)$ has a colimit cone of morphisms $\mathrm{id}_S : (S, a_i) \to (S, \bigvee_i a_i)$. Products with a singleton (S, b) preserve this colimit cone iff

$$b \wedge (\bigvee_i a_i) = \bigvee_i (b \wedge a_i)$$

in L. Thus Fzs L can only be cartesian closed if L is a complete Heyting algebra.

71.5. Crisp and void fuzzy sets. We say that a fuzzy set (A, α) is *crisp* if $\alpha(x) = \top$ for all $x \in A$, and *void* if $\alpha(x) = \bot$ for all $x \in |A|$. Crisp H-fuzzy sets are the coarse objects, and void fuzzy sets the discrete objects, of Fzs H.

A. Pultr [88] observed that assigning to every H-fuzzy set (A, α) the set of all $x \in A$ with $\alpha(x) = \top$ defines a functor from Fzs H to sets, acting on morphisms by restriction. It is easily seen that this functor is right adjoint to the crisp fuzzy set functor from sets to Fzs H.

71.6. Reduced fuzzy sets. An L-fuzzy set (A, α) may have members $x \in A$ with $\alpha(x) = \bot$ in L, i.e. members which are definitely not members of the fuzzy set. Some authors have argued that this situation should be excluded, by requiring that $\alpha(x) \neq \bot$ for all $x \in A$. We say in this case that (A, α) is

a *reduced L-fuzzy set*, and we denote by Fzr L the full subcategory of Fzs L
with reduced L-fuzzy sets as its objects.

Proposition. *For every complete lattice L, partial morphisms in Fzr L
are represented, and Fzr L is a coreflective full subcategory of Fzs L.*

PROOF. For a fuzzy set (A, α), let $r(A, \alpha)$ be the fuzzy set obtained
by removing all elements x with $\alpha(x) = \perp$, and restricting α accordingly.
This clearly defines a functor from fuzzy sets to reduced fuzzy sets, obtained
for morphisms by domain and codomain restrictions. If (B, β) is a reduced
fuzzy set, then every morphism $f : (B, \beta) \to (A, \alpha)$ in Fzs L maps (B, β) into
$r(A, \alpha)$; thus the functor r is right adjoint to the inclusion functor.

Strong monomorphisms for Fzr L are embeddings, as for Fzs L, and partial morphisms of Fzr L are represented in the same way as for Fzs L; see the
proof of 71.3.

71.7. We may call a complete Heyting algebra H *irreducible* if $x \wedge y = \perp$
in H, for x, y in H, only if $x = \perp$ or $y = \perp$. In particular, a complete chain
is an irreducible complete Heyting algebra.

Proposition. *For a complete lattice L, the category Fzr L is cartesian
closed if and only if L is an irreducible complete Heyting algebra.*

PROOF. If L is an irreducible complete Heyting algebra, then finite limits
in Fzr L are limits in Fzs L, and the proof of 71.4 shows that Fzr L is cartesian closed. The only change is that the function space $(C^B, [\beta, \gamma])$ must be
restricted to mappings g with $[\beta, \gamma](g) \neq \perp$.

If L is not irreducible, consider a singleton (S, a), and a set (B, β) with
$B = \{0, 1\}$ and $\beta(0) = \beta(1) = \top$. The coequalizer of the two morphisms from
(S, a) to (B, β) then identifies 0 and 1. If we take products with (S, b), where
$a \wedge b = \perp$, then $(S, a) \times (S, b)$ is empty in Fzr L. It follows that the coequalizer
of the two mappings is not preserved. For L irreducible, the second part of the
proof of 71.4 remains valid for Fzr L.

71.8. Let $i :$ Fzr $L \to$ Fzs L be the full and faithful inclusion functor,
and $r :$ Fzs $L \to$ Fzr L its right adjoint, with $r(A, \alpha)$ obtained by removing
from A all $x \in A$ with $\alpha(x) = \perp$. We note the following result.

Proposition. *If L is an irreducible complete Heyting algebra, then Fzr L
is quasitopological over sets, and the adjunction $i \dashv r$ is a surjective geometric
morphism.*

PROOF. Quotient sinks in Fzs L are collectively surjective. Thus the final
lift for a quotient sink is a reduced L-fuzzy set if the domain objects for the

sink are reduced fuzzy sets, and $\operatorname{Fzr} L$ admits final lifts for quotient sinks. If L is irreducible, then finite sources with reduced fuzzy sets as codomains have reduced initial lifts. Thus in this case, $\operatorname{Fzr} L$ has initial lifts for finite sources, and the functor i preserves them. It follows that i preserves finite limits, so that $i \dashv r$ is a geometric morphism. The functor i clearly reflects isomorphisms; thus $i \dashv r$ is surjective.

72. H-valued Sets and Relations

72.1. H-sets. From now on, H will always denote a complete Heyting algebra. We define an *H-valued set*, or briefly an *H-set*, as a pair $A = (|A|, \delta_A)$, consisting of a set $|A|$ and a mapping $\delta_A : |A| \times |A| \to H$, subject to the following two conditions.

(1) $\delta_A(y, x) = \delta_A(x, y)$,
(2) $\delta_A(x, y) \wedge \delta_A(y, z) \leqslant \delta_A(x, z)$,

for all elements x, y, z of $|A|$. An H-set is also called an H-valued *totally fuzzy set*.

We interpret δ_A as fuzzy equality with truth-values in H, with $\delta_A(x, y)$ the truth-value of $x =_A y$ for x, y in $|A|$.

By the usual type of *abus de langage*, we may use the same symbol for an H-set and its underlying ordinary set, and we may omit the subscript of δ_A if the context permits it.

72.2. Membership degree and fuzzy sets. The two conditions in 72.1 state that fuzzy equality $x =_A y$ is symmetric and transitive, in a world with truth values in H. Instead of reflexivity $\delta_A(x, x) = \top$, we ask for validity of a statement $x =_A x \iff x \in A$, by defining *degree* or *extent of membership* ε_A in an H-set A by $\varepsilon_A(x) = \delta_A(x, x)$, for $x \in |A|$. It follows from the conditions of 72.1 that

$$\delta_A(x, y) \leqslant \varepsilon_A(x) \wedge \varepsilon_A(y),$$

for all x, y in $|A|$.

We shall consider H-fuzzy sets to be special H-sets, with equality made crisp by putting $\delta_A(x, x') = \bot$ for distinct elements x, x' of $|A|$.

72.3. Fuzzy relations. For H-sets A and B, we define an *H-valued fuzzy relation* $f : A \to B$ as a mapping $f : |A| \times |B| \to H$ which satisfies the following conditions.

(1) $f(x, y) \leqslant \varepsilon_A(x) \wedge \varepsilon_B(y)$,
(2) $f(x, y) \wedge \delta_A(x, x') \leqslant f(x', y)$,

(3) $f(x, y) \wedge \delta_B(y, y') \leqslant f(x, y')$,

for all members x, x' of $|A|$ and y, y' of $|B|$.

These conditions state that fuzzy relations must be extensional. We note that fuzzy equality in A is a fuzzy relation $\delta_A : A \to A$.

We compose fuzzy relations $f : A \to B$ and $g : B \to C$ by putting

$$(g \circ f)(x, z) = \bigvee_y (f(x, y) \wedge g(y, z))$$

for $x \in |A|$ and $z \in |C|$, with the supremum taken over all $y \in |B|$.

Using the infinite distributive law of a complete Heyting algebra, one sees easily that this composition is a relation $g \circ f : A \to C$, and that H-sets and H-valued relations are the objects and morphisms of a category, with identity relations δ_A as identity morphisms.

72.4. Proposition. *For H-sets A and B, relations $f : A \to B$ with pointwise ordering form a complete Heyting algebra. Composition preserves suprema; we have*

$$g \circ \left(\bigvee_i f_i \right) = \bigvee_i (g \circ f_i) \quad \text{and} \quad \left(\bigvee_i g_i \right) \circ f = \bigvee_i (g_i \circ f)$$

if the compositions are defined.

PROOF. Pointwise ordering means that we put $f \leqslant g$ iff $f(x, y) \leqslant g(x, y)$ for all $x \in |A|$ and $y \in |B|$. For relations $f_i : A \to B$, it is easily seen that

$$\left(\bigvee f_i \right)(x, y) = \bigvee f_i(x, y),$$

for $x \in |A|$ and $y \in |B|$ defines an H-valued relation from A to B; this is clearly the pointwise supremum. Pointwise meets

$$(f \wedge g)(x, y) = f(x, y) \wedge g(x, y)$$

are obtained similarly, and

$$\top_{A,B}(x, y) = \varepsilon_A(x) \wedge \varepsilon_B(y)$$

defines a full relation $\top_{A,B}$. With these constructions, the formal laws of a complete Heyting algebra are easily verified. It is also easy to verify that composition $g \circ f$ preserves suprema.

72.5. Dual relations. Putting $f^{\mathrm{op}}(y, x) = f(x, y)$ assigns to every H-valued relation $f : A \to B$ a *dual relation* $f^{\mathrm{op}} : B \to A$. This construction is clearly involutive, *i.e.* $(f^{\mathrm{op}})^{\mathrm{op}} = f$. We have

$$(g \circ f)^{\mathrm{op}} = f^{\mathrm{op}} \circ g^{\mathrm{op}}$$

if the compositions are defined, and identity relations δ_A are self-dual. Duality clearly preserves order.

Lemma. *Every H-valued relation $f : A \to B$ satisfies*

$$(1) \qquad\qquad f \leqslant f \circ f^{\mathrm{op}} \circ f.$$

PROOF. We have

$$(f \circ f^{\mathrm{op}} \circ f)(x, y) = \bigvee_u \bigvee_v (f(x, u) \wedge f^{\mathrm{op}}(u, v) \wedge f(v, y))$$

$$\geqslant f(x, y) \wedge f^{\mathrm{op}}(y, x) \wedge f(x, y) = f(x, y)$$

for $x \in |A|$ and $y \in |B|$.

72.6. Some types of relations. For H-sets A and B, we say that an H-valued relation $f : A \to B$ is:

single-valued if always $f(x, y) \wedge f(x, y') \leqslant \delta_B(y, y')$,
injective if always $f(x, y) \wedge f(x', y) \leqslant \delta_A(x, x')$,
total if always $\bigvee_{y \in |B|} f(x, y) = \varepsilon_A(x)$,
surjective if always $\bigvee_{x \in |A|} f(x, y) = \varepsilon_B(y)$.

These concepts are dual in pairs; a relation f is single-valued or total iff f^{op} is injective or surjective respectively. If we translate the conditions into first-order H-valued logic, they become

$$f(x, y) \wedge f(x, y') \implies y = y'$$

for being single-valued, and

$$x \in A \iff (\exists y \in B)(f(x, y))$$

for totality, with dual statements for injectivity and surjectivity.

Identity relations δ_A clearly satisfy all four conditions.

72.7. Proposition. *An H-valued relation $f : A \to B$ is single-valued if and only if $f f^{\mathrm{op}} \leqslant \delta_B$, injective if and only if $f^{\mathrm{op}} f \leqslant \delta_A$, total if and only if $f^{\mathrm{op}} f \geqslant \delta_A$, and surjective if and only if $f f^{\mathrm{op}} \geqslant \delta_B$.*

Corollary. *Single-valued, injective, total and surjective relations define subcategories of the category of H-sets and H-valued relations, with $f = f f^{\mathrm{op}} f$ if f is single-valued or injective.*

PROOF. The first two claims follow immediately from the definitions. If $f^{\mathrm{op}} f \geqslant \delta_A$, then in particular

$$(f^{\mathrm{op}} f)(x, x) = \bigvee_y f(x, y) \geqslant \varepsilon_A(x)$$

for $x \in A$, and with 72.3.(1), f is total.

Conversely, if f is total, then

$$\delta_A(x, x') = \delta_A(x, x') \wedge \left(\bigvee_y f(x, y)\right) = \bigvee_y (\delta_A(x, x') \wedge f(x, y) \wedge f(x, y))$$

$$\leqslant \bigvee_y (f(x, y) \wedge f(x', y)) = (f^{\mathrm{op}} f)(x, x')$$

for x, x' in A.

The proof for surjective relations is dual to this.

For the Corollary, we have already noted that identity relations are in each of the proposed subcategories. If $f : A \to B$ and $g : B \to C$ are single-valued, then we have

$$gf(gf)^{\mathrm{op}} = gff^{\mathrm{op}}g^{\mathrm{op}} \leqslant g\delta_B g^{\mathrm{op}} = gg^{\mathrm{op}} \leqslant \delta_C ;$$

thus gf is single-valued. Closure under composition is obtained similarly for the other three conditions. If f is single-valued or injective, then $ff^{\mathrm{op}}f \leqslant f$, and equality results from 72.5.

73. Categories Set H and Mod H of H-valued sets.

73.1. The category Set H. Let H again be a complete Heyting algebra. We say that an H-valued relation $f : A \to B$ of H-sets is *functional* if f is single-valued and total. By 72.7, functional H-valued relations are the morphisms of a category with H-sets as objects. We denote this category by Set H. Morphisms of Set H could be called H-valued *fuzzy mappings*. The category Set H was introduced and studied by D. HIGGS in a never published but widely distributed preprint [49]. Many results of [49] and some additional results can be found in [32]. The category Set H is clearly the appropriate category for H-valued fuzzy sets with fuzzy equality and fuzzy mappings.

73.2. We show first that the ordering of relations in 72.4 becomes discrete when restricted to functional relations.

Proposition. *If $f : A \to B$ and $g : A \to B$ in Set H and $f \leqslant g$, then $f = g$.*

PROOF. We have $\delta_B \leqslant f^{\mathrm{op}} f$ and $gg^{\mathrm{op}} \leqslant \delta_A$ by 72.7, and $f^{\mathrm{op}} \leqslant g^{\mathrm{op}}$ if $f \leqslant g$. Thus $g \leqslant gf^{\mathrm{op}} f \leqslant gg^{\mathrm{op}} f \leqslant f$ if $f \leqslant g$.

73.3. Induced morphisms. We say that a morphism $f : A \to B$ of Set H is *induced* by a mapping $\bar{f} : |A| \to |B|$ of the underlying sets if

(1) $$f(x, y) = \varepsilon_A(x) \wedge \delta_B(\bar{f}(x), y)$$

for all $x \in |A|$ and $y \in |B|$.

The identity morphism δ_A is an example; it is clearly induced by the identity mapping $\mathrm{id}_{|A|}$.

Proposition. *For H-sets A and B, a mapping $\bar{f} : |A| \to |B|$ induces a morphism $f : A \to B$ of Set H if and only if*

$$(2) \qquad\qquad \delta_A(x, x'), \leqslant \delta_B(\bar{f}(x), \bar{f}(x'))$$

for all x, x' in $|A|$.

PROOF. The mapping f defined by (1) clearly satisfies conditions (1) and (3) of 72.3, and f is single-valued if f is a relation. We have

$$\bigvee_y f(x, y) = \varepsilon_A(x) \wedge \delta_B(\bar{f}(x), \bar{f}(x))$$

for $x \in |A|$. Thus f is total iff always $\varepsilon_A(x) \leqslant \varepsilon_B(\bar{f}(x))$; this is a special case of (2). Condition 72.3.(2) for f defined by (1) above means that always

$$\delta_A(x, x') \wedge \delta_B(\bar{f}(x), y) \;\leqslant\; \delta_B(\bar{f}(x'), y).$$

This is clearly the case if (2) above is satisfied. Conversely, if f defined by (1) is a total relation, then 72.3.(2) for f becomes (2) above for $y = \bar{f}(x)$.

73.4. If $f : A \to B$ in Set H is induced by a mapping $\bar{f} : |A| \to |B|$, then $\bar{f}$ need not be determined uniquely by f. The following result makes clear what happens.

Proposition. *If a morphisms $f : A \to B$ of Set H is induced by a mapping $\bar{f} : |A| \to |B|$, then the following are equivalent for a mapping $g : |A| \to |B|$.*

(i) $\varepsilon_A(x) \wedge \delta_B(\bar{f}(x), y) \leqslant \delta_B(g(x), y)$ *for all $x \in |A|$ and $y \in |B|$.*
(ii) $\varepsilon_A(x) \leqslant \delta_B(\bar{f}(x), g(x))$ *for all $x \in |A|$.*
(iii) *g induces the same morphism of Set H as $\bar{f}$.*

PROOF. (i) follows immediately from (iii) and the definitions.
We get (ii) immediately from 73.3.(2), with $x' = x$, and from (i) with $y = \bar{f}(x)$.
If (ii) is valid, then

$$\varepsilon_A(x) \wedge \delta_B(\bar{f}(x), y) = \varepsilon_A(x) \wedge \delta_B(\bar{f}(x), y) \wedge \delta_B(\bar{f}(x), g(x))$$
$$= \varepsilon_A(x) \wedge \delta_B(g(x), y) \wedge \delta_B(\bar{f}(x), g(x)) = \varepsilon_A(x) \wedge \delta_B(g(x), y),$$

and (iii) is valid.

73.5. Composition with induced morphisms. Compositions gf with g induced by a mapping cannot be simplified much, but we note the following result.

Proposition. *If* $f : A \to B$ *is induced by* $\bar{f} : |A| \to |B|$, *then*

$$(gf)(x, z) = g(\bar{f}(x), z)$$

for $g : B \to C$, *and for* $x \in |A|$ *and* $z \in |C|$. *In particular,* gf *is induced by* $\bar{g}\bar{f}$ *if* g *is induced by* $\bar{g} : |B| \to |C|$.

PROOF. The composition is given by

$$(gf)(x, z) = \bigvee_y (\varepsilon_A(x) \wedge \delta_B(\bar{f}(x), y) \wedge g(y, z)) = \varepsilon_A(x) \wedge g(\bar{f}(x), z).$$

If g is induced by $\bar{g}$, this becomes

$$(gf)(x, z) = \varepsilon_A(x) \wedge \delta_C(\bar{g}(\bar{f}(x)), z),$$

since $\varepsilon_A(x) \leqslant \varepsilon_B(\bar{f}(x))$ by 73.3.

73.6. Crisp mappings of H-sets. We cannot simply define a crisp mapping $f : A \to B$ of H-sets as a mapping $f : |A| \to |B|$ of the underlying sets. First, the mapping should preserve equality:

$$x =_A x' \quad \Longrightarrow \quad f(x) =_B f(x').$$

This means that f should satisfy 73.3.(2), and therefore induce a morphism of Set H. Second, mappings should be extensional, *i.e.*

$$(\forall x \in A)(f(x) =_B g(x)) \quad \Longrightarrow \quad f = g.$$

This means that mappings $f : |A| \to |B|$ and $g : |A| \to |B|$ should define the same mapping $A \to B$ if they satisfy 73.4.(ii) and thus induce the same morphism of H-sets.

The result of these considerations is that we define the category $\mathrm{Mod}^0 H$ of H-*valued models* as the subcategory of Set H with all H-sets as objects, but with morphisms $f : A \to B$ restricted to morphisms induced by some mapping $\bar{f} : |A| \to |B|$. This category was introduced in a different form by G.P. MONRO [75]; we use his notation. We denote by $\mathrm{Mod}\, H$ the full subcategory of $\mathrm{Mod}^0 H$ with objects A with $|A|$ not empty; see 75.8.

The categories $\mathrm{Mod}\, H$ and $\mathrm{Mod}^0 H$ are not isomorphic (see 75.8), but have very similar properties. We often prove a common property of both categories only for one of them but use it for both.

73.7. Morphisms of totally fuzzy sets. We recall that H-valued totally fuzzy sets are the same as H-sets. The considerations outlined above led D. Ponasse [86] to define the following category which he denoted by JTF. Objects are J-sets, or totally fuzzy sets, with fuzzy equality values in a complete Heyting algebra J. A morphism $f : A \to B$ is a binary relation $f : |A| \to |B|$, which must satisfy the following conditions, for elements of $|A|$ and of $|B|$.

(i) For every x in $|A|$, there is y in $|B|$ such that xfy.

(ii) If xfy and $x'fy'$, then $\delta_A(x, x') \leqslant \delta_B(y, y')$.

(iii) If xfy and $\varepsilon_A(x) \leqslant \delta_B(y, y')$, then xfy'.

A composition gf in JTF is defined as the compositon gf in the category of sets and relation, saturated so that (iii) is satisfied for the composition.

Somewhat surprisingly, the following result seems to be new. Its proof uses the Axiom of Choice.

Proposition. *For a complete Heyting algebra J, the categories* $\mathrm{Mod}^0 J$ *and* JTF *are isomorphic.*

PROOF. If $\bar{f} : |A| \to |B|$ is a choice function for a morphism $f : A \to B$ of JTF, then $\bar{f}$ induces a morphism of Set H by (ii) and 73.3, and f is the union (or supremum) of all functions $\bar{g} : |A| \to |B|$ which induce the same morphism of Set H, by (iii) and 73.4.

Conversely, if $\bar{f} : |A| \to |B|$ induces a morphism of Set H and we put xfy for $y \in |B|$ iff $\varepsilon_A(x) \leqslant \delta_B(\bar{f}(x), y)$, then it follows easily from 73.3 and 73.4 that f satisfies conditions (i), (ii) and (iii), with $\bar{f}$ as a choice function.

Composition of morphisms in JTF and in $\mathrm{Mod}^0 J$ is obtained by composition of choice functions; thus the bijections described above define inverse isomorphisms of categories.

73.8. Discussion. We can assign to every morphism $f : A \to B$ of Set H a relation $\bar{f} : |A| \to |B|$ by putting $x\bar{f}y$, for (x, y) in $|A| \times |B|$, iff $f(x, y') = \varepsilon_A(x) \wedge \delta_B(y, y')$ for every $y' \in |B|$. It is easily verified that this relation always satisfies conditions (ii) and (iii) of 73.7, and f is induced by any choice function of $\bar{f}$ if $\bar{f}$ also satisfies 73.7.(i). The relation $\bar{f}$ is then a morphism of PONASSE's category; we say that $\bar{f}$ *induces* the morphism f of $\mathrm{Mod}^0 H$.

More generally, we say that a relation $f_0 : |A| \to |B|$ *induces* a morphism $f : A \to B$ of $\mathrm{Mod}^0 H$ if f_0 satisfies conditions (i) and (ii) of 73.7, and a choice function of f_0 induces f. Putting xfy if xf_0y_1 and $\varepsilon_A(x) \leqslant \delta_B(y_1, y)$, for some $y_1 \in |B|$ then defines the morphism $\bar{f}$ of PONASSE's category which induces f.

74. H-valued Subsets

74.1. H-valued subsets. We define an H-subset structure α of an H-set A as a mapping $\alpha : |A| \to H$ which must satisfy:

(1) $\alpha(x) \leqslant \varepsilon_A(x)$,

(2) $\alpha(x) \wedge \delta_A(x, x') \leqslant \alpha(x')$,

for all x, x' in $|A|$, and the H-subset A_α of A with this structure as the H-set $(|A|, \delta_\alpha)$ with the same underlying set as A, with structure δ_α given by

$$\delta_\alpha(x, x') = \alpha(x) \wedge \delta_A(x, x') = \alpha(x') \wedge \delta_A(x, x'),$$

for x, x' in $|A|$.

We omit the straightforward proof of the following result.

Proposition. *If A_α is an H-subset of an H-set A, then $\mathrm{id}_{|A|}$ induces an injective morphism $j_\alpha = \delta_\alpha : A_\alpha \to A$ of Set H. Every H-set A is an H-subset of itself, with structure $\alpha = \varepsilon_A$, and if A_α is an H-subset of A and A_β an H-subset of H_α, then A_β is an H-subset of A.*

We call $j_\alpha = \delta_\alpha : A_\alpha \to A$ an *inclusion morphism* of H-sets.

74.2. We order H-subset structures of an H-set A pointwise, putting $\beta \leqslant \alpha$ iff $\beta(x) \leqslant \alpha(x)$ for all $x \in |A|$. This condition is satisfied iff A_β is an H-subset of A_α. It is easily verified that H-subset structures of A with this order form a complete Heyting algebra, with ε_A as largest element, and with $j_\alpha = \delta_A = \mathrm{id}_A$ for $\alpha = \varepsilon_A$. The mapping $\mathrm{id}_{|A|}$ induces a morphism $j : A_\beta \to A_\alpha$ in Set H iff $\beta \leqslant \alpha$.

The following result follows immediately from the definitions; we omit its proof.

Proposition. *If $f : A \to B$ is an H-valued relation of H-sets, then every mapping $y \mapsto f(x, y) : |B| \to H$, for $x \in |A|$, is an H-subset structure of B.*

74.3. Direct and inverse images. For $f : A \to B$ in Set H and an H-subset structure α of A, we define the *image* or *direct image* of α by f as the H-subset structure $f^{\to}\alpha$ of B given by

$$(f^{\to}\alpha)(y) = \bigvee_{x \in |A|} (\alpha(x) \wedge f(x, y)),$$

for $y \in |B|$. For an H-subset structure β of B, we put

$$(f^{\leftarrow}\beta)(x) = \bigvee_{y \in |B|} (\beta(y) \wedge f(x, y))$$

for $x \in |A|$ to define the *inverse image* of β by f. It is easily verified that $f^{\rightarrow}\alpha$ and $f^{\leftarrow}\beta$ are H-subset structures of B and of A respectively.

Lemma. *If $f : A \to B$ is induced by a mapping $\bar{f} : |A| \to |B|$, then $(f^{\leftarrow}\beta)(x) = \varepsilon_A(x) \wedge \beta(\bar{f}(x))$ for $x \in |A|$.*

PROOF. This follows easily from the definitions.

74.4. Proposition. *If $f : A \to B$ in Set H, then the following are equivalent for H-subset structures α of A and β of $|B|$.*

(i) $f^{\rightarrow}\alpha \leqslant \beta$.

(ii) $\alpha \leqslant f^{\leftarrow}\beta$.

(iii) $f j_\alpha = j_\beta g$ for a morphism $g : A_\alpha \to B_\beta$ of Set H.

If f is induced by a mapping $\bar{f}$, then g in (iii) is induced by $\bar{f}$.

PROOF. Let x, x' be in $|A|$ and y, y' in $|B|$. If (i) is valid, then

$$\bigvee_y (\beta(y) \wedge f(x,y)) \geqslant \bigvee_y (f(x,y) \wedge \bigvee_{x'}(\alpha(x') \wedge f(x',y)))$$

$$= \bigvee_{x'} (\alpha(x') \wedge \bigvee_y (f(x,y) \wedge f(x'y)))$$

$$\geqslant \bigvee_{x'} (\alpha(x') \wedge \delta_A(x,x')) = \alpha(x),$$

using $f^{op}f \geqslant \delta_A$ from 72.7, and (ii) is valid.

It is easily seen that

(1)　　　$(f j_\alpha)(x,y) = f(x,y) \wedge \alpha(x)$　　　and　　　$(j_\beta g)(x,y) = g(x,y)$

if the compositions are defined, with $g(x,y) \leqslant \beta(y)$. We have

$$\alpha(x) \wedge f(x,y) \leqslant \bigvee_{y'} (f(x,y) \wedge f(x,y') \wedge \beta(y'))$$

$$\leqslant \bigvee_{y'} (\delta_B(y,y') \wedge \beta(y')) = \beta(y),$$

if (ii) is valid, and it follows easily that $g(x,y) = f(x,y) \wedge \alpha(x)$ defines a morphism $g : A_\alpha \to B_\beta$ with $j_\beta g = f j_\alpha$.

If (iii) is valid, then $g(x,y) = f(x,y) \wedge \alpha(x) \leqslant \beta(y)$ by (1) above, and (i) follows. If f is induced by $\bar{f}$, it also follows that g is induced by $\bar{f}$.

74.5. Factorizations. The following result and 74.6.1 below show that H-subset inclusions and surjective morphisms define a diagonal polarity in Set H and in Mod H (see Section 10).

Proposition. *Every commutative square*

$$
\begin{array}{ccc}
A & \xrightarrow{\;e\;} & C \\
{\scriptstyle f}\downarrow & & \downarrow{\scriptstyle g} \\
B_\beta & \xrightarrow{\;j_\beta\;} & B
\end{array}
$$

(1)

in Set H, with e surjective and j_β an H-subset inclusion, has a unique diagonal $u : C \to B_\beta$ in Set H, with $f = ue$ and $g = j_\beta u$. If g is induced by a mapping $\bar{g}$, then u is also induced by $\bar{g}$.

PROOF. We have $ee^{\mathrm{op}} = \delta_C$ for e surjective and total, by 72.7, and then $u = fe^{\mathrm{op}}$ if $f = ue$. This relation u satisfies

$$
j_\beta u = j_\beta fe^{\mathrm{op}} = gee^{\mathrm{op}} = g,
$$

and u is total since f and e^{op} are total. By a dual argument, $u = j_\beta{}^{\mathrm{op}}g$, and it follows that u is single-valued and satisfies $ue = f$. Thus the factorization exists in Set H. For $(z, y) \in |C| \times |B|$, we have

$$
u(z, y) = \bigvee_{y' \in |B|} (g(z, y') \wedge \delta_B(y', y) \wedge \beta(y)) = g(z, y) \wedge \beta(y)
$$

since $u = j_\beta{}^{\mathrm{op}}g$. Thus u is induced by $\bar{g}$ if g is induced by $\bar{g}$.

74.6. Proposition. *For a commutative square*

$$
\begin{array}{ccc}
A_\alpha & \xrightarrow{\;g\;} & B_\beta \\
{\scriptstyle j_\alpha}\downarrow & & \downarrow{\scriptstyle j_\beta} \\
A & \xrightarrow{\;f\;} & B
\end{array}
$$

(1)

in Set H or $\mathrm{Mod}^0 H$, the morphism g is surjective if and only if $\beta = f^{\to}\alpha$, and (1) is a pullback square if and only if $\alpha = f^{\leftarrow}\beta$.

74.6.1. Corollary. *Every morphism f of Set H or $\mathrm{Mod}^0 H$ factors $f = me$ with e surjective and m an H-subset inclusion.*

PROOF. We have $g(x, y) = \alpha(x) \wedge f(x, y)$ for $x \in |A|$ and $y \in |B|$, by the proof of 74.4; it follows immediately that g is surjective iff $\beta = f^{\to}\alpha$. For $\alpha = \varepsilon_A$ and $j_\alpha = \mathrm{id}_A$, the Corollary follows.

Now assume $\alpha = f^{\leftarrow}\beta$. If $fu = j_\beta v$, then u factors $u = j_{\alpha'}e$ for a subset structure α' of A by the Corollary, with e surjective. But then we can factor $v = he$ and $fj_{\alpha'} = j_\beta h$ for a morphism h, by 74.5, and $\alpha' \leqslant \alpha$ follows from 74.4. Thus $j_{\alpha'} = j_\alpha t$ for a morphism t, and $h = gt$ follows as j_β is injective. Since $u = j_\alpha te$ and $v = gte$, we have the desired pullback. Conversely, $\alpha = f^\leftarrow \beta$ if (1) is a pullback, since pullbacks are determined up to equivalence and equivalent H-subset inclusions are equal.

74.7. Universal quantifiers. We recall from 20.2 that a universal quantifier functor $\forall_f$ along a morphism f is right adjoint to an inverse image functor $f^\leftarrow$ along f.

Proposition. *Set H and Mod H have universal quantifiers, with*

$$(1) \qquad (\forall_f \alpha)(y) = \varepsilon_B(y) \wedge \bigwedge_{x \in |A|} (f(x, y) \to \alpha(x))$$

for $f : A \to B$ in Set H, $y \in |B|$, and an H-subset structure α for A.

PROOF. For an H-subset structure β of B, we have $f^\leftarrow \beta \leqslant \alpha$ iff $f(x, y) \wedge \beta(y) \leqslant \alpha(x)$ for every $x \in |A|$ and $y \in |B|$, hence iff always $\beta(y) \leqslant f(x, y) \to \alpha(x)$, and hence iff $\beta \leqslant \forall_f \alpha$, as given by (1). It is easily verified that (1) defines an H-subset structure of B; we omit the details.

75. Constructions for H-valued Sets

75.1. Isomorphisms. These are easily described.

Proposition. *A morphism f of Set H is an isomorphism of Set H if and only if f is injective and surjective, and an isomorphism of $\mathrm{Mod}^0 H$ if and only if f is surjective and equivalent in $\mathrm{Mod}^0 H$ to an H-subset inclusion. In either case, the inverse isomorphism of f is the dual relation f^{op}.*

PROOF. The result for $\mathrm{Mod}^0 H$ follows immediately from the diagonal polarity, obtained in 74.5 and 74.6.1, between surjective morphisms and H-subset inclusions. If a morphism $f : A \to B$ of Set H is injective and surjective, then f^{op} is a morphism of Set H, with $ff^{\mathrm{op}} = \delta_B$ and $f^{\mathrm{op}} f = \delta_A$ by 72.7. Conversely, if $gf = \delta_A$, then $f^{\mathrm{op}} = gff^{\mathrm{op}} \leqslant g$, and dually $f^{\mathrm{op}} \geqslant g$ if $fg = \delta_B$. Thus $f^{-1} = f^{\mathrm{op}}$ for an isomorphism f of Set H.

75.2. Terminal objects. It is easily seen that if $|T|$ is non-empty, with $\delta_T(x, x') = \top$ for all x, x' in $|T|$, then T is a terminal object of Set H and of Mod H. The only morphism $f : A \to T$, for an H-set A, is then induced by every mapping $|A| \to |T|$.

If H is not trivial, *i.e.* not a singleton, then not all terminal objects of Set H or of Mod H are of this form. It is easily seen that the set H with equality structure $\delta_H(a,b) = a \wedge b$ is a terminal object of Set H and of Mod H, with the unique morphism $f : A \to H$ given by $f(x,a) = \varepsilon_A(x) \wedge a$ for $x \in |A|$ and $a \in H$, and hence induced by the mappings $x \mapsto \varepsilon_A(x)$ and $x \mapsto \top$.

If H is trivial, then every set has exactly one H-set structure, and it is easily seen that any two sets are isomorphic in Set H, with any two non-empty sets isomorphic in Mod^0 H and in Mod H.

75.3. Finite Products. For H-sets A and B, we form the H-set $A \times B$ as the product $|A| \times |B|$ of the underlying sets, with fuzzy equality

$$\delta((x_1, y_1), (x_2, y_2)) = \delta_A(x_1, x_2) \wedge \delta_B(y_1, y_2).$$

It is easily verified that this defines an H-set $A \times B$, and that the projections of $|A| \times |B|$ induce morphisms $p : A \times B \to A$ and $q : A \times B \to B$ of H-sets. The following result shows that these morphisms are the projections of a product.

Proposition. *For H-sets A, B, C, and for morphisms $f : C \to A$, $g : C \to B$, and $h : C \to A \times B$ of Set H, we have $f = ph$ and $g = qh$, for the projections p and q of $A \times B$, if and only if*

$$h(z, x, y) = f(z, x) \wedge g(z, y)$$

for all $x \in |A|$, $y \in |B|$, and $z \in |C|$, with h induced by a mapping if and only if f and g are induced by mappings.

PROOF. We have $ph = f$ and $qh = g$ iff always

$$h(z, x, y) = \bigvee_{x', y'} (h(z, x', y') \wedge \delta_A(x, x') \wedge \delta_B(y, y'))$$

$$= \bigvee_{x', y'} (h(z, x, y') \wedge h(z, x', y) \wedge \delta_A(x, x') \wedge \delta_B(y, y'))$$

$$= \bigvee_{y'} (h(z, x, y') \wedge \delta_B(y', y)) \wedge \bigvee_{x'} (h(z, x', y) \wedge \delta_A(x', x))$$

$$= (qh)(z, y) \wedge (ph)(z, x) = f(z, x) \wedge g(z, y),$$

with suprema over all $x' \in |A|$ and $y' \in |B|$.

If h is induced by a mapping $\bar{h}$, then ph and qh are induced by mappings. Conversely, if f and g are induced by mappings $\bar{f}$ and $\bar{g}$, then it is easily seen that h is induced by the mapping $\langle \bar{f}, \bar{g} \rangle$.

75.4. Equalizers. We construct equalizers in Set H and Mod H as H-subset inclusions; all H-subset inclusions are equalizers by 75.6 below.

Proposition. *For a pair* $A \overset{f}{\underset{g}{\rightrightarrows}} B$ *of morphisms of* Set H, *the inclusion* $j_\alpha : A_\alpha \to A$ *with*

$$\alpha(x) = \bigvee_{y \in |B|} (f(x,y) \wedge f(x,y)),$$

for $x \in |A|$, *is an equalizer of* f *and* g.

PROOF. For the given α and $(x,y) \in |A| \times |B|$, we have

$$(f j_\alpha)(x,y) = f(x,y) \wedge \alpha(x) = \bigvee_{y'} (f(x,y) \wedge f(x,y') \wedge g(x,y'))$$

$$= \bigvee_{y'} (f(x,y) \wedge \delta_B(y,y') \wedge g(x,y')) = f(x,y) \wedge g(x,y),$$

and $f j_\alpha = g j_\alpha$ follows. If $fu = gu$ for $u : C \to A$, then

$$u(z,x) \wedge f(x,y) = u(z,x) \wedge f(x,y) \wedge \bigvee_{x'} (u(z,x') \wedge g(x',y))$$

$$= f(x,y) \wedge \bigvee_{x'} (u(z,x) \wedge u(z,x') \wedge g(x',y))$$

$$= f(x,y) \wedge \bigvee_{x'} (u(z,x) \wedge \delta_A(x,x') \wedge g(x',y))$$

$$= u(z,x) \wedge f(x,y) \wedge g(x,y)$$

for $(x,y,z) \in |A| \times |B| \times |C|$. Taking suprema $\bigvee_y$ on both sides, we get $u(z,x) = u(z,x) \wedge \alpha(x)$. Thus u factors through j_α.

75.5. Monomorphisms and epimorphisms. We note the following result.

Proposition. *A morphism* $f : A \to B$ *of* Set H *or of* Mod H *is monomorphic if and only if* f *is injective, and epimorphic if and only if* f *is surjective.*

PROOF. If f is injective, then $f^{op}f = \delta_A$ in the category of H-valued relations since $f^{op}f \geqslant \delta_A$ in any case, and thus f is monomorphic not only in Set H, but also in the category of H-valued relations. Dually, f is epimorphic if f is surjective.

Now consider the H-subset structure

$$\mu(x_1, x_2) = \bigvee_{y \in |B|} (f(x_1,y) \wedge f(x_2,y))$$

of $A \times A$. By 73.5, we have

$$(fp)(x_1, x_2, y) = f(x_1, y) \wedge \varepsilon_A(x_2)$$

and
$$(fq)(x_1, x_2, y) = f(x_2, y) \wedge \varepsilon_A(x_1)$$

for the projections p and q of $A \times A$; thus the subset inclusion j_μ of $A \times A$ is an equalizer of fp and fq by 75.4. If f is monomorphic, then p and q induce the same morphism $pj_\mu = qj_\mu$ from $(A \times A)_\mu$ to A, and it follows from 73.4 that always $\mu(x_1, x_2) \leqslant \delta_A(x_1, x_2)$ for x_1, x_2 in $|A|$. This is the case iff f is injective.

We shall see in 75.6.1 that H-subset inclusions are strong monomorphisms, in Set H and in Mod H. Thus if f is epimorphic and $f = me$ by 74.6.1, with e surjective and m an H-subset inclusion, then m is an isomorphism, and $f = m^{\text{op}}e$ surjective.

75.6. The subobject classifier. We have noted that $T = \{*\}$, with $\varepsilon_T(*) = \top$, is a terminal object of Set H and of Mod H. We define Ω by $|\Omega| = H$, with $\delta_\Omega(a, b) = a \leftrightarrow b$ for a, b in H. The mapping $* \mapsto \top$ induces a morphism $\top : T \to \Omega$ of Set a, with $\top(*, a) = a \leftrightarrow \top = a$ for a in H.

Proposition. *The morphism* $\top : T \to \Omega$ *is a classifier for H-subset inclusions in* Set H *and in* Mod H, *with* ch j_α *induced by* $\alpha : |A| \to H$ *for an H-subset structure α of an H-set A.*

75.6.1. Corollary. *Every H-subset inclusion is an equalizer, and the strong monomorphisms of* Set H *and of* Mod H *are the monomorphisms equivalent to H-subset inclusions.*

PROOF. If α is an H-subset structure of an H-set A, then $\delta_A(x, x') \leqslant \alpha(x) \leftrightarrow \alpha(x')$ for x, x' in $|A|$, by 74.1.(2). Thus

$$(1) \qquad (\text{ch } j_\alpha)(x, a) = \varepsilon_A(x) \wedge (a \leftrightarrow \alpha(x))$$

defines a morphism ch $j_\alpha : A \to \Omega$ in Mod H, induced by α.

Conversely, a square

$$(2) \qquad \begin{array}{ccc} A_\alpha & \longrightarrow & T \\ {\scriptstyle j_\alpha}\downarrow & & \downarrow{\scriptstyle \top} \\ A & \xrightarrow{\ \varphi\ } & \Omega \end{array}$$

in Set H is a pullback square iff j_α is an equalizer of φ and $\top t$, for the morphism $t : A \to T$, hence iff

$$(3) \qquad \alpha(x) = \bigvee_{a \in H} (\varphi(x, a) \wedge \top(*, a)) = \bigvee_{a \in H} (\varphi(x, a) \wedge a)$$

for x in $|A|$.

If $\varphi = \operatorname{ch} j_\alpha$, then $\varphi(x, a) \wedge a \leqslant \alpha(x)$, with equality for $a = \top$; thus (2) is a pullback square. Conversely, if (2) is a pullback square, then (3) holds, and

$$\varphi(x, a) \wedge \alpha(x) = \bigvee_{a'} (\varphi(x, a) \wedge \varphi(x, a') \wedge a')$$

$$= \bigvee_{a'} (\varphi(x, a) \wedge (a \leftrightarrow a') \wedge a') = \varphi(x, a) \wedge a$$

follows for $(x, a) \in |A| \times H$. But then $\varphi \leqslant \operatorname{ch} j_\alpha$, and $\varphi = \operatorname{ch} j_\alpha$ by 73.2.

By 75.4 and the Proposition, every H-subset inclusion is an equalizer, in Set H and in $\operatorname{Mod}^0 H$. If a strong monomorphism f factors $f = j_\beta e$ by 74.6.1, with e epimorphic by 75.5, it follows that e is an isomorphism, and f equivalent to j_β.

75.7. Reduced and void H-sets. For an H-set A, we say that A is *void* if $\varepsilon_A(x) = \bot$ for all $x \in |A|$. Every H-set A has a void H-subset A_ζ with $\zeta(x) = \bot$ for all $x \in |A|$, and the unique H-set Z with $|Z| = \emptyset$ is void.

For an H-set A, we obtain a *reduced H-set* rA by deleting all elements of A with $\varepsilon_A(x) = \bot$ and restricting δ_A to the remaining elements. If A is void, then $rA = Z$ with $|Z| = \emptyset$.

Proposition. *Every H-set A is isomorphic in* Set H *to the reduced set* rA. *In particular, all void H-sets are isomorphic in* Set H, *and initial objects of* Set H.

PROOF. Restriction of $\delta_A(x, x')$ to $x \in |rA|$ defines a morphism $d : rA \to A$ of Set H which is clearly injective and surjective, and thus an isomorphism by 75.1. If A is void, then rA is the initial object Z of Set H.

75.8. Reduced and void H-sets in $\operatorname{Mod} H$. For an H-set A with reduced set rA, the isomorphism $d : rA \to A$ of 75.7 is clearly induced by the set inclusion mapping. If $|rA|$ is non-empty, then this mapping has a left inverse which induces the inverse morphism of d. Thus A and rA are isomorphic in $\operatorname{Mod}^0 H$ if A is not void in Set H.

The void sets A_ζ with $|A|$ non-empty are isomorphic in $\operatorname{Mod}^0 H$; the unique morphism $f : A_\zeta \to B_\zeta$ is induced by every mapping from $|A|$ to $|B|$. However, if Z is the empty set with its unique H-set structure and $|A|$ is non-empty, then the unique morphism $f : Z \to A_\zeta$ in $\operatorname{Mod}^0 H$ is injective and surjective, but not an isomorphism: the inverse morphism f^{op} in Set H cannot be induced by a mapping.

If T is the standard terminal object of $\operatorname{Mod}^0 H$, with $|T| = \{*\}$ and $\varepsilon_T(*) = \top$, then T_ζ is the object 0^* of 46.1, and there is a unique morphism

$f : T_\zeta \to A$ for every H-set A with $|A|$ non-empty. Thus T_ζ is an initial object for the full subcategory $\text{Mod}\, H$ of $\text{Mod}^0\, H$ with objects A for which $|A|$ is not empty. It follows with 77.7 and 75.6.1 that $\text{Mod}\, H$ is a solid quasitopos, the solidification of the quasitopos $\text{Mod}^0\, H$.

76. Sheaves and Presheaves in Set H

76.1. H-sets induced by presheaves. We obtain a functor Φ from set-valued presheaves on H to Set H as follows. We assign to a presheaf $A : H \to \text{SET}$ the set $\coprod_i Ai$ of all pairs (x, i) with $i \in H$ and $x \in Ai$, and we put

$$\delta_{\Phi A}((x, i), (y, j)) = \bigvee \{ h \leqslant i \wedge j : x \mid h = y \mid h \}$$

for pairs (x, i) and (y, j) in this disjoint union. It is easily verified that this defines an H-set ΦA, with $\varepsilon_{\Phi A}(x, i) = i$ for $x \in Ai$; we say that ΦA is the *induced H-set* of the presheaf A.

A morphism $f : A \to B$ of presheaves induces a mapping $\Phi f = \coprod f_i$ of the associated coproducts. For $x \in Ai$ and $y \in Aj$, we have

$$f_i(x) \mid h = f_h(x \mid h) = f_h(y \mid h) = f_j(y) \mid h$$

if $x \mid h = y \mid h$, with $h \leqslant i \wedge j$. It follows from this and the definition of $\delta_{\Phi A}$ and $\delta_{\Phi B}$ that Φf satisfies the condition of 73.3. Thus Φf induces a morphism of H-sets which we also denote by Φf, and it follows immediately from 73.5 that we have obtained a functor Φ, from the functor category $[H^{\text{op}}, \text{SET}]$ to Set H or to $\text{Mod}^0\, H$.

76.2. Restrictions. We can order an H-set A by putting $x \leqslant y$, for elements x, y of $|A|$, if $\delta_A(x, y) = \varepsilon_A(x)$. It is easily verified that this relation is reflexive and transitive. We say that x is a *restriction* of y if $x \leqslant y$.

If $A = \Phi A_1$ for a presheaf $A_1 : H^{\text{op}} \to \text{SET}$, then for $x \in A_1 i$ and $j \leqslant i$, we have

$$\varepsilon_{\Phi A}(x \mid j, j) = j = \delta_{\Phi A}((x \mid j, j), (x, i))$$

by the definitions. Thus restrictions in A_1 become restrictions in ΦA_1.

Lemma. *If $f : A \to B$ is an H-valued relation and $y' \leqslant y$ in B, then*

$$f(x, y') = f(x, y) \wedge \delta_B(y, y') = f(x, y) \wedge \varepsilon_B(y')$$

for all x in $|A|$.

PROOF. We have $f(x, y') \wedge \delta_B(y, y') = f(x, y) \wedge \delta_B(y, y')$, and the result follows since $f(x, y') \leqslant \varepsilon_B(y') = \delta_B(y, y')$.

76.3. Separated H-sets and presheaves. We say that an H-set A is *separated* if the restriction preorder of A is an order, *i.e.* if $x \leqslant y$ and $y \leqslant x$ only for $x = y$.

Proposition. *A presheaf $A : H^{\mathrm{op}} \to \mathrm{SET}$ is separated if and only if the induced H-set ΦA is separated.*

PROOF. We have $(x, i) \leqslant (y, j)$ for elements of ΦA iff $i \leqslant j$, and $x \,|\, h = y \,|\, h$ for elements h of H with supremum i in H. If A is separated, this is the case iff $x = y \,|\, i$; thus ΦA is separated. Conversely, if x, y are in Ai and $x \,|\, h = y \,|\, h$ for elements h of H with supremum i, then

$$\varepsilon(x, i) = \delta((x, i), (y, i)) = \varepsilon(y, i)$$

in ΦA. Thus $x = y$, and A is separated, if ΦA is separated.

76.4. Separated presheaves in Set H. We say that an H-set A is a *presheaf* if for $x \in |A|$ and $i \leqslant \varepsilon_A(x)$ in H, there is always a restriction $y \leqslant x$ of x in A with $\varepsilon_A(y) = i$. We may denote such a restriction y by $x \,|\, i$. The H-set ΦA induced by a presheaf functor $A : H^{\mathrm{op}} \to \mathrm{SET}$ is always a presheaf in this sense, with restrictions $x \,|\, i$ inherited from the functor A. These restrictions are coherent, *i.e.* $(x \,|\, i) \,|\, j = x \,|\, j$ for $j \leqslant i \leqslant \varepsilon_A(x)$, but we may not be able to obtain coherent restrictions in an H-set which is a presheaf in Set H, but not separated.

We now have separated presheaves as functors $A : H^{\mathrm{op}} \to \mathrm{SET}$ and as special H-sets. By 76.3, every separated presheaf functor induces a separated presheaf in Set H. We show that the two separated presheaf concepts are equivalent by proving the following converse.

Proposition. *If an H-set A is a separated presheaf, then restrictions $x \,|\, i$ in A are unique and satisfy*

$$(1) \qquad\qquad x \,|\, i = y \,|\, i \quad \Longleftrightarrow \quad i \leqslant \delta_A(x, y),$$

for x, y in A and $i \leqslant \varepsilon_A(x) \wedge \varepsilon_A(y)$ in H. Thus A is induced by a presheaf functor, up to a bijection which preserves all values $\delta_A(x, y)$.

PROOF. If $y \leqslant x$ and $y' \leqslant x$ with $\varepsilon_A(y) = \varepsilon_A(y')$, then

$$\varepsilon_A(y) = \delta_A(x, y) \wedge \delta_A(x, y') \leqslant \delta_A(y, y') \leqslant \varepsilon_A(y),$$

and thus $y \leqslant y'$ and $y' \leqslant y$. If A is separated, then $y = y'$ follows.

If we put $Ai = \varepsilon_A^{\leftarrow}(\{i\})$ for $i \in H$, then we may regard A as a coproduct of the disjoint subsets Ai. For $i \leqslant j$ in H, restrictions $x \,|\, i$ define a mapping from Aj to Ai. Since restrictions in A are unique, restriction mappings $Aj \to Ai$ define a functor $H^{\mathrm{op}} \to$ SET. If (1) is valid, then A is the H-set induced by this functor.

If $h \leqslant \delta_A(x, y)$ in H, for x, y in A, then

$$\delta_A(x \mid h, y \mid h) \geqslant \delta_A(x, x \mid h) \wedge \delta_A(x, y) \wedge \delta_A(y, y \mid h) = h\,,$$

and $\varepsilon_A(x \mid h) = \delta_A(x \mid h, y \mid h) = \varepsilon_A(y \mid h)$ follows. But then $x \mid h = y \mid h$ since A is separated. Conversely, if $x \mid h = y \mid h$, then

$$h = \delta_A(x, x \mid h) \wedge \delta_A(y, y \mid h) \leqslant \delta_A(x, y)\,;$$

thus (1) is valid and we are done.

76.5. Sheaves. We say that two elements x and y of an H-set A are *compatible* if

$$\delta_A(x, y) = \varepsilon_A(x) \wedge \varepsilon_A(y)\,.$$

If A is a separated presheaf, this means by 76.4 that $x \mid i \wedge j = y \mid i \wedge j$ for $i = \varepsilon_A(x)$ and $j = \varepsilon_A(y)$. Now we can define a sheaf exactly as in 28.2; a presheaf A is a *sheaf* if for every family of pairwise compatible elements x_i of A, there is exactly one element x of A with $\varepsilon_A(x) = \sup \varepsilon_A(x_i)$, for which each x_i is a restriction of x.

It is easily seen that a presheaf functor $A : H^{\mathrm{op}} \to$ SET is a sheaf if and only if the induced H-set ΦA is a sheaf.

76.6. Proposition. *If an H-set F is a sheaf, then every morphism $f : A \to F$ of Set H is induced by a unique mapping $f_0 : |A| \to |F|$ such that $\varepsilon_F(f_0(x)) = \varepsilon_A(x)$ for every $x \in |A|$. This mapping f_0 preserves restrictions $x \mid i$ if A is a presheaf.*

76.6.1. Corollary. *The restriction of the functor $\Phi : [H^{\mathrm{op}}, \mathrm{SET}] \to$ Set H to sheaves is an equivalence of categories.*

PROOF. For a single-valued relation $f : A \to F$ and $x \in |A|$, all restrictions $y \mid f(x, y)$ are compatible by 76.4. For a sheaf F, it follows that $y \mid f(x, y) = f_0(x) \mid f(x, y)$, for all y and a unique $f_0(x)$, with

$$\varepsilon_F(f_0(x)) = \bigvee_y f(x, y) = \varepsilon_A(x)$$

if f is total. With 76.2, we have

$$f(x, y) = f(x, y \mid f(x, y)) = f(x, f_0(x) \mid f(x, y)) \leqslant f(x, f_0(x)) \leqslant \varepsilon_A(x)$$

for all $y \in |F|$, and $f(x, f_0(x)) = \varepsilon_A(x)$ follows. Now

$$f(x, y) \wedge \delta_F(f_0(x), y) = f(x, f_0(x)) \wedge \delta_F(f_0(x), y) = \delta_F(f_0(x), y),$$

and $\qquad\qquad f(x, y) = f(x, y) \wedge f(x, f_0(x)) \leqslant \delta_F(f(x_0), y);$

thus f is induced by f_0. If also $f(x, y) = \delta_F(y_1, y)$ for all $y \in |F|$, with $\varepsilon_F(y_1) = \varepsilon_A(x)$, then $\varepsilon_F(y_1) = \delta_F(f_0(x), y_1) = \varepsilon_F(f_0(x))$, and $y_1 = f_0(x)$ since F is separated.

By 76.2 and its dual, and by what we already proved, we have

$$f(x \mid i, \ f_0(x) \mid i) = f(x, f_0(x)) \wedge i = i = \varepsilon_A(x \mid i) = \varepsilon_B(f_0(x) \mid i)$$

if $x \mid i$ is defined. It follows by unicity of $f_0(x)$ that $f_0(x \mid i) = f_0(x) \mid i$. Thus f_0 preserves restrictions.

If $A = \Phi A_1$ and $F = \Phi F_1$, for a presheaf functor A_1 and a sheaf functor F_1, then for every $i \in H$ there is a unique mapping $f_i : A_1 i \to F_1 i$ such that $f_0(x, i) = (f_i(x), i)$ for $x \in A_1 i$. By the preceding paragraph, the mappings f_i preserve restrictions; thus they define a morphism $\hat{f} : A_1 \to F_1$ of presheaves, with $f = \Phi \hat{f}$ by the construction of f_0. In particular, the functor Φ restricted to sheaves is full and faithful; this proves the Corollary.

76.7. Proposition. *For a separated presheaf B in* Set H, *every morphism $f : A \to B$ in* $\mathrm{Mod}^0 H$ *is induced by a unique mapping $f_0 : |A| \to |B|$ such that $\varepsilon_B(f_0(x)) = \varepsilon_A(x)$ for every $x \in |A|$. If A is a presheaf in* Set H, *then f_0 preserves restrictions.*

76.7.1. Corollary. *The restriction of the functor $\Phi : [H^{\mathrm{op}}, \mathrm{SET}] \to \mathrm{Mod}^0 H$ to separated presheaves is an equivalence of categories.*

PROOF. Put $f_0(x) = \bar{f}(x) \mid \varepsilon_A(x)$ if f is induced by $\bar{f}$. This satisfies the condition, and

$$\delta_B(f_0(x), \bar{f}(x)) = \varepsilon_A(x).$$

Thus f_0 also induces f, by 73.4.

If also $x f y_1$ with $\varepsilon_B(y_1) = \varepsilon_A(x)$, then

$$\varepsilon_A(x) \leqslant \delta_B(f_0(x), y_1) \leqslant \varepsilon_B(y_1) = \varepsilon_A(x).$$

Thus $y_1 = f_0(x)$ since B is separated, and $f_0(x)$ is unique.

The second part of the Proposition and the Corollaru are proved as in 76.6.

77. Set H is a Topos and $\operatorname{Mod} H$ a Quasitopos

77.1. Singletons. We define a *singleton* in an H-set A as an H-subset structure σ of A such that

$$(1) \qquad\qquad \sigma(x) \wedge \sigma(x') \leqslant \delta_A(x, x')$$

for all x, x' in $|A|$. We put in particular

$$s_x(x') = \delta_A(x, x')$$

for x, x' in $|A|$. This clearly assigns a singleton s_x in A to every member x of $|A|$.

We denote by SA the H-set of all singletons in A, with fuzzy equality defined by

$$\delta_{SA}(\sigma, \tau) = \bigvee_{x \in |A|} (\sigma(x) \wedge \tau(x))$$

for singletons σ and τ in A. It is easily verified that this fuzzy equality satisfies the two conditions of 72.1.

Singletons were introduced in [49] and named in [32]. We observe that the unique morphism $A_\sigma \to T$, from an H-subset A_σ of A to a terminal object T of Set H, is injective if and only if σ is a singleton in A, but we also observe that the void H-subset structure of A (75.7) is a singleton in A by our present definition. If $|A|$ is empty, then SA is a void H-set with $|SA|$ a singleton. For a terminal object T of Set H, ST is isomorphic to the set H with fuzzy equality $\delta(a, b) = a \wedge b$ for elements a, b of H.

77.2. Proposition. *For members x, x' of an H-set A, and for a singleton σ in A, we have*

$$\delta_{SA}(\sigma, s_x) = \sigma(x) \qquad \text{and} \qquad \delta_{SA}(s_x, s_{x'}) = \delta_A(x, x').$$

For singletons σ, τ in A, we have $\sigma \leqslant \tau$ in SA if and only if $\sigma(x) \leqslant \tau(x)$ for every $x \in |A|$.

PROOF. With 74.1.(2), the first equation follows immediately from the definitions, and the second equation is a special case.

If $\sigma(x) \leqslant \tau(x)$ for every $x \in |A|$, then clearly $\sigma \leqslant \tau$. Conversely, if $\sigma \leqslant \tau$, then

$$\sigma(x) = \bigvee_{x'}(\sigma(x) \wedge \sigma(x') \wedge \tau(x')) \leqslant \bigvee_{x'}(\delta_A(x, x') \wedge \tau(x')) = \tau(x)$$

for every $x \in |A|$.

77.3. Proposition. *For an H-set B, the H-set SB of singletons in B is a sheaf in Set H, and the mapping $y \mapsto s_y : |B| \to |SB|$ induces an isomorphism $s : B \to SB$ of Set H, given by $s(y,\sigma) = \sigma(y)$ for $y \in |B|$ and a singleton σ in B. If $f : A \to B$ is a morphism of Set H, then $f(x,\text{---})$ is a singleton in B for every element x of $|A|$, and the morphism $sf : A \to SA$ is induced by the mapping $x \mapsto f(x,\text{---})$.*

PROOF. By 73.3 and 77.2, the mapping $y \mapsto s_y$ induces a morphism $s : B \to SB$ of Set H, with

$$s(y,\sigma) = \varepsilon_B(y) \wedge \delta_{SB}(s_y,\sigma) = \sigma(y)$$

for $y \in |B|$ and a singleton σ in B. This morphism is clearly surjective, and injective by 77.1.(1). Thus s is an isomorphism in Set H by 75.1. It follows immediately from the last part of 77.2 that SB is separated. We obtain the sheaf structure of SB in 77.4.

For $f : A \to B$ in Set H and $x \in |A|$, the mapping $y \mapsto f(x,y) : |B| \to H$ is a singleton in B, by 74.2 and the single-valuedness of f. We have

$$(sf)(x,\sigma) = \bigvee_y (f(x,y) \wedge \sigma(y)) = \delta_{SB}(f(x,\text{---}),\sigma)$$

for a singleton σ in B; thus sf is induced by the mapping $x \mapsto f(x,\text{---})$.

77.4. For a singleton σ in B and $i \leqslant \varepsilon_{SB}(\sigma)$ in H, we put

$$(\sigma \mid i)(y) = i \wedge \sigma(y)$$

for every $y \in |B|$. This defines a singleton $\sigma \mid i$ in B with $\varepsilon_{SB}(\sigma \mid i) = i$, and $\sigma \mid i$ is a restriction of σ by 77.2.

For a family (σ_μ) of compatible singletons, put

$$\sigma(y) = \bigvee_\mu \sigma_\mu(y)$$

for $y \in |B|$. This clearly defines an H-subset structure of B. For y, y' in $|B|$, we have

$$\sigma(y) \wedge \sigma(y') = \bigvee_\mu \bigvee_\nu (\sigma_\mu(y) \wedge \sigma_\nu(y')) = \bigvee_\mu \bigvee_\nu (\delta_{SB}(s_y,\sigma_\mu) \wedge \delta_{SB}(\sigma_\nu,s_{y'}))$$

$$= \bigvee_\mu \bigvee_\nu (\delta_{SB}(s_y,\sigma_\mu) \wedge \delta_{SB}(\sigma_\mu,\sigma_\nu) \wedge \delta_{SB}(\sigma_\nu,s_{y'}))$$

$$\leqslant \delta_{SB}(s_y,s_{y'}) = \delta_B(y,y'),$$

by 77.2 and compatibility of the σ_μ. Thus σ is a singleton, and the supremum of the σ_μ in SB.

77.5. Theorem. *For a complete Heyting algebra H, the category* Set H *is a topos, equivalent to the topos* Sh H *of set-valued sheaves for the canonical Grothendieck topology of H.*

PROOF. By 76.6.1, sheaves in Set H define a full subcategory of Set H which is equivalent to Sh H. Since every H-set B is isomorphic in Set H to the sheaf SB, the category Set H is also equivalent to the topos Sh H.

77.6. The isomorphism $s : B \to SB$ of Set H is always a morphism of $\mathrm{Mod}^0\, H$, but in general not an isomorphism of $\mathrm{Mod}^0\, H$ because s^{op} is not always induced by a mapping. The remedy for this is to introduce the H-set $\bar{S}B$ with all singletons of the form $s_y \mid i$ as elements, for $y \in |B|$ and $i \leqslant \varepsilon_B(y)$, with $\delta_{\bar{S}B}$ obtained by restriction of δ_{SB} to singletons $s_y \mid i$. This H-set is clearly a separated presheaf, and we have the following result.

Proposition. *The morphism $\bar{s} : B \to \bar{S}B$ of $\mathrm{Mod}^0\, H$ induced by the mapping $y \mapsto s_y$ is an isomorphism in $\mathrm{Mod}^0\, H$, and a morphism $f : A \to B$ of Set H is induced by a mapping $\bar{f} : |A| \to |B|$ if and only if every singleton $f(x, -)$, for $x \in |A|$, is an element of $\bar{S}B$. If this is the case, then the morphism $\bar{s}f : A \to \bar{S}B$ of $\mathrm{Mod}^0\, H$ is induced by the mapping $x \mapsto f(x, -)$.*

PROOF. It is clear that $y \mapsto s_y$ induces a morphism $\bar{s} : B \to \bar{S}B$ of $\mathrm{Mod}^0\, H$ as shown. In the other direction, put $\sigma t y$, for a singleton σ in B and $y \in |B|$, if σ is a restriction $s_y \mid i$. If $\sigma t y$, then

$$\varepsilon_{\bar{S}B}(\sigma) \wedge \delta_B(y, y') = \delta_B(y, y') \wedge i = \sigma(y')$$

for $y' \in |A|$. Thus the relation t induces the morphism $\bar{s}^{\mathrm{op}} : \bar{S}B \to B$, and $\bar{s}$ and $\bar{s}^{\mathrm{op}}$ are inverse isomorphisms in $\mathrm{Mod}^0\, H$.

If $f : A \to B$ in Set H is induced by a mapping $\bar{f}$, then

(1) $$f(x, -) = s_y \mid \varepsilon_A(x),$$

for $x \in |A|$ and $y = \bar{f}(x)$, by the definitions. Conversely, the relation $x \bar{f} y$ of 73.8 is valid iff (1) holds. Thus f is a morphism of $\mathrm{Mod}^0\, H$ iff each singleton $f(x, -)$ in B is an element of $\bar{S}B$, and it follows as in the proof of 77.3 that the mapping $x \mapsto f(x, -)$ induces the morphism $\bar{s}f : A \to \bar{S}B$ of $\mathrm{Mod}^0\, H$.

77.7. Theorem. *For a complete Heyting algebra H, the category $\mathrm{Mod}^0\, H$ is a quasitopos, equivalent to the quasitopos of separated set-valued presheaves for the canonical Grothendieck topology of H. The inclusion functor from $\mathrm{Mod}^0\, H$ to Set H is the inverse image part of an injective geometric morphism $I \dashv S$, with SB the sheaf of singletons in B for an H-set B.*

PROOF. By 76.7 and 77.6, every object of $\text{Mod}^0 H$ is isomorphic in $\text{Mod}^0 H$ to a separated presheaf, and for separated presheaves A, B in Set H, every morphism $f : A \to B$ is induced by a unique morphism of presheaves. On the other hand, a morphism of presheaves induces a morphism in $\text{Mod}^0 H$; this proves the first part of the Theorem.

For H-sets A and B, the isomorphism $s : B \to SB$ of 77.3 induces a natural bijection between morphisms $f : A \to B$ of Set H and morphisms $sf : A \to SB$; with all morphisms sf in $\text{Mod}^0 H$ by 76.6. This defines an adjunction $I \dashv S$ with I the inclusion functor. By 75.3 and 75.4, finite products and equalizers in $\text{Mod}^0 H$ are also limits in Set H; thus the functor I preserves finite limits. The functor S is full and faithful since each counit $s^{\text{op}} : ISB \to B$ is an isomorphism in Set H.

78. The Topos Structure of H-Sets

78.1. Remarks. Using the isomorphisms $s : B \to SB$ of 77.3 and the results of Chapter 2 and Section 76, we can obtain the topos structure of Set H from that of Sh H, and similarly the quasitopos structure of Mod H from that of Sep H, but direct constructions are simpler. Powerset objects have been constructed by D. HIGGS [49]; the other constructions may be new.

Virtually the same constructions can be used for Mod H and $\text{Mod}^0 H$ as for Set H; all morphisms of Set H with a codomain of the form PB or $\tilde{B}$ or $[C, B]$ are induced by mappings.

The constructions of objects PB, $\tilde{B}$, and $[C, B]$ are quite similar. We present the construction of powerset objects with full details, and the constructions of objects $\tilde{B}$ and $[C, B]$ with less detail.

78.2. Powerset objects. For an H-set B, we define an H-set PB as the set of all H-subset structures of B, with fuzzy equality given by

$$\delta_{PB}(\beta, \gamma) = \bigwedge_{y \in |A|} (\beta(x) \leftrightarrow \gamma(x)).$$

It is easily seen that this defines an H-set, with $\varepsilon_{PB}(\beta) = \top$ for every H-subset structure β of B. Transitivity of δ_{PB} follows from the formal law

$$a \wedge (a \leftrightarrow b) = b \wedge (a \to b) = a \wedge b$$

for Heyting algebras. We show in 78.4 below that this works.

78.3. Characteristic morphisms. We note that relations $f : A \to B$ of H-sets, as defined in 72.3, are the same as H-subset structures of the product H-set $A \times B$.

A relation $f : A \to B$ assigns to every $x \in |A|$ an H-subset structure $f(x, —)$ of B. For x, x' in $|A|$, we have

$$\delta_A(x, x') \leqslant \bigwedge_{y \in |B|} (f(x, y) \leftrightarrow f(x', y)),$$

by 72.3.(2) and the definitions. Thus mapping x to $f(x, —)$ induces a morphism $\chi f : A \to PB$, the *characteristic morphism* of f, with

$$(\chi f)(x, \beta) = \varepsilon_A(x) \wedge \bigwedge_y (f(x, y) \leftrightarrow \beta(y))$$

for $x \in |A|$ and $\beta \in |PB|$. In terms of H-valued logic, we have

$$\beta = (\chi f)(x) \quad \Longleftrightarrow \quad x \in A \wedge (\forall y \in B)(f(x, y) \Longleftrightarrow y \in B_\beta).$$

for $x \in |A|$ and $\beta \in |PB|$.

78.4. Theorem. *For an H-set B, the H-set PB is a powerset object in Set H and in $\mathrm{Mod}^0 H$, with the membership relation $PB \to B$ given by $\ni_B (\beta, y) = \beta(y)$, for $y \in |B|$ and an H-subset structure β of B, and with*

$$\varphi = \chi f \quad \Longleftrightarrow \quad f = (\varphi \times \delta_B)^{\leftarrow} \ni_B$$

for a relation $f : A \to B$ and a morphism $\varphi : A \to PB$.

PROOF. It is readily verified that $\ni_B$ is an H-valued relation as shown. If $f = (\varphi \times \delta_B)^{\leftarrow} \ni_B$, then always

$$f(x, y) = \bigvee_{\beta, y'} (\varphi(x, \beta) \wedge \delta_B(y, y') \wedge \beta(y')) = \bigvee_\beta (\varphi(x, \beta) \wedge \beta(y)).$$

It follows that always $\varphi(x, \beta) \wedge \beta(y) \leqslant f(x, y)$, and

$$\varphi(x, \beta) \wedge f(x, y) = \bigvee_{\beta'} (\varphi(x, \beta) \wedge \varphi(x, \beta') \wedge \beta'(y))$$

$$= \bigvee_{\beta'} (\varphi(x, \beta \wedge \delta_{PB}(\beta, \beta') \wedge \beta'(y))) = \varphi(x, \beta) \wedge \beta(y).$$

But then $\varphi \leqslant \chi f$, and $\varphi = \chi f$ by 73.2.

If $\varphi = \chi f$, then always $\varphi(x, \beta) \wedge \beta(y) \leqslant f(x, y)$. For $\beta = f(x, —)$, we have $\varphi(x, \beta) = \varepsilon_A(x)$, and $\varphi(x, \beta) \wedge f(x, y) = f(x, y)$. Thus $(\varphi \times \delta_B)^{\leftarrow} \ni_B = f$ for $\varphi = \chi f$.

78.5. Representing partial morphisms. Up to equivalence, partial morphisms in Set H and in Mod H are single-valued relations; we factor a single-valued relation $f : A \to B$ as $A \xleftarrow{j\alpha} A_\alpha \xrightarrow{f} B$ with $\alpha(x) = \bigvee_y f(x, y)$ for $x \in |A|$.

The second parts of 77.3 and 77.7 clearly remain valid for partial morphisms. We utilize this by letting SB, for an H-set B, be the object of all singletons in B if we deal with Set H, and the object $\bar{S}B$ of 77.6 when dealing with Mod H or $\text{Mod}^0 H$.

With this convention, let $\tilde{B}$ be the H-set with the same elements as SB, but with fuzzy equality obtained by restricting δ_{PB} in 78.2 to singletons in SB. We then define $\vartheta_B : B \to \tilde{B}$ by

$$\vartheta_B(y, \sigma) = \sigma(y),$$

for $y \in |B|$ and a singleton σ in SB.

78.6. Lemma. SB *is an H-subset of $\tilde{B}$, with*

(1)
$$\bigvee_y (\sigma(y) \wedge \tau(y)) = \bigvee_y \sigma(y) \wedge \bigwedge_y (\sigma(y) \leftrightarrow \tau(y))$$

for singletons σ and τ in SB, and with $\vartheta_B = js$ for the isomorphism $s : B \to SB$ and the H-subset inclusion $j : SB \to \tilde{B}$.

PROOF. The righthand side of (1) is the largest $t \leqslant \bigvee_y \sigma(y)$ in H with

$$t \wedge \sigma(y) = t \wedge \tau(y) = t \wedge \sigma(y) \wedge \tau(y)$$

for all $y \in |B|$. Taking suprema in the complete Heyting algebra H, we get

$$t = t \wedge \bigvee_y \sigma(y) = t \wedge \bigvee_y (\sigma(y) \wedge \tau(y))$$

On the other hand, we have

$$\sigma(y) \wedge \bigvee_{y'} (\sigma(y') \wedge \tau(y')) = \bigvee_{y'} (\sigma(y) \wedge \sigma(y') \wedge \tau(y'))$$

$$= \bigvee_{y'} (\sigma(y) \wedge \delta_B(y, y') \wedge \tau(y')) = \sigma(y) \wedge \tau(y)$$

for $y \in |B|$, and $\bigvee_y (\sigma(y) \wedge \tau(y)) \leqslant t$ follows. This proves (1), and $\vartheta_B = js$ follows easily.

78.7. Theorem. *For an H-set B, the morphism $\vartheta_B : B \to \tilde{B}$ represents partial morphisms with codomain B.*

PROOF. If $f : A \to B$ is a single-valued relation, then the mapping $x \mapsto f(x, -)$ induces a morphism $\varphi : A \to \tilde{B}$, as in 78.3, given by

$$(1) \qquad \varphi(x, \sigma) = \varepsilon_A(x) \wedge \bigwedge_y (f(x, y) \leftrightarrow \sigma(y))$$

for $x \in |A|$ and a singleton σ. Conversely, if $\varphi : A \to \tilde{B}$ is given, then $f = \vartheta_B{}^{op}\varphi$ defines a single-valued relation $f : A \to B$, with

$$(2) \qquad f(x, y) = \bigvee_\sigma (\varphi(x, \sigma) \wedge \sigma(y))$$

for x, y in $A \to B$, and we have pullback squares

$$(3) \qquad \begin{array}{ccccc} A_\alpha & \xrightarrow{f_0} & B & \xrightarrow{s} & SB \\ \downarrow{\scriptstyle j_\alpha} & & \downarrow{\scriptstyle \vartheta_B} & & \downarrow{\scriptstyle j} \\ A & \xrightarrow{\varphi} & \tilde{B} & \xrightarrow{id} & \tilde{B} \end{array},$$

with $\alpha = \varphi^{\leftarrow} \varepsilon_{SB}$. Thus we have

$$\alpha(x) = \bigvee_\sigma (\varphi(x, \sigma) \wedge \varepsilon_{SB}(\sigma)) = \bigvee_\sigma (\varphi(x, \sigma) \wedge \bigvee_y \sigma(y))$$

$$= \bigvee_{\sigma, y} (\varphi(x, \sigma) \wedge \sigma(y)) = \bigvee_y f(x, y)$$

if $f(x, y)$ is given by (2), and f in (2) is the partial morphism $(j_\alpha, f_0) : A \to B$ in (3). As in the proof of 78.4, (1) and (2) define a bijection between morphisms $\varphi : A \to \tilde{B}$ and single-valued relations $f : A \to B$; this completes the proof.

78.8. Function H-sets. For H-sets C and B, we denote by $[C, B]$ the set of all H-valued relations $h : C \to B$ with each $h(z, -)$ in SB for $z \in |C|$, where SB is defined by the convention of 78.5. Thus each relation h in $[C, B]$ is single-valued. We define fuzzy equality in $[C, B]$ by putting

$$\delta_{[C,B]}(h, h') = \bigwedge_{z \in |C|} \left(\varepsilon_C(z) \to \bigvee_{y \in |B|} (h(z, y) \wedge h'(z, y)) \right)$$

for relations h, h' in $[C, B]$. This measures the degree to which the singletons $h(z, -)$ and $h'(z, -)$ are equal for each $z \in |C|$, and $\varepsilon_{[C,B]}(h)$ measures the degree to which h has a value at every $z \in |C|$. Fuzzy equality $\delta_{[C,B]}$ is clearly symmetric; it is easily seen that it is also transitive.

For H-sets A, B, C, a morphism $f : A \times C \to B$ of Set H or Mod H assigns to every $x \in |A|$ a single-valued relation $\hat{f}(x) = f(x, -, -)$, with each $\hat{f}(x)(z, -) = f(x, z, -)$ a singleton in SB.

Lemma. *For a morphism* $f : A \times C \to B$, *the mapping* $\hat{f}$ *induces a morphism* $\varphi : A \to [C, B]$.

PROOF. By 72.3 and the definitions, we must show that always

$$\delta_A(x, x') \wedge \varepsilon_C(z) \leqslant \bigvee_y (f(x, z, y) \wedge f(x', z, y)),$$

for x, x' in $|A|$ and $z \in |C|$. We have

$$\delta_A(x, x') \wedge \bigvee_y (f(x, z, y) \wedge f(x', z, y)) = \bigvee_y (\delta_A(x, x') \wedge f(x, z, y) \wedge f(x', z, y))$$

$$= \bigvee_y (\delta_A(x, x') \wedge f(x, z, y)) = \delta_A(x, x') \wedge \varepsilon_A(x) \wedge \varepsilon_C(z),$$

and the desired inequality follows.

78.9. Evaluation. We define evaluation $\text{ev} : [C, B] \times C \to B$ by

$$\text{ev}(h, z, y) = h(z, y),$$

for $(z, y) \in |C| \times |B|$ and h in $[C, B]$. This works by 77.3 and 77.6, due to the following result.

Lemma. *The mapping* $(h, z) \mapsto h(z, -)$, *for* h *in* $[C, B]$ *and* $z \in |C|$, *induces a morphism* $s \cdot \text{ev} : [C, B] \times C \to SB$.

PROOF. We must show that always

$$\delta_{[C,B]}(h, h') \wedge \delta_C(z, z') \leqslant \bigvee_y (h(z, y) \wedge h'(z', y)).$$

The lefthand side of this inequality is the largest $t \leqslant \delta_C(z, z')$ with

$$t \wedge \varepsilon_C(z_1) \leqslant \bigvee_y (h(z_1, y) \wedge h'(z_1, y))$$

for all $z_1 \in |C|$. It follows for $z_1 = z'$ that

$$t \leqslant \bigvee_y (\delta(z, z') \wedge h(z', y) \wedge h(z', y)) \leqslant \bigvee_y (h(z, y) \wedge h'(z', y)).$$

78.10. Theorem. *The categories* Set H *and* Mod H *are cartesian closed with function* H-*sets* $[C, B]$.

PROOF. For $f : A \times C \to B$, we have by 78.8 a morphism $\varphi : A \to [C, B]$ given by

$$(1) \qquad \varphi(x, h) = \varepsilon_A(x) \wedge \bigwedge_z (\varepsilon_C(z) \to \bigvee_y (f(x, z, y) \wedge h(z, y)))$$

for $x \in |A|$ and h in $[C, B]$. For $\varphi : A \to [C, B]$, it is easily seen that the morphism $\mathrm{ev} \cdot (\varphi \times \delta_C) : A \times C \to B$ is given by

$$(2) \qquad f(x, z, y) = \bigvee_h (\varphi(x, h) \wedge h(z, y)).$$

We omit the fairly straightforward proof that (1) and (2) are equivalent.

79. Fuz H and Related Categories

79.1. Eytan's category Fuz H **of fuzzy sets.** If we identify H-valued fuzzy sets with H-sets A such that $\delta_A(x, x') = \perp$ for all pairs of distinct elements x, x' of $|A|$, as in 72.2, then the category Fuz H of fuzzy sets of M. EYTAN [28] is the full subcategory of Set H with H-fuzzy sets as its objects. In Set H, finite products and H-subsets of H-fuzzy sets are again H-fuzzy sets. Thus Fuz H is closed under finite limits in Set H, and it follows that monomorphisms are the same as injective morphisms. Except for the subobject classifier $\top : 1 \to \Omega$, all constructions of Sections 73–75 can be carried out in Fuz H; thus this category has a first-order logic with truth values in H.

Contrary to EYTAN's assertion in [28], Fuz H is usually not a topos, but not much is known about this. We state the two known results.

79.2. Proposition [84]. *If H is a complete Boolean algebra, then* Fuz H *is a Boolean topos, equivalent to* Set H *and to* Sh H. *Conversely, if* Fuz H *is a topos, then H is Boolean.*

PROOF. For an H-set A, let A^* denote the set A with equality made crisp, by putting $\varepsilon_{A^*}(x) = \varepsilon_A(x)$ for $x \in |A|$, and $\delta_{A^*}(x, x') = \perp$ for distinct elements x, x' of $|A|$. Then $\mathrm{id}_{|A|}$ induces a morphism $e : A^* \to A$ with $e(x, x') = \delta_A(x, x')$ for x, x' in $|A|$. This morphism e is clearly surjective. If H is Boolean, then e has a right inverse $m : A \to A^*$ by Theorem 5.39 of [58]; thus A is isomorphic to an H-subset of A^*. This subset is again an object of Fuz H; thus Fuz H is equivalent to the topoi Set H and Sh H.

We refer to [84] or [85] for the proof of the converse.

79.3. The following result applies in particular to $H = [0, 1]$, the real unit interval.

Proposition [15]. *If H is irreducible (71.7), then* Fuz H *is a quasitopos, equivalent to the quasitopos* Fzr H *of reduced fuzzy sets.*

PROOF. The reduced H-set rA of an H-fuzzy set A is an H-fuzzy set, isomorphic to A in Fuz H by 75.7. Thus Fuz H is equivalent to the full subcategory of Set H with reduced fuzzy sets as its objects.

If $f : A \to B$ is a morphism in this category, and $x \in |A|$, then $f(x, y) \neq \perp$ for some $y \in |B|$ since f is total, and $f(x, y)$ for at most one $y \in |B|$ if H is irreducible, since f is single-valued. Thus $f(x, y) \neq \perp$ for exacly one element $y = f_0(x)$ of $|B|$, with $f(x, y) = \varepsilon_A(x) \leqslant \varepsilon_B(y)$. This shows that f is induced by exactly one morphism $f_0 : A \to B$ of Fzr H. Conversely, a morphism $f_0 : A \to B$ of Fzr H clearly induces a morphism $f : A \to B$ of Set H, and the desired equivalence of categories follows.

79.4. Fuzzy sets with crisp mappings. The category Fzs H, studied in Section 71, is one category of fuzzy sets with crisp equality and crisp mappings. We obtain other categories of this kind, which we shall denote by $\text{Fuz}_m H$ and $\text{Fuz}_m^0 H$, by intersecting the subcategory Fuz H of Set H with the subcategories Mod H (see 75.8) and $\text{Mod}^0 H$. These categories are vertices in a diagram

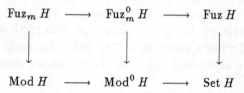

of subcategories and inclusion functors. In this diagram, the lower righthand inclusion has a right adjoint by 77.7, and the two horizontal left-hand inclusions have left adjoints. We do not know whether other functors in the diagram have left or right adjoints.

79.5. Reduced and void sets in Fuz H. Reduced fuzzy sets (71.6) are a special case of reduced H-sets, as defined in 75.7. Thus every fuzzy set A is isomorphic in Fuz H to its reduced fuzzy set rA, obtained by deleting all elements x with $\varepsilon_A(x) = \perp$. In particular, all fuzzy sets A with rA empty are isomorphic in Fuz H, and initial objects of Fuz H.

The situation changes for $\text{Fuz}_m^0 H$. In this category, A and rA still are isomorphic if rA is not empty, and the void fuzzy set Z with $|Z|$ empty is an initial object. If A is void, with rA empty but $|A|$ not empty, then there is no morphism $A \to Z$ in $\text{Fuz}_m^0 H$, but a unique morphism $f : A \to B$ for every fuzzy set B with $|B|$ non-empty.

The category $\text{Fuz}_m H$ is the full subcategory of $\text{Fuz}_m^0 H$ with the empty object Z deleted. In this category, every void object A with rA empty (and hence not an object of $\text{Fuz}_m H$) is an initial object; the unique morphism $f : A \to B$ for an object B is induced by any mapping from $|A|$ to $|B|$.

79.6. Discussion. A mapping $\bar{f} : |A| \to |B|$ induces a morphism $f : A \to B$ of $\text{Fuz}_m^0 H$ iff $\varepsilon_A(x) \leqslant \varepsilon_B(\bar{f}(x))$ for all $x \in |A|$, by 73.3, and thus iff $\bar{f} : A \to B$ in $\text{Fzs}\, H$. This defines a functor from $\text{Fzs}\, H$ to $\text{Fuz}_m^0 H$ which is clearly full, but not faithful, and hence functors from $\text{Fzs}\, H$ to $\text{Fuz}\, H$ and to $\text{Set}\, H$. Claims have been made, without valid proofs, that some of these functors have left or right adjoints.

By 73.4, morphisms $f : A \to B$ and $g : A \to B$ of $\text{Fzs}\, A$ induce the same morphism of $\text{Fuz}_m^0 H$ iff $\varepsilon_A(x) \leqslant \delta_B(f(x), g(x))$ for all $x \in |A|$. If $\varepsilon_A(x) \neq \perp$, this means that $f(x) = g(x)$, but if $\varepsilon_A(x) = \perp$, then $g(x)$ can be any element of $|B|$. Thus morphisms $A \overset{f}{\underset{g}{\rightrightarrows}} B$ of $\text{Fzs}\, H$ induce the same morphism of $\text{Fuz}_m^0 H$ iff they have the same restriction to a morphism $h : rA \to rB$ of reduced fuzzy sets (see 71.6 and 75.7). This leads to the following result.

79.7. Theorem. *For a complete Heyting algebra H, the reduction functor $r : \text{Fuz}_m H \to \text{Fzr}\, H$ is an equivalence of categories.*

PROOF. By the preceding discussion, a morphism $f : A \to B$ of $\text{Fuz}_m H$ is induced by a unique morphism $rf : rA \to rB$ of $\text{Fzr}\, H$, obtained by restricting to rA and rB a mapping $\bar{f} : |A| \to |B|$ which induces f. Thus reduction defines a full and faithful functor r. Every initial object of $\text{Fuz}_m H$ has the same reduction Z, and every non-initial object A of $\text{Fuz}_m H$ is isomorphic in $\text{Fuz}_m H$ to the reduced fuzzy set rA. Thus r is an equivalence of categories.

79.8. Subset structures in $\text{Fzs}\, H$ and $\text{Fzr}\, H$. If we use 72.2 to regard H-fuzzy sets as special H-sets, then all definitions and results of Section 74 can be carried over to $\text{Fzs}\, H$, with the following changes.

In $\text{Fzs}\, H$, direct and inverse images for $f : A \to B$, and quantifiers $\exists_f$ and $\forall_f$, are given by

$$(f^{\to}\alpha)(y) = (\exists_f \alpha)(y) = \bigvee_{f(x)=y} \alpha(x),$$

$$(f^{\leftarrow}\beta)(x) = \varepsilon_A(x) \wedge \beta(f(x)),$$

$$(\forall_f \alpha)(y) = \varepsilon_B(y) \wedge \bigwedge_{f(x)=y} (\varepsilon_A(x) \to \beta(y)),$$

for H-subset structures α of A and β of B, and for $x \in |A|$ and $y \in |B|$. 74.4 and 74.7 are easily verified for these structures.

For the factorization of 74.5 and 74.6, we must replace surjective morphisms by *fine* morphisms $e : A \to C$, with $\varepsilon_C(e(x)) = \varepsilon_A(x)$ for $x \in |A|$, and $\varepsilon_C(y) = \bot$ for y in $|C|$ not in the range of f. Then $ge = f$ in 74.5.(1) at the set level, with $u = g : C \to B_\beta$ as the diagonal, and $g = f$ at the set level in 74.6.

We define subset inclusions $j : S \to A$ in Fzr H as morphisms with $|S|$ a subset of $|A|$ and $j : |S| \to |A|$ the subset inclusion. There is then a natural isomorphism between the complete lattices of subsets A_α of A in Fzs H and subsets S of A in Fzr H, given by $S = rA_\alpha$. We use this isomorphism to obtain direct and inverse images, and quantifiers $\exists_f$ and $\forall_f$, in Fzr H. For the factorization of 74.5 and 74.6, surjective morphisms must be replaced by quotient morphisms, *i.e.* fine morphisms which are surjective at the set level.

80. First Order Fuzzy Logic

80.1. The language of fuzzy logic. Let H be a complete Heyting algebra, as before. For the categories Set H, Mod H and Mod^0 H, the internal logic of Chapter 3 is an H-valued fuzzy logic. If **E** is one of the other categories discussed in the present Chapter, then we restrict the language. We leave out the higher-order terms of 34.1.(7) and 34.2.(7), and we admit only atomic statements $t =_A u$ and $t \in A_\alpha$ in 34.1.(4) and 34.2.(4), for terms t and u of type A and an H-subset structure α of an object A of **E**. All other ingredients in 34.1 and rules in 34.2 remain unchanged. We replace A_α in (4) by A for the coarse H-subset structure $\alpha = \varepsilon_A$ of an object A.

Higher-order constructions can be added to this language, for categories **E** which have them in an appropriate form.

80.2. Interpretations and validity of formulas. Terms in the language of fuzzy logic are interpreted by morphisms of the given category **E**, exactly as in Section 35, by the rules 35.2.(1)–(3). Interpretations $|\Phi|_L$ of a statement Φ, for a list L of distinct variables which includes all variables in Φ, are H-subset structures of P_L, defined by the following rules.

(4) $|t \in A_\alpha|_L = |t|_L \leftharpoonup \alpha$, and $|t =_A u|_L = \langle |f|_L, |g|_L \rangle \leftharpoonup \delta_A$, for terms t, u of type A with $|t|_L$ and $|u|_L$ defined.

(5) Interpretations $|\Phi|_L$ for propositi ̄al connectives of statements are defined by the corresponding propositional connectives in the complete Heyting algebra of H-subset structures of P_L.

(6) If L does not contain x and $|\Phi|_{Lx}$ is defined, then

$$|(\forall x)\Phi|_L = \forall_p |\Phi|_{Lx} \quad \text{and} \quad |(\exists x)\Phi|_L = \exists_p |\Phi|_{Lx},$$

for the projection $p = \pi_L^{Lx} : P_{Lx} \to P_L$.

A statement Φ in the language of fuzzy logic is valid if $|\Phi|_L = \varepsilon_{P_L}$ for every interpretation of Φ, or equivalently for an internal interpretation of Φ. All applicable results of Chapter 3 remain valid for fuzzy logic.

80.3. Discussion. In order to define fuzzy logic for a category $\mathbf{E}$ of fuzzy sets, we have assumed the following for $\mathbf{E}$.

(1) Finite limits and colimits.

(2) A complete Heyting algebra of fuzzy subset structures of A, for every objet A of $\mathbf{E}$.

(3) A diagonal polarity $(\mathcal{E}, \mathcal{M})$ in $\mathbf{E}$, with $\mathcal{M}$ the class of fuzzy subset inclusions in $\mathbf{E}$ and $\mathcal{E}$ preserved by pullbacks.

(4) Direct images $f^{\to}\alpha = \exists_f \alpha$, inverse images $f^{\leftarrow}\beta$, and universal quantifiers $\forall_f \gamma$ for every morphism f of $\mathbf{E}$, obtained by Galois connections $\exists_f \dashv f^{\leftarrow}$ and $f^{\leftarrow} \dashv \forall_f$.

These conditions are satisfied for every category of fuzzy sets discussed so far. In Set and $\mathrm{Mon}\,H$, the monomorphisms equivalent to H-subset inclusions are the strong monomorphisms, and fuzzy logic is equivalent to the first order part of the internal logic. In $\mathrm{Fzs}\,H$, the monomorphisms equivalent to H-subset inclusions are the morphisms which are bijective at the set level. In $\mathrm{Fzr}\,H$, every monomorphism is equivalent to a fuzzy subset inclusion. For $\mathrm{Fuz}\,H$, we do not have a good characterization of H-subset inclusions.

For $\mathrm{Set}\,H$ and $\mathrm{Mod}\,H$, the class $\mathcal{E}$ in (3) consists of all epimorphisms. For $\mathrm{Fzs}\,H$, we get the class of fine morphisms, and for $\mathrm{Fzr}\,H$ the class of strong epimorphisms or quotient morphisms. In these examples, $\mathcal{E}$ is preserved by pullbacks.

With inverse images $f^{\leftarrow}\beta$ in $\mathbf{E}$ defined as pullbacks, a direct image $f^{\to}\alpha = \exists_f \alpha$ is obtained from an $(\mathcal{E}, \mathcal{M})$-factorization of $f j_\alpha$. If $\mathbf{E}$ is cartesian closed and $\mathcal{E}$ is preserved by pullbacks, then every pullback functor p^*, for projections p of products, has a right adjoint p_* by 18.5, and universal quantifiers $\forall_p \alpha$ are equivalent to morphisms $p_* j_\alpha$. The example $\mathrm{Fzr}\,H$, for a complete Heyting algebra H which is not irreducible in the sense of 71.7, shows that $\mathbf{E}$ can have universal quantifiers $\forall_f \alpha$ without being cartesian closed.

80.4. Łukasiewicz logic. For the real unit interval $[0, 1]$, internal conjunction is given by $a \wedge b = \min(a, b)$, and internal implication by the formula

$$a \to b = \begin{cases} 1, & \text{if } a \leqslant b, \\ b, & \text{if } b < a. \end{cases}$$

As a function of real variables a, b, implication is thus discontinous at points (a, a) with $a \neq 1$. Practitioners of fuzzy sets have avoided these discontinuities by using LUKASIEWICZ implication, given by

$$a \Rightarrow b = \min(1, 1 + b - a)$$

for a, b in $[0, 1]$.

Using a non-standard implication means using a logic in which many familiar laws are no longer valid. In order to minimize this inconvenience, it has been proposed to change conjunction as well as implication, using a non-standard conjunction $a * b$ for which the adjunction (or Galois connection)

(1) $a * b \leqslant c \quad \Longleftrightarrow \quad a \leqslant b \Rightarrow c$

remains valid. For Lukasiewicz implication, we get

$$a * b = \max(0, a + b - 1);$$

this is known as Lukasiewicz conjunction. We observe, however, that this does not solve the problem; other laws become invalid. For example, the only idempotents a for Lukasiewicz conjunction, with $a * a = a$, are 0 and 1.

80.5. Non-standard conjunctions and implications. The connectives of 80.4 can be generalized by considering a complete lattice L with a binary operation $*$ obeying the following formal laws:

(i) $a * b = b * a$,

(ii) $(a * b) * c = a * (b * c)$,

(iii) $a * \top = a = \top * a$,

(iv) $a * (\bigvee_i b_i) = \bigvee_i (a * b_i)$,

with additional laws which may be needed. Thus L with $*$ is a commutative monoid with neutral element $\top$. It follows from these laws that $*$ is monotone with respect to the order of L, with

$$a * b \leqslant a \wedge b$$

for a, b in L, and that 80.4.(1) defines a binary operation $\Rightarrow$ on L, right adjoint to $*$.

A non-standard conjunction is usually not idempotent, *i.e.* the formal law $a * a = a$ is not valid. If we have the formal law

(v) If $a \leqslant b$, then $a = b * c$ for some c in L,

and $e * e = e$, then $a * e = a \wedge e$ for every a in L.

Motivating examples are complete Heyting algebras with $* = \wedge$, and the real unit interval with Lukasiewicz conjunction and implication. These examples satisfy (v).

80.6. Non-standard fuzzy logic. Operations $*$ and $\Rightarrow$ are easily defined for fuzzy subset structures α and β of an object A of Fzs L, by putting

$$(\alpha * \beta)(x) = \alpha(x) * \beta(x) \quad \text{and} \quad (\alpha \Rightarrow \beta)(x) = \varepsilon_A(x) \wedge (\alpha(x) \Rightarrow \beta(x))$$

for $x \in |A|$. Operations $*$ and $\Rightarrow$ for subset structures in Fzr L are obtained from this by reduction (71.6). Thus propositional connectives $*$ and $\Rightarrow$ can be interpreted for the fuzzy logic of Fzs L or Fzr L. The interpretation of quantifiers depends on the interpretation of substitution; I have not seen this discussed in the literature.

80.7. $(L, *)$-sets. We assume now that a complete lattice L is given, with a binary operation $*$ which satisfies conditions 80.4.(i)–(iv). For this structure, we can obtain categories of fuzzy sets by replacing conjunction $\wedge$ and implication $\rightarrow$, in the definitions of this Chapter and their interpretations, by the operations $*$ and $\Rightarrow$. Since $*$ is in general not idempotent, some conditions in the definitions may have to be sharpened.

In [56] and [57], an $(L, *)$-*valued set* is defined as a pair (A, α) consisting of a set A and a fuzzy equality $\alpha : A \times A \rightarrow L$, subject to the following conditions.

(1) $\alpha(x, y) \leqslant \alpha(x, x) \wedge \alpha(x, y)$,

(2) $\alpha(y, x) = \alpha(x, y)$,

(3) $\alpha(x, y) * (\alpha(y, y) \Rightarrow \alpha(y, z)) \leqslant \alpha(x, z)$,

for all elements x, y, z of A. The value $\alpha(x, x)$ is then the extent of membership of $x \in A$ in (A, α).

In [16] and [57], fuzzy equality is subject to stricter conditions. With the stricter conditions, all values $\varepsilon_A(x)$ become idempotent, with the result that equality is fuzzy, but membership is crisp, if Łukasiewicz conjunction is used.

80.8. Categories of $(L, *)$-sets. There are several possible definitions of morphisms of $(L, *)$-sets. One way is to define a morphism $f : A \rightarrow B$ of $(L, *)$-sets as a mapping $f : |A| \times |B| \rightarrow H$, with the following conditions.

(1) $\varepsilon_A(x) * f(x, y) = f(x, y) = f(x, y) * \varepsilon_B(y)$,

(2) $f(x, y) * \delta_A(x, x') \leqslant f(x', y)$, and $f(x, y) * \delta_B(y, y') \leqslant f(x, y')$,

(3) $f(x, y) * f(x, y') \leqslant \delta_B(y, y')$,

(4) $\bigvee_y f(x, y) = \varepsilon_A(x)$,

for members x, x' of $|A|$ and y, y' of $|B|$. Composition of $f : A \rightarrow B$ and $g : B \rightarrow C$ is then defined by

$$(g \circ f)(x, z) = \bigvee_{y \in |B|} (f(x, y) * g(y, z)),$$

for (x, z) in $|A| \times |C|$. Other categories of $(L, *)$-sets are obtained in similar fashion.

80.9. Higher-order structures. In a category of L-fuzzy sets with a non-standard conjunction, we can define a *tensor product*

$$(A, \alpha) \otimes (B, \beta) = (A \times B, \alpha \otimes \beta)$$

of objects (A, α) and (B, β) by putting

$$(\alpha \otimes \beta)(x, y) = \alpha(x) * \beta(y)$$

for (x, y) in $|A| \times |B|$. This can also be done for $(L, *)$-sets, putting

$$(\alpha \otimes \beta)((x, y), (x', y')) = \alpha(x, x') * \beta(y, y')$$

for x, x' in A and y, y' in B. This defines a monoidal structure, and right adjoints of tensor product functors $(A, \alpha) \otimes -$ then define symmetric monoidal closed structures of categories of L-fuzzy sets or of $(L, *)$-sets. The tensor product $(A, \alpha) \otimes (B, \beta)$ is usually not the ordinary product; thus we do not get a cartesian closed structure. U. HÖHLE and L.N. STOUT have investigated such structures in their articles cited in the Bibliography of these Notes.

BIBLIOGRAPHY

[1] JIŘI ADÁMEK, Classification of concrete categories. *Houston Jour. of Math.* **12** (1986), 305–326.

[2] J. ADÁMEK and H. HERRLICH, Cartesian closedness, quasitopoi and topological universes. *Comm. Math. Univ. Carolinae* **27** (1986), 235–257.

[3] J. ADÁMEK and H. HERRLICH, A characterization of concrete quasitopoi by injectivity. To appear in: *Jour. Pure Applied Algebra* (1990).

[4] J. ADÁMEK, H. HERRLICH, G.E. STRECKER, *Abstract and Concrete Categories.* Wyley, New York etc. (1990).

[5] J. ADÁMEK and J. REITERMAN, The quasitopos hull of the category of uniform spaces. *Topology Appl.* **27** (1987), 97–104.

[6] J. ADÁMEK and J. REITERMAN, The quasitopos hull of the category of uniform spaces – a correction. To appear in: *Topology Appl.* .

[7] J. ADÁMEK, J. REITERMAN, F. SCHWARZ, On universally topological hulls and quasitopos hulls. *Seminarberichte Fernuniversität Hagen* **34** (1989), 1–11.

[8] P. ANTOINE, Extension minimale de la catégorie des espaces topologiques. *C.R. Acad. Sci. Paris* **262** (1966) Sér. A, 1389–1392.

[9] P. ANTOINE, Étude élémentaire des catégories d'ensembles structurés. *Bull. Sci. Math. Belgique* **18** (1966), 142–166, 387–414.

[10] M.A. ARBIB and E.G. MANES, *Arrows, Structures and Functors. The categorical imperative.* Academic Press, 1975.

[11] MICHAEL BARR, Fuzzy set theory and topos theory. *Canad. Math. Bull.* **29** (1986), 501–508.

[12] M. BARR and C. WELLS, *Toposes, Triples and Theories.* Grundlehren der Math. Wissensch. **278**, Springer–Verlag (1985).

[13] ANDREAS BLASS, The interaction between category theory and set theory. *Mathematical Applications of Category Theory.* Contemporary Mathematics **30** (1984), pp. 5–29.

[14] N. BOURBAKI, *Éléments de mathématique I. Livre I, Théorie des ensembles*. Actualités Scientifiques et Industrielles, **1141, 1212, 1243, 1258**. Hermann, Paris, 1948–1957.

[15] J.C. CARREGA, The categories Set$H$ and Fuz$H$. *Fuzzy Sets and Systems* **9** (1983), 227–332.

[16] U. CERRUTI and U. HÖHLE, Categorical foundations of fuzzy set theory with applications to algebra and topology. *The Mathematics of Fuzzy Systems*, ed. by ANTONIO DI NOLA, ALDO G.S. VENTRI, Verlag TÜV Rheinland, Köln (1986), pp. 51–86.

[17] G. CHOQUET, Convergences. *Ann. Univ. Grenoble, Sect. Sci. Math. Phys.* (N.S.) **23** (1948), 57–112.

[18] J. COULON – J.-L. COULON, Remarques sur certaines catégories d'ensembles totalement flous. *BUSEFAL* **21, 23, 24** (1985).

[19] J. COULON; J.L. COULON, Classification de morphismes partiels dans la catégorie JTF^{00} d'ensembles totalement flous. *BUSEFAL* **29** (1986).

[20] J. COULON and J.-L. COULON, Is the category JTF a topos? *Fuzzy Sets and Systems* **27** (1988), 31–44.

[21] J. COULON and J.-L. COULON, Morphisms in the category JTF^{00}. Weak classifier of monomorphisms. *Portugaliae Math.* **45** (1988), 29–47.

[22] BRIAN DAY, A reflection theorem for closed categories. *Jour. Pure Applied Algebra* **2** (1972), 1–11.

[23] BRIAN DAY, Limit spaces and closed span categories. *Category Seminar, Sydney 1972/73*. Lecture Notes in Math. **420** (1974), pp. 65–74.

[24] B.J. DAY and G.M. KELLY, On topological quotient maps preserved by pullbacks or products. *Proc. Camb. Phil. Soc.* **67** (1970), 553–558.

[25] D. DOITCHINOV, A unified theory of topological, proximal and uniform spaces. *Doklady Akad. Nauk SSSR* **156** (1964), 21–24. Translated in: Soviet Mathematics (Doklady) **5** (1964), 595–598.

[26] EDUARDO DUBUC, Adjoint Triangles. *Reports of the Midwest Category Seminar II*. Lecture Notes in Math. **61** (1968), pp. 69–91.

[27] EDUARDO DUBUC, Concrete quasitopoi. *Applications of Sheaves. Proceedings of the Durham Conference*. Lecture Notes in Math. **753** (1979), pp. 239–254.

[28] M. EYTAN, Fuzzy sets, a topos-logical point of view. *Fuzzy Sets and Systems* 5 (1981), 47–67.

[29] H.R. FISCHER, Limesräume. *Math. Annalen* 137 (1959), 269–303.

[30] MICHAEL FOURMAN, *Connections between Category Theory and Logic.* Ph.D. Thesis, University of Oxford, 1974.

[31] MICHAEL FOURMAN, The logic of topoi. *Handbook of Mathematical Logic*, pp. 1033–1090. Studies in Logic and the Foundations of Mathematics 90, North Holland, Amsterdam, 1977.

[32] M.P. FOURMAN and D.S. SCOTT, Sheaves and logic. *Applications of Sheaves. Proceedings of the Durham Conference.* Lecture Notes in Math. 753 (1979), pp. 302–401.

[33] PETER FREYD, *Abelian Categories. An introduction to the theory of functors.* New York, 1964.

[34] PETER FREYD, Aspects of topoi. *Bull. Austral. Math. Soc.* 7 (1972), 1–76.

[35] G. GIERZ, K.H. HOFMANN, K. KEIMEL, J.D. LAWSON, M. MISLOVE, D.S. SCOTT, *A Compendium of Continuous Lattices.* Springer–Verlag, Berlin, 1980.

[36] ROGER GODEMENT, *Topologie algébrique et théorie des faisceaux.* Act. Sci. et Ind. 1252, Hermann, Paris (1958).

[37] J.A. GOGUEN, *L*-Fuzzy Sets. *Jour. Math. Anal. Appl.* 18 (1967), 145–174.

[38] J.A. GOGUEN, Categories of *V*-sets. *Bull. Amer. Math. Soc.* 75 (1969), 622–624.

[39] J.A. GOGUEN, Concept representation in natural and artificial languages: Axioms, extensions and applications for fuzzy sets. *Internat. Jour. Man–Machine Studies* 6 (1974), 513–561.

[40] ROBERT GOLDBLATT, *Topoi : the Categorical Analysis of Logic.* Studies in Logic and the Foundations of Mathematics 98. Rev. ed. 1984.

[41] SIEGFRIED GOTTWALD, Fuzzy set theory wih *t*-norms and φ-operators. *The Mathematics of Fuzzy Systems*, ed. by ANTONIO DI NOLA, ALDO G.S. VENTRI, Verlag TÜV Rheinland, Köln (1986), pp. 143–195.

[42] F. HAUSDORFF, Gestufte Räume. *Fund. Math.* 25 (1935), 486–502.

[43] HORST HERRLICH, Topological structures. *Math. Centre Tracts* **52** (1974), pp. 59–122.

[44] HORST HERRLICH, Topological functors. *Gen. Topology Appl.* **4** (1974), 125–142.

[45] HORST HERRLICH, Universal Topology. *Categorical Topology.* Sigma Series in Pure Math. 5, pp. 223–281. Heldermann–Verlag, Berlin (1984).

[46] HORST HERRLICH, Topological improvements of categories of structured sets. *Topology Appl.* **27** (1987), 145–155.

[47] HORST HERRLICH and HARTMUT EHRIG, The construct PRO of projection spaces: its internal structure. *Categorical Methods in Computer Science with Aspects from Topology.* Lecture Notes in Computer Science **393** (1989), pp. 286–293.

[48] HORST HERRLICH and G.E. STRECKER, *Category Theory*, 2nd Ed. Heldermann–Verlag, Berlin, 1979.

[49] DENIS HIGGS, *A category approach to boolean–valued set theory.* Preprint, 1973.

[50] DENIS HIGGS, Injectivity in the topos of complete Heyting algebra valued sets. *Canad. Jour. Math.* **36** (1984), 550–568.

[51] RUDOLF–E. HOFFMANN, *Die kategorielle Auffassung der Initial- und Finaltopologie.* Dissertation, Universität Bochum, 1972.

[52] RUDOLF–E. HOFFMANN, Topological functors and factorizations. *Archiv der Math. (Basel)* **31** (1975), 1–7.

[53] H. HOGBE–NLEND, *Théorie des Bornologies et Applications.* Lecture Notes in Math. **213** (1971).

[54] ULRICH HÖHLE, Fuzzy sets and subobjects. *Fuzzy Set Theory and its Applications (Louvain-la-Neuve 1985)*, pp. 69 - 76. Kluwer-Nijhoff, Boston (1985).

[55] ULRICH HÖHLE, Monoidal closed categories, weak topoi, and generalized logics. *To appear in*: Fuzzy Sets and Systems, 1990.

[56] ULRICH HÖHLE, M-valued sets and sheaves over integral commutative CL-monoids. *To appear in*: Applications of Category Theory to Fuzzy Sets, Kluwer, 1990.

[57] U. HÖHLE and L.N. STOUT, *Foundations of Fuzzy Sets*, Preprint 1990.

[58] P.T. JOHNSTONE, *Topos Theory.* Academic Press, 1977.

[59] P.T. JOHNSTONE, On a topological topos. *Proc. London Math. Soc.* (3) **38** (1979), 237–271.

[60] D. KAN, Adjoint functors. *Trans. Amer. Math. Soc.* **87** (1958), 294–329.

[61] MIROSLAV KATĚTOV, On continuity structures and spaces of mappings. *Comm. Math. Univ. Carolinae* **6** (1965), 257–278.

[62] D.C. KENT, Convergence functions and their related topologies. *Fund. Math.* **54** (1964), 125–133.

[63] A. KLEIN, Relations in categories. *Illinois Jour. of Math.* **14** (1970), 536–550.

[64] A. KOCK and C.J. MIKKELSEN, *Non–standard extensions in the theory of toposes.* Aarhus Univ. Preprint Series 1971/2, No. 25.

[65] A. KOCK and G.C. WRAITH, *Elementary Toposes.* Lecture Notes Series, no. 30. Mat. Inst. Aarhus Univ. 1971.

[66] H.-J. KOWALSKY, Limesräume und Komplettierung. *Math. Nachrichten* **12** (1954), 301–340.

[67] J. LAMBEK and P.J. SCOTT, *Introduction to higher order categorical logic.* Cambridge studies in higher mathematics 7, Cambridge University Press (1986).

[68] R. S. LEE, The category of uniform convergence spaces is cartesian closed. *Bull. Austral. Math. Soc.* **15** (1976), 461–465.

[69] E. LOWEN and R. LOWEN, A quasitopos containing CONV and MET as full subcategories. *Internat. Jour. Math. Sci.* **11** (1988), 417–438.

[70] E. LOWEN and R. LOWEN, Topological quasitopos hulls of categories containing topological and metric objects. *Cahiers Topologie Géom. Différentielle Catégoriques* **30** (1989), 213–228.

[71] ROBERT LOWEN, Mathematics and fuzziness. *Fuzzy Sets Theory and Applications*, Reidel, Dordrecht, Boston, Lancaster, Tokyo (1986), pp. 3–38.

[72] S. MACLANE, *Categories for the Working Mathematician.* Springer-Verlag, 1971.

[73] M.M. MAWANDA, On a categorical analysis of Zadeh generalized subsets of sets I. *Categorical Algebra and its Applications.* Lecture Notes in Math. **1348** (1988), 257–269.

[74] WILLIAM MITCHELL, Categories of boolean topoi. *Jour. Pure Applied Algebra* **3** (1973),193–201.

[75] G.P. MONRO, Quasitopoi, logic and Heyting-valued models. *Jour. Pure Applied Algebra*, **42** (1986), 141–164.

[76] G.P. MONRO, A category-theoretic approach to Boolean-valued models of set theory. *Jour. Pure Applied Algebra*, **42** (1986), 245–274.

[77] C.V. NEGOIŢĂ, Fuzzy sets in topoi. *Fuzzy Sets and Systems* **8** (1982), 93–99.

[78] C.V. NEGOIŢĂ, D.A. RALESCU, *Applications of Fuzzy Sets to Systems Analysis.* Wiley, New York and Toronto, 1975.

[79] LOUIS NEL, Topological universes and smooth Gelfand–Naimark duality. *Mathematical Applications of Category Theory.* Contemporary Mathematics **30** (1984), pp. 244–271.

[80] GERHARD OSIUS, Logical and set–theoretical tools in elementary topoi. *Model Theory and Topoi.* Lecture Notes in Math. **445** (1975), pp. 297–346.

[81] ROBERT PARÉ, Colimits in topoi. *Proc. Amer. Math. Soc.* **80** (1974), 556–561.

[82] J. PENON, Quasi–topos. *C.R. Acad. Sci. Paris* **276** (1973), Sér. A, 237–240.

[83] J. PENON, Sur les quasi–topos. *Cahiers Topologie Géom. Différentielle* **18** (1977), 181–218.

[84] ANDREW M. PITTS, Fuzzy sets do not form a topos. *Fuzzy Sets and Systems* **8** (1982), 101–104.

[85] D. PONASSE, Some remarks on the category Fuz(H) of M. Eytan. *Fuzzy Sets and Systems* **9** (1983), 199–204.

[86] D. PONASSE, Une nouvelle conception des ensembles flous. *BUSEFAL* **17** (1984), 4–9.

[87] D. PONASSE, Categorical studies of fuzzy sets, *Fuzzy Sets and Systems* **28** (1988), 235–244.

[88] ALEŠ PULTR, Fuzzy mappings and fuzzy sets. *Acta Univ. Carolinae— Math. et Phys.* **17** (1976), 441–459.

[89] HELENA RASIOWA and ROMAN SIKORSKI, *The Mathematics of Metamathematics.* Monografie Matematyczne **41**, Warszawa 1963.

[90] CLAUS MICHAEL RINGEL, Diagonalisierungspaare, I, II. *Math. Zeitschr.* **117** (1970), 249–266, **122** (1971), 10–32.

[91] M. SCHRODER, Solid convergence spaces. *Bull. Austral. Math. Soc.* **8** (1973), 443–459.

[92] FRIEDHELM SCHWARZ, Description of the topological universe hull. *Categorical Methods in Computer Science with Aspects from Topology.* Lecture Notes in Computer Science **393** (1989), pp. 325–339.

[93] E. SPANIER, Quasi–topologies. *Duke Math. Jour.* **30** (1963), 1–14.

[94] LAWRENCE N. STOUT, *General Topology in an Elementary Topos.* Ph.D. Thesis, University of Illinois, 1974.

[95] L.N. STOUT, Topoi and categories of fuzzy sets. *Fuzzy Sets and Systems* **12** (1984), 169–184.

[96] L.N. STOUT, *A Survey of Fuzzy Set and Topos Theory.* Preprint, 1988.

[97] L.N. STOUT, *The Logic of Unbalanced Subobjects in a Category with Two Closed Structures.* Preprint, 1989.

[98] GAISI TAKEUTI and SATOKO TITANI, Global intuitionistic fuzzy set theory. *The Mathematics of Fuzzy Systems,* ed. by ANTONIO DI NOLA, ALDO G.S. VENTRI, Verlag TÜV Rheinland, Köln (1986), pp. 51–86.

[99] HERBERT TOTH, Categorial properties of *f*-set theory. *Fuzzy Sets and Systems* **33** (1989), 99–109.

[100] A. TOZZI and O. WYLER, On categories of supertopological spaces. *Acta Univ. Carolinae—Math. et Phys.* **28** (1987), 137–149.

[101] OSWALD WYLER, On the categories of general topology and topological algebra. *Archiv der Math. (Basel)* **22** (1971), 7–17.

[102] OSWALD WYLER, Top categories and categorical topology. *Gen. Topology Appl.* **1** (1971), 17–28.

[103] OSWALD WYLER, *Elementary Topoi.* Lecture Notes, Carnegie–Mellon University, 1973/4.

[104] OSWALD WYLER, Are there topoi in topology? *Categorical Topology, Mannheim 1975.* Lecture Notes in Math. **540** (1976), pp. 699–719.

[105] OSWALD WYLER, Solid convergence structure equals pseudotopology. *Bull. Austral. Math. Soc.* **15** (1976), 273–275.

[106] OSWALD WYLER, Function spaces in topological categories. *Categorical Topology, Proc. Berlin 1978.* Lecture Notes in Math. **719** (1979), pp. 411–420.

[107] OSWALD WYLER, *Quasitopos Hulls.* Unpublished manuscript, 1986.

[108] OSWALD WYLER, On convergence of filters and ultrafilters to subsets. *Categorical Methods in Computer Science with Aspects from Topology.* Lecture Notes in Computer Science **393** (1989), pp. 340–350.

[109] L.A. ZADEH, Fuzzy sets. *Information Control* **8** (1965), 338–353.

[110] L.A. ZADEH, Fuzzy logic and approximate reasoning. *Synthese* **30** (1975), 407–428.

INDEX

U